THE IDEA

OF BIODIVERSITY

THE IDEA

OF BIODIVERSITY

PHILOSOPHIES

OF PARADISE

DAVID TAKACS

· · · · · · · · · · · · · · · · · · ·

The Johns Hopkins University Press

Baltimore and London

05 04 03 02 01 00 99 98 97 96   5 4 3 2 1

The Johns Hopkins University Press
2715 North Charles Street
Baltimore, Maryland 21218-4319
The Johns Hopkins Press Ltd., London

Library of Congress Cataloging-in-Publication Data
will be found at the end of this book.
A catalog record for this book is available from
the British Library.

ISBN 0-8018-5400-8

For Jae Wise, and Larry Carbone

All we can do is gaze in wonderment at the diversity of discursive species, just as we do at the diversity of plant or animal species.

JEAN-FRANÇOIS LYOTARD, *The Postmodern Condition: A Report on Knowledge*

 "Each individual, each element of the universe is an infinitely faceted mirror that reflects all other individuals and entities in the universe. We're all constantly reflecting everything else. So it's like a hologram. Each part contains the whole. That's very mystical, but—speaking as a scientist—our genetic material remembers the days in the organic soup when we were basically a few simple biochemical pathways—they're still there. They haven't changed hardly at all from the bacterial days. So in a sense, even a scientist has to admit that we still are a part and parcel of everything else, everything else organic on the planet. Regarding the question of 'What is intrinsic value?': I'm not a philosopher and I haven't figured it out. But intuitively, when I'm asked, you know, 'Should we save this species or that species, or this place or that place?' the answer is always 'Yes!' with an exclamation point. Because it's *obvious*. And if you ask me to justify it, then I switch into a more cognitive consciousness and can start giving you reasons, economic reasons, aesthetic reasons. They're all dualistic in a sense. But the feeling that underlies it is that 'Yes!' And that 'Yes!' comes out of the affirmation of being part of it all, being part of this whole evolutionary processes. And agreeing with Arne Naess that each species, each entity should be allowed to continue its evolution and to live out its destiny—It's not ordained or anything, but just do its thing, as we say. Why not? And the 'why not' is there's too many people."

MICHAEL E. SOULÉ, interview

CONTENTS

• • • • • • • • • • • • •

We need a more sophisticated, more creative, more critical, more joyous understanding of wild nature, human nature, and the bonds between them. Our lives depend on it.

I aim to make a small contribution. This book took root in 1989 when, during my second year of graduate school at Cornell, I was mired in a sophomore slump, flailing about for direction and purpose. Although I had signed up for a degree in the history and philosophy of science and technology, I was member of, and housed in, the section of ecology and systematics, of which the graduate field of ecology and evolutionary biology is a part. Don't ask; one needs a flashlight, a guidebook, and a week's supply of food and water to wend one's way through Cornell's baroque and Byzantine administrative structure. Anyway, I was truly surprised to find that what the vast majority of people did in Corson Hall had little to do with pressing environmental problems. The knowledge they generated just did not seem applicable to anything in "the real world." I heard virtually no discussion of environmental problems; no one talked about conservation biology.

At the time, I found this fairly appalling. How could they—and by "they" I mean both faculty members and the graduate students they supervised—choose their research agendas without attempting to answer some of the pressing questions needed to inform conservation work? I wrote a paper about this, foamed at the mouth a bit, and decided that rather than blathering about it, I should be doing the kind of work I was advocating. I considered transferring to a university friendlier to conservation biology. But I was torn, because I enjoyed contemplating the dialectic between science and society.

Then Daniel Janzen gave a talk at Cornell.[1] He explained his work as champion of Guanacaste National Park in Costa Rica. Through ample quantities of charisma, charm, and chutzpah, Janzen raised the money to buy out a moribund ranching industry in order to grow a huge national park of largely regenerated tropical dry forest. Furthermore, he was instrumental in establishing Costa

Rica's Instituto Nacional de Biodiversidad (INBio), which aims to catalogue each and every one of the nation's half million (give or take a few) species, and to figure out how to extract as many Costa Rican colones as possible from them. Janzen was training rural folks—former bartenders, housewives, poachers, preachers—to become parataxonomists, to collect and prepare the specimens that would become part of the national data base.

When he showed a slide of Oscar Arias (Costa Rica's Nobel Peace Prize–winning president) giving out diplomas to the first graduating class of parataxonomists, I got all choked up. I also knew I had a book topic. How did this idea of biodiversity get to be a force that moved societies here and abroad? What happens when scientists like Dan Janzen step outside of "value-neutral" science to become advocates on behalf of biodiversity? Why do they care so much about biodiversity in the first place, anyway? And why do I?

I have divided the partial responses to these and many other questions that have emerged along the way into eight interrelated chapters. First, I introduce the various tensions that this book chronicles, and that I wish to help resolve. Chapter 2 gives a selective history of some representative biologists who, over the past half century, have sought to remake the way we see the natural world so that we will not unduly remake the natural world. I had originally planned to provide a thoroughgoing history of the evolution of the concept of biodiversity, along with a comprehensive history of biologists advocating conservation. This project would have taken me well into my retirement years; neither my common sense nor my funding sources thought this advisable. Suffice it to say that Chapter 2 might provide grist for several other books, none of which I plan to write.

In Chapter 3, I analyze why biodiversity rhetoric surplanted previous conservation paradigms focused on the terms *nature, wilderness,* and *endangered species.* Chapter 4 uses a synergistic pairing of the disciplines of environmental history and science studies to examine how and why biologists proselytize on behalf of biodiversity. In Chapter 5, I provide a detailed look at the multiplicity of values biologists find in biodiversity. Chapters 6 and 7 are case studies: I view Costa Rica's National Institute of Biodiversity as a strategic research site where the supposed boundaries between science, society, and nature blur, and I show how the biologists running it are reshaping how an entire nation relates to nature. In Chapter 7, I examine the intricate web of resources spun by E. O. Wilson on behalf of biodiversity and his own position to speak for it. Finally, Chapter 8 returns to the tangled skein of values imbued in the idea of biodiversity, with an eye to extricating the hidden core.

To decide who to interview for this work, I listed the U.S. biologists I encountered in my reading who had promoted the values of biodiversity outside traditional scientific forums. They are not meant to represent Everybiologist; rather, they stand on the cutting edge—albeit not in the sense we traditionally give that phrase when applied to science—of their field. Although I had been granted generous time and money to do this research, I was not quite able to go everywhere to talk to everyone. Inasmuch as I picked these people because they had added public advocacy of conservation to their previously full menu of activities, scheduling meeting times became somewhat complicated. I could not interview some biologists because they were in out-of-the-way places; others were abroad when I was to be in their part of this country. Eventually, I did interview twenty-three biologists, whose time and cooperation in my research I greatly appreciate.

All interviews with U.S. biologists were conducted in person during 1992. I had a core of questions I wished to ask each biologist. The interviewees' own time constraints made some interviews shorter than others. Some had left paper trails that allowed me to skip some questions and delve into others more deeply. What I am saying is that each interview was different. This variability correlates with non-statisticity; this is a qualitative research project, which means I weave my own story from an embarrassment of riches I have to work with. My Costa Rican methodology took even freer form. I had prearranged my visit to INBio in May–June 1993 with director Rodrigo Gámez. Once there, I took advantage of opportunities as they arose. I send my appreciation, too, to all those at INBio and in Costa Rica who gave generously of their time. The people I interviewed are listed in the Appendix.

It will be obvious to some readers that I respect and admire many

of my interviewees. Other readers may perhaps think me disrespect-
ful of the scientists I interviewed because of my attempts to analyze
or deconstruct what they say and how they say it. Do these observations invalidate this study because they reveal my lack of value-
neutrality, my lack of "objectivity"? I don't believe so. Two decades
of science studies scholarship have peeled away the veneer of value-
neutrality in academic research (although I get the uncomfortable
feeling that some science studiers asymmetrically exempt themselves from this general principle). Qualitative methodological paradigms accept that value-neutral research is a myth.[2] As conservation
biologist Michael Soulé observed during our interview, "[W]hat is
'objectivity'? What is 'subjectivity'? What does *value-neutral* mean?
If you try to define *value-neutral*, you quickly find you can't. It's a
meaningless term, because we all have values." This book is about
biologists, who, for better or worse, often wear their values on their
sleeves. Surely I should not exempt myself from doing the same.

Nonobjective studies may still command assent. Of course they
do! In fact, you've never read any other kind of study. Let my readers
be wary when reading this—as they should be when digesting and
assessing any tidbit or mountain of information anywhere. I believe
recognition and acknowledgment of the objectivity dilemma is a
strength of this work; with this as a starting point, I have striven to
be as open-minded, fair, and balanced as possible.

I shall try to bare my biases for you as I go, so that you do not
have to read between the lines for them. Perhaps the most important one is that, like many of the biologists I profile here, I am engaged in the search for enduring, deep-rooted, universal, accurate,
honest, socially just arguments, schemes, and ethics for conservation. Another bias is that I am not scientistic; I feel no special reverence for science or scientists. However, I was trained as a scientist;
I am housed in a department of science; many of my best friends
are scientists. I believe I have a balanced view of what goes on in
the esoteric world of science, while not having been captured by the
normative forces exerted on its practitioners.

I hurl at the reader of this book a barrage of quotations, because I
think they make the story richer and more interesting. For the sake
of completeness and texture, at various points I include lengthy se-

quences of quotations from my interviews with the biologists who inhabit this story; you should skip these if not interested in such detail. I have arranged each such sequence alphabetically, which means that Peter Brussard will bear the brunt of scrutiny. Do him a favor: if you are merely perusing some of the responses to a given question, sometimes start somewhere in the middle. Not all my interviewees are represented in each sequence, as I did not always pose a given question to each biologist I interviewed. An ellipsis means I have edited out less pertinent information; where an ellipsis might normally be called for in the context of a quotation, as where a speaker pauses or has a change of mind in midstream, I use a dash (—). All quotations for which the source is not otherwise indicated derive from my interviews.

If these quotations from my interviews lack polish, it is because they were spontaneous conversational responses. In most cases, I have made the decision not to edit to smooth things out, as it is difficult to discern where felicitous changes in grammar segue into infelicitous changes in meaning. The drawback is that the reader may judge harshly respondents who may have temporarily drowned in their own syntax. Cut them some slack: this happens to all of us in conversation, particularly when we are attempting to express intricate thoughts or emotions.

I was fortunate during my Cornell career to teach freshman (freshperson?) writing seminars. I insist on gender-neutral language from my students. Alas, I have not always found an elegant and foolproof way of using gender-neutral pronouns in my own work. Here I alternate between "his" and "her" when discussing a generic "biologist" or other person. Keep in mind that while I try to use "his" and "her" equally, many more men than women work in conservation biology; grammar and real life do not always map neatly onto each other.

Part of the point of this work is that it is difficult to distinguish biodiversity, a socially constructed idea, from biodiversity, some concrete phenomenon. My editors ask that I put *biodiversity* and its close relative, *nature*, in italics whenever I discuss these words without direct reference to the outside world to which these terms relate. But the terms are always informed by that tangible world;

and we have no direct access to that world without our conceptual processes. The decision whether to italicize has come down to a coin flip in a few cases. I believe any confusion this creates will encourage you to ponder the flexible boundaries between the real world and our depictions of it.

ACKNOWLEDGMENTS

Are facts socially constructed? I don't know. Are values socially constructed? I'm not sure. Is this work a social construction? You bet it is.

A Cornell University A. D. White graduate fellowship; Cornell University travel grants; the Organization for Tropical Studies; a Cornell John S. Knight Writing Program Buttrick-Crippen fellowship; a National Science Foundation training grant to Cornell's Department of Science and Technology Studies; and a National Science Foundation dissertation improvement grant (DIR-9121580) provided the funds that made this project possible.

An army of superwomen run Cornell, somehow making sense of senseless bureaucratic impediments and staggering personal incompetence (ours not theirs). Alberta Jackson, LuAnn Kenjerska, DeeDee Albertsman, Sue Drake, Sarah Albrecht, Beth Marks, Amy Devaul, Louise Harrington, Louise Gunn, Debbie VanGalder, and Lillian Isacks have corrected my numerous blunders. I thank them for their grace, wisdom, and unfailing good cheer.

Reaching back to the past, Greg Mertz introduced me to the wonders of biodiversity long before the term existed. Jeanne Palm taught me to write. Glenn Hausfater, Carol Saunders, and Michael Pereira took an overexcited kid to Kenya and turned me on, not only to baboons, but also to the splendors and heartbreaks of conservation, development, and Africa. Claire de Boer Van der Ven showed me that anything is possible, and the people of Keur Seny Gueye, Senegal, taught me the meaning of life.

I was able to stretch my research grants much further than I could have imagined thanks to the hospitality of the following people, who offered me food, shelter, and good company during my interview travels: Majken Ryherd Keira and Mamadou Keira, Sally Scott and Jimmy Potash, Greg Case, Nadine and George Haydachuk, Peter Zahler and Chantal Dietemann, Peter and Wynne Wimberger, Linda Beck and Mark Pires, Matt Jensen, Pat DeMaio, Soren Ryherd and Vida Jakabhazy, Ylva Hernlund, Ron Scott, and Maya

Hernlund-Scott, Phil and Danielle Eidenberg-Noppe, Nancy Clum, Leslie Welch and Hannah Sidibé, Theresa and Bobby Ayers, Steven Sather, David Antman, Dan Peck, la familia Mora-Zamora, and Sarah Townsend, who was there once again to lead me out of the woods.

Robert Harington, my editor at Johns Hopkins University Press, was unfailingly enthusiastic and helpful. Peter Dreyer provided astute, necessary copyediting.

The committee of professors who advised me at Cornell—Will Provine, David Pimentel, Sunny Power, and Peter Taylor—provided me with moral support, intellectual fodder, and good friendship. Additional intellectual stimulus and crucial feedback was offered by Douglas Allchin, Thom Boyce, Sharon Brisolara, Peter Dear, Tom Dunlap, Stephanie Fried, Tom Gieryn, Sheila Jasanoff, Bruce Lewenstein, Laura Murray, Kavita Philip, Majken Ryherd, Sergio Sismondo, Michael Soulé, Sarah Townsend, Edward O. Wilson, Carol Yoon, and an ornery (but helpful) anonymous reviewer.

My students are a constant source of instant gratification and deep inspiration to me. Thanks to (nearly) all of them. And thanks to the folks at the John S. Knight Writing Program for their continued nurturance.

Shortly after my arrival in Ithaca eighteen years ago, I was blessed when two families adopted me. I have cherished their friendship, support, and love over the long haul. I am grateful to all de Boer and Zahler friends, parents, children, spouses, spouse equivalents (past and present), pets, and other symbionts for enriching my life.

The Wolofs have a proverb: "Ku am nit, ñakkul dara"—who has people lacks nothing. Some of what little sanity I have maintained while writing this book has been due to the companionship afforded by these people while sweating: Dan Shapiro, Dan Peck, Josh Nowlis, Michael Nachman, and Yanek Mieczleowski. Most especially, Thom Boyce and Yvette de Boer kept me inspired, laughing, and drenched. In drier times, I have lacked nothing in Ithaca while working on this project due to Carol Yoon, Merrill Peterson, Becky Sladek, Laura Katz, Dan Berger, Ben Normark, Roxanna Normark, Susi Remold, Renée Perry, Becky Jarrell, Claudia Coen, Jordan West, Peter Wimberger, Michele McClure, Nancy Clum, Pat Doak, Tully,

Hester Parker, Sue Reed, Evan Bloom, Jennifer Astone, Sharon Briso-lara, Luis Acebal, Sergio Sismondo, Laura Murray, Liz Davey, Kavita Philip, Stephanie Fried, Jerry Shing, Phil Kogan, Byron Suber, Bob Marra, Connie Adams, Maya the Dog, and, finally, Steven Sather and Majken Ryherd Keira. Now I expect you all to buy several copies of this book.

Larry Carbone has read this work closely and provided crucial intellectual and editorial assistance. His work on laboratory animal veterinarians and changing standards of ethical treatment of animals has strongly influenced my ideas here. Together we have attempted to figure out how to balance the content of our work between rigorous adherence to disciplinary imperatives and dedicated service to the human and nonhuman constituents we study. And Larry has proven repeatedly and remarkably distractable when the need to procrastinate arises.

Even those listed above who know me really well cannot imagine the nightmare I can sometimes be. For the length of this project, I was blessed with a kind and caring companion whose patience and support I could not have imagined nor deserved. Jae Wise not only kept me happy; he has also read every word of this work and heard about it until I'm sure he doesn't want to know from biodiversity anymore ever. Jae's love and wisdom can be found on every page herein.

THE IDEA

OF BIODIVERSITY

TENSIONS AT THE
CROSSROADS
OF SCIENCE, NATURE,
& CONSERVATION

Around you, life pulses and throbs: vibrant, gorgeous, teeming, unutterably fecund, unimaginably complex, inextricably interconnected. Inside your head, too, this extravagant glut of life plays out its drama. I write *nature*; neurons network; you conjure up images of a personalized nature, yours uniquely. When I write *biodiversity*, perhaps other images appear to your mind's eye. Take notice: How do these differ from the images evoked by the word *nature*?

Conservation biologists have generated and disseminated the term *biodiversity* specifically to change the terrain of your mental map, reasoning that if you were to conceive of nature differently, you would view and value it differently. As a result of a determined and vigorous campaign by a cadre of ecologists and biologists over the past decade, biodiversity has become a focal point for the environmental movement. In this work, I analyze what biodiversity represents to the biologists who operate in broader society on its behalf.

At places distant from where you are, but also uncomfortably close, a holocaust is under way. People are slashing, hacking, bulldozing, burning, poisoning, and otherwise destroying huge swaths of life on Earth at a furious pace. The term *biodiversity* is a tool for a zealous defense of a particular social construction of nature

· · · · · · · · · · · ·

1

that recognizes, analyzes, and rues this furious destruction of life on Earth. When they deploy the term, biologists aim to change science, conservation, cultural habits, human values, our ideas about nature, and, ultimately, nature itself.

As a result, anyone interested in the dwindling resources *biodiversity* represents must turn to conservation biologists for guidance. In the name of biodiversity, biologists hope to increase their say in policy decisions, to accrue resources for research, gain a pivotal position in shaping our view of nature, and, ultimately, stem the rampant destruction of the natural world. To accomplish all this, biologists speak for an array of values that go far beyond what one might think of as falling within the ambit of their expertise. Their factual, political, emotional, aesthetic, ethical, and spiritual feelings about the natural world are embodied in the concept of *biodiversity*; so packaged, *biodiversity* is used to shape public perceptions of, feelings about, and actions toward that world. Biologists hope to have a say in forging a new ethics, new moral codes, even new faiths. By staking out new sources of power for themselves, they ultimately hope to gain control over nature—over how and where and even why wild organisms and natural processes are allowed to endure. By altering our mental configurations of nature, biologists seek to alter the geographical configurations of nature.

Biodiversity lies at the heart of a complex web; strands radiate outward, taut with tensions.

Any phenomenon in the natural world results from the tensions played out in the evolutionary drama. The genes, individuals, ecosystems, and processes—even the human values—subsumed under the term *biodiversity* result, at least in part, from the incalculable set of tensions and struggles we now call "natural selection." The ceaseless skirmishes of each living organism (and of each of its ancestors throughout the eons of evolution) to cope with the vicissitudes of the environment have created all that now concerns us when we worry about biodiversity. This evolutionary struggle, according to some of the biologists I portray here, may also have produced the ecological and emotional affiliations with nature that make us fearful for the natural matrix within which we evolved.

That is to say, manifold, exquisite evolutionary tensions helped fashion everything in nature that conservation biologists fight for; these tensions also produced the biologists themselves and the urges that prompt them to fight for nature.

That many feel the need to fight for the natural world speaks to perhaps the most crucial and important tension human beings must resolve as we careen into the next millennium. We constantly struggle for self-aggrandizement or survival at the expense of the natural world. People attempting to secure even the barest necessities of food and shelter for themselves and their kin must perforce engage in some level of destruction of the natural matrix that allows for the continued unfurling of human existence. And most humans struggle not just for the daily necessities; they struggle for a richer life, for wealth mined from the bounty that nature has been producing and storing for billions of years.

However, biologists warn, this struggle by ever-increasing numbers of people for ever-increasing wealth threatens the species that have resulted from the evolutionary process, the ecosystems they inhabit, the natural cycles that provide the water we drink and the air we breathe. To speak at all of a threat to biodiversity is to acknowledge the conflict among humans who are engaged in destroying the natural world and those who seek to preserve it in some of its integrity. Battles over biological resources rage a few miles from where you sit, as they do in every remote corner of the Earth. These battles pit the wants of wealthy humans against the needs of poor humans. They set at odds the perceived needs of humans and those of many millions of other species, and of the natural processes that nourish them and us. Scientists who love the natural world forged the term *biodiversity* as a weapon to be wielded in these battles.

"Scientists who love"—the contradiction in terms this image may conjure up suggests another tension chronicled here. Science is commonly thought of by the public and portrayed by its practitioners as an objective, cold, nonpartisan, value-neutral enterprise. Scientists discover facts, mediate truths about nature: on this image their continued prosperity is thought to ride. Yet a group of biologists have been as partisan as can be in their attempts to preserve biodiversity. Biologists speak for it in Congress and on the *Tonight*

Show. They whisper it into the ears of foreign leaders. They extol its virtues to the Harvard Divinity School. They transport 10 percent of the U.S. Senate to spend nights in the heart of the Amazon so that biodiversity will work its persuasive charms firsthand. They weave sensuous word tapestries in books meant to seduce readers to love biodiversity and therefore join biologists in attempts to sculpt the political, physical, and normative landscape to its needs. They profess to be experts on an array of economic, ecological, and even aesthetic and spiritual values of biodiversity that would seem to stretch the limits of what we normally consider to lie within scientists' expertise.

Hence the tension: in so doing, biologists jeopardize the societal trust that allows them to speak for nature in the first place. Of course, this also creates tensions between biologists who feel that their love for biodiversity compels them to argue for its preservation—and who feel others ought to pull more of their fair share of the weight—against those other biologists who feel that such advocacy jeopardizes the very image that allows scientists their resources and power in society.

Within the community of biologists who advocate for biodiversity, we find other sets of tensions. Where and how is it appropriate to be an advocate? What are the values of biodiversity that should be promoted? Why should any of us care? These tensions do not originate with biologists who promote biodiversity. Since the early days of conservation in the United States (and certainly before that), crusaders for nature have disagreed about strategy and tactics. However, in attempting to make us care about biodiversity, biologists consciously seek to bridge the ideological dichotomies that divide them.

Who will speak for nature? Who should be recognized as expert on the values of the natural world? Who best knows what is good for nature, and implicitly, therefore, what is good for humans? By attempting to become spokespersons for biodiversity, biologists create further tensions around these questions.

While they whip up public concern over the diminution of biodiversity, biologists simultaneously garner the resources that go hand in hand with such increased concern. Biodiversity is a formidable constituency; its representatives wield significant power in a society that cares to preserve it. Biologists, who have been called upon to provide "facts" about the natural world, now clear space to speak of nature's "values." That they should be experts on all aspects of the natural world is sometimes disputed by politicians, religious folk, economists, corporate executives, citizens whose land rights are jeopardized, indigenous people who claim to be guardians of knowledge about and protectors of biodiversity, and a host of other constituencies who have a stake in nature. Tensions are guaranteed to arise through this struggle to speak for nature.

The term *biodiversity* also arises from tensions over official environmental protection policies. Concern for biodiversity and concern for endangered species—hitherto the strongest weapon in U.S. conservationists' arsenal—lead policymakers down different paths. Yet protection of endangered species sometimes creates more problems than it solves. Biodiversity preservation creates its own conundrums, causing controversy over how we should go about protecting the natural world, and on what levels of the biological hierarchy we should focus.

Around the time the term *biodiversity* made its appearance, another shibboleth, *sustainable development*, was also gaining clout. In graduate school, I earned an official minor in conservation and sustainable development. Contemplate this for a moment and you see another tension: how can we simultaneously conserve and develop? The growing needs of a growing population would seem to clash with the needs of those who want unsullied natural refugia. These forces clash in titanic battles over economic development, social justice, and long-term survival.[1]

Away from the front lines, the battle over biodiversity also reflects a battle within the academy and in society at large over relativism and realism vis-à-vis both facts and values. At the ivy-covered humanities end of campus, "truth" is out; all is relative. Values do not inhere in the natural world; they are subjective. Even "facts" are social constructions, representing whatever we all agree to, without validifying referents in nature. Yet to speak for biodiversity, a realist's worldview seems a priori necessary. Not only must biologists (and lay environmentalists) produce usable facts about the natural world; they also choose to argue for the overwhelming "truth" of the values that fall under biodiversity's aegis. These values are promoted, not as relative and shifting, but as enduring, and even biologically based. In fact, for some biologists, these values would seem to fall as ineluctably within the realm of biologists' expertise as do facts. I discuss this more fully throughout. I hope to offer some resolution to this tension for the purposes of my work and for other work in the humanities and social sciences that looks at environmentalism, and other questions of social justice or ethics, and seeks to do so in a way that lends itself to change in the real world.

This work also reflects tensions in the three academic disciplines with which I am most closely affiliated: conservation biology, environmental history, and science studies. Conservation biologists do not merely seek to study and document the ecological and evolutionary phenomena most crucial to informed conservation policy; they are "mission-oriented," and the mission is to conserve the very objects they study. Embodied in conservation biology are a set of normative principles as inextricable from the science as the formulae, theories, and models they use or the entities they investigate.[2] How to make it so that the facts appear credible, unshaped by the advocacy or the norms?

Environmental historian Barbara Liebhardt writes that those who practice her discipline aim to "discern the threads that weave people and their environments together in particular patterns." Donald Worster urges that environmental historians "must go wherever the human mind has grappled with the meaning of nature."[3] Where better to focus this book's analytical gaze, then, than on on the *ideas* scientists have generated, and actively promoted to society, to change the way we treat the Earth? Ideas may produce ecological change, and environmental historians may highlight the dialectic, and the tensions, between our treatment of the Earth and our ideas of it.

Conservation biologists have set themselves the mission of crafting and selling the tools that will make the Earth sustainable for biodiversity. According to Roderick French, environmental historians are concerned with "this most basic of existential concerns: can the human species learn so to modify its behavior that earth history will have a future?"[4] Here, then, we find tensions overlapping be-

tween environmental history and conservation biology. Practitioners of the two disciplines share the plight of attempting to be at once
dispassionate chroniclers and sometimes vocal partisans of a set of
goals inextricable from what is being chronicled. Both engage in a
kind of boundary redefinition, attempting to expand outward so that
their words will be taken seriously in society, but simultaneously
risking loss of credibility. Both are concerned about the relation between nature and our ideas of nature, while trapped in a vortex that
makes discriminating between the two nearly impossible. Both attempt to change their audiences' ideas of nature, one by creation of
the term *biodiversity* and promotion of the intellectual, normative,
and emotional connotations subsumed under it, and the other by
heightening awareness of the role of nature in culture and history
and of the role of culture and history in nature. And neither group
feels they can afford to fail in their self-defined mission.

Science studies descends from programs that sprang up at universities in the 1960s to study relations between science, technology,
and society; to channel science down paths to the public good; and
to make more concrete applications for the feeder disciplines of history, philosophy, sociology, and politics of science. Practitioners of
science studies are currently attempting to solidify the subject as a
recognized academic discipline, with its own norms, theories, acceptable matters of study and ways of thinking. It is often fiercely
relativistic, taking a constructivist standpoint. As sociologist of science Brian Martin points out, science studies has transformed the
discourse from a critique of science for scientists to a critique of
scientists for other science studies practitioners.[5] It is hard to reconcile a relativist viewpoint with my desire to have biologists—who
often hold entrenched ideas about the nature and practice of science and conservation—read and reflect on my work. Hence a tension in science studies between those who would toe the academic
line to further the relativist theorizing of the discipline—no matter
how alienating to practicing scientists—and those, like myself, who
would bring science studies to the people we analyze.

And so to the tensions within myself. I make no bones about it:
I am an environmentalist, although that term is amorphous enough
to have lost some currency and precision. To put it another way: I

love the natural world. I believe my existence and all our existences are horribly diminished as humans encroach on the domains of wild creatures. I love roaming natural landscapes, studying other organisms. Nature's many creations, its unfathomable complexity, give me constant aesthetic enjoyment. I have a reverence for the natural world akin to what others in some contexts call spiritual. I think it hubris for humans wantonly to destroy other creatures and the natural matrix in which they live. By my reckoning, this feeling for life must also include reverence for my fellow humans; solutions to environmental crises must be as generous and protective toward men and women as they are to the natural world.

Yet I am a scholar with obligations to objectivity. My disciplinary homes lie at the boundaries of social sciences and the humanities, areas where traditional notions of truth currently hold scant sway. To "deconstruct" an idea is not difficult, and I attempt to do that here with biodiversity. Yet how does one deconstruct constructively? How to make it so that others do not misuse my analyses to obstruct those biologists attempting to stem the destruction of biological diversity? How does one bring advocacy to scholarship while remaining far enough removed from the events one chronicles to make some stab at objectivity? How can one feel about the natural world as strongly as I do, and as do the biologists whose exploits I narrate, and not believe that those feelings approach the truth in some sense? How can I balance my healthy skepticism about conservation biologists' proselytizing on behalf of biodiversity against my fervent hope that they succeed? In reporting the tensions at whose nexus biodiversity is located, I hope to resolve some of them in myself.

2

By activism on behalf of what they call *biodiversity*, conservation biologists seek to redefine the boundaries of science and politics, ethics and religion, nature and our ideas about it. They believe that humans and the other species with which we share the Earth are imperiled by an unparalleled ecological crisis, whose roots lie in an unheeded ethical crisis. *Biodiversity* is the rallying cry currently used by biologists to draw attention to this crisis and to encapsulate the Earth's myriad species and biological processes, as well as a host of values ascribed to the natural world. An elite group of biologists aims to forge a new ethic, in which biodiversity's multiplicity of values will be respected, appreciated, and perhaps even worshiped.

Yet conservation activism by biologists is not something new under the sun. Today's conservation biologists had forerunners from whom they could learn lessons about how to put ecological science to work to change the way we view the natural world, and thus make us wish to preserve it.[1] I wish here to avoid the perils of Whiggism—selectively depicting a strand leading to the phenomenon one is extolling today, as if history had aimed an arrow at one's target. Rather, when considering the rise to prominence of the term *biodiversity*, we may consider four timelines: biologists' ideas of biological diversity; biologists' promotion of biological diversity to the public; the public's actions and values vis-à-vis biological diversity; and changes in diversity itself. Events along each timeline affect and constrain the others.

· · · · · · · · · · · ·

For centuries, biologists and their intellectual antecedents viewed flora and fauna as objects to be collected, dissected, and classified. As human activities increasingly penetrated diversity's domains, this notion of life's richness jumped the bounds of science and took on broader social meaning. Although others had preceded them in sounding alarms, around the middle part of this century, biologists began in earnest to reify biological diversity as a threatened entity and intensified activities to convince the public to conserve this new commodity.

From the nineteenth-century conservationist-naturalists George Perkins Marsh and John Muir, through the mid-twentieth-century ecological biologists Aldo Leopold, Charles S. Elton, and Rachel Carson, and on to a cadre of 1990s conservation biologists, American naturalist-ecologists have proclaimed nature's diverse values. They have shared a conundrum: how do you convince others to care about what you love? Since the early days of conservation in the United States, those who have sought to safeguard parts of the natural world from despoilment have disagreed as to philosophy, strategy, and tactics.

Scholars have documented the tensions between seemingly antithetical ways of valuing, and thus arguing for, nature. Stephen Fox writes of the emotional/aesthetic "radical amateur" descendants of John Muir, who see nature as a sacred source imbued with spiritual value and seek its eternal "preservation." This view is juxtaposed with the cold economic spirit of the "scientific professionals" descended from the first head of the U.S. Forest Service, Gifford Pinchot, from whom we have inherited a forest system and a philosophy of "conservation" in which nature is to be used wisely as a resource for human welfare.[2] Others have pinpointed similar dichotomies and attributed different names to those who hold what seem like opposing beliefs. Max Oelschlaeger rues the banishment by the Cartesian-Newtonian scientific paradigm of the aesthetic-religious approach to knowing the natural world. Similar dualities are constructed between the arcadian and the imperialist views of nature by Donald Worster; between moralists and economic aggregators by Bryan Norton; between the naturalist's intimate, aestheticized "look" and the scientist's omniscient, controlling "gaze" by

Gregg Mitman; and between intuitive/emotive ecological Diony-
suses and analytical/rational Apollos by Daniel Botkin. They are
reflected when deep ecologists pose their own biocentric beliefs
against "shallow ecologists" who continue to insist that the Earth
exists for human use and human valuers.[3]

Modern-day conservation biologists reiterate these operational
and ideological dualities, while at the same time consciously attempting to bridge them—to solve what has been called the "conservation dilemma" or "environmentalist's dilemma"—in an attempt to
make us care about biodiversity.[4] Throughout the twentieth century,
biologists have wrestled with how best to present the natural world
to us so that we may share their apprehension about its fate. Often
they have privately held a set of most unscientific, biocentric, arcadian beliefs and devotion toward nature while publicly promoting
more anthropocentric, utilitarian, scientifically respectable rationales. At times, however, they have openly promoted a set of values,
a worldview, that may strike us as most unscientific. For those who
would preserve diversity, and who understand and *feel* the full range
of values diversity holds for humans, these choices are deliberate
and crucial for shaping an idea of nature and a corresponding ethic
that will nurture and protect the entities and processes they have
held most dear. Today, we call these entities and processes *biodiversity*, but they have gone by other names: *natural variety*, *flora and
fauna*, *wildlife*, *fellow creatures*, *wilderness*, or, simply, *nature*.

ALDO LEOPOLD

Aldo Leopold (1887–1948) was a forester, game manager, wildlife ecologist, hunter, philosopher, and writer.[5] Prominent environmental historians and philosophers have mined Leopold's writings,
which offer a rich lode for (sometimes contradictory) exegesis.[6] Here
I wish to portray Leopold as a twentieth-century forerunner of
today's conservation biologists. Leopold's *A Sand County Almanac*
(in later years published with additional essays from his collection
Round River) takes several cues from his fellow Wisconsinite John
Muir.[7] In prose that often soars, Leopold brings together natural
history observations, ecological science, and homespun wilderness
philosophy. He hopes that once seared into our brain, heart, and

soul, the land and the organisms that inhabit it will become seared into our ethical code: "We can be ethical only in relation to something we can see, feel, understand, love, or otherwise have faith in," he observes.[8]

Much of *A Sand County Almanac* is an elegy for the world as it was and might still be; it is a paean to the values that humans could delight in if only we would preserve enough wild places and open up our spirits to the lessons wild places can teach us. The volume does not merely put forward Leopold's philosophy and prescriptions. It is also a manual on how to combine science and literature, rational analysis and emotional musings.[9] Leopold paves the way for others to amalgamate diverse values in the service of forging a new ethic based on biologists' expertise, where expertise is to be understood as deep knowledge of and *feeling for* the array of values in nature.

When we examine the public defense of biodiversity by today's conservation biologists, we find that Leopold was often there first. He writes of restoration ecology. He recognizes the connections between cultural and biological diversity. He sees that economic valuations of biological species are often cheap rationalizations, yet he nonetheless presents nature as a multifarious resource for humankind.[10] He lacks a mechanism, but presages E. O. Wilson's contention that a connection to other forms of life is rooted in our genes, and that in eliminating wilderness, we deny our descendants their right to be fully human: "A man may not care for golf and still be human, but the man who does not like to see, hunt, photograph, or otherwise outwit birds or animals is hardly normal," he asserts. "Opportunity for exercise of all the normal instincts has come to be regarded more and more as an inalienable right. The men who are destroying our wildlife are alienating one of these rights, and doing a thorough job of it. More than that, they are doing a permanent job of it."[11]

Leopold also foresaw the argument that organisms provide unknown ecosystem services: "What of the vanishing species, the preservation of which we now regard as an esthetic luxury? They helped build the soil; in what unsuspected ways may they be essential to its maintenance?" This foreshadows what I call the "argument from ignorance" that lurks in the biodiversity concept:

The outstanding scientific discovery of the twentieth century is not television, or radio, but rather the complexity of the land organism. Only those who know the most about it can appreciate how little is known about it. The last word in ignorance is the man who says of an animal or plant: "What good is it?" If the land mechanism as a whole is good, then every part is good, whether we understand it or not. If the biota, in the course of aeons, has built something we like but do not understand, then who but a fool would discard seemingly useless parts? To keep every cog and wheel is the first precaution of intelligent tinkering.[12]

Science depicts the complexity of what Leopold refers to as "the land organism"—today we might call it the ecosystem—and tells us we know little about it. The policy prescription that follows is: save it all, for you know not what you do. The whole may or may not be worth more than the sum of its parts—Leopold is agnostic about that—but since we do not know what the parts do, or even what they are, it behooves us to save everything.

Susan Flader describes Leopold's "romance of diversity"—not just of organisms, but of landscapes, of climates, of cultures: "Valuing diversity as he did, it was natural that he should ponder its relationship to the processes he was observing."[13] She traces the way Leopold's general, aesthetic preoccupation with diversity evolved into a specific, scientific concept of it as an ecological property and commodity. As his predecessors would, Leopold reified diversity of all kinds as a commodity to be cherished—and possibly lost.

The ecologist understands, deeply, what it means to tinker with the diversity that comprises the ecosystem—what it might mean when a species-component of that diversity disappears from the landscape:

It is easy to say that the loss [of grouse in the North Woods] is all in our mind's eye, but is there any sober ecologist who will agree? He knows full well that there has been an ecological death, the significance of which is inexpressible in terms of contemporary science. A philosopher has called this imponderable essence the *numenon* of material things. It stands in contradistinction to *phenomenon*, which is ponderable and predictable, even to the tossings and turnings of the remotest star.[14]

Ecologists step in as experts on the ineffable. They guard the numinous, something essential that science cannot quite express as fact, yet that only the ecological scientist can feel or know in some way. As they would for his late-twentieth-century descendants, ecology and evolution offered Leopold scientific prescriptions for both land management and ethical development. He saw ecology as "an infant just learning to talk, and [that], like other infants, is engrossed with its own coinage of big words. Its working days lie in the future. Ecology is destined to become the lore of Round River, a belated attempt to convert our collective knowledge of biotic materials into a collective wisdom of biotic navigation."[15] Jargon did not impress him, and he rued the fact that "the construction of instruments is the domain of science, while the detection of harmony is the domain of poets."[16] Ecology could navigate between these poles, could provide guideposts to help humans belong to, rather than dominate, the Earth.

A Sand County Almanac is an attempt to use ecological and evolutionary precepts as such a guide. Evolutionary biology teaches lessons about our place in the universe:

> It is a century now since Darwin gave us the first glimpse of the origin of species. We know now what was unknown to all the preceding caravan of generations: that men are only fellow-voyagers with other creatures in the odyssey of evolution. This new knowledge should have given us, by this time, a sense of kinship with fellow-creatures; a wish to live and let live; a sense of wonder over the magnitude and duration of the biotic enterprise.
>
> Above all we should, in the century since Darwin, have come to know that man, while now captain of the adventuring ship, is hardly the sole object of its quest, and that his prior assumptions to this effect arose from the simple necessity of whistling in the dark.[17]

Evolutionary biology thus reveals one of the tenets of an environmental ethic: it must avoid the hubris of unthinking anthropocentrism. To truly appreciate the natural world and understand why it must be conserved, the evolutionary perspective is irreplaceable: "Only those able to see the pageant of evolution can be expected to value its theater, the wilderness, or its outstanding achievement, the grizzly," for example.[18] Only those who can think "like a mountain" can begin to transcend their own immediate interests to grasp the exquisite and venerable interconnections between all organisms.[19]

As additional maxims to be heeded in his new creed, Leopold proposed such nascent ecological and evolutionary revelations as the workings of the biotic pyramid and the interdependence of all things. He cites as a rationale for conservation "the tacit evidence of evolution, in which diversity and stability are so closely intertwined as to seem two names for one fact." Leopold observes that "science has given us many doubts, but it has given us at least one certainty: the trend of evolution is to elaborate and diversify the biota."[20]

Herein lies a problem with relying on scientific facts for the tenets of an environmental ethic: science is sometimes wrong. The facts of ecology and evolution on which to base this ethic may turn out to be the wishes of the ethic makers. The ecological theory that correlates species diversity with ecological stability seems to have been widely violated. Leopold's "one certainty" of science has been recently made somewhat less certain, if we accept Stephen Jay Gould's contention that the "cone of diversity" is just another ecological myth, and that evolution produced its maximum diversity over 500 million years ago.[21]

How can one ever extricate what the natural world really tells us from what we wish the natural world told us? Years after Leopold's death, the creators of the term *biodiversity* would come to recognize the skein of human desires wrapped around nature. They, like Leopold before them, would derive a set of values from ecological science, just as they have derived facts. Both facts and values are open to debate; both lie within the ecologist's ambit.

A Sand County Almanac may be confused or ambiguous at times, but some of the book's brilliance arises from its ambiguities. Like the natural world it chronicles, Leopold's work is rich enough for each reader to interpret it as he or she likes. Max Oelschlaeger contrasts authors who revere Leopold as a pioneer biocentrist (as Oelschlaeger does) to others who classify him as a thoroughgoing anthropocentrist. Some, like Roderick Nash, believe Leopold discovered an inevitable truth: ethical regard must be extended to the Earth and its nonhuman denizens. Bryan Norton, however, says Leopold believed nothing that metaphysical: he merely offered as many pragmatic arguments as he could, while straddling the line between pragmatic and more "ecstatic" valuations of the land and its inhabitants.[22]

I believe Norton comes close to the truth. Leopold knew that his was not the final word, but merely the prolegomena to a new synthesis. Like many who came after him, Leopold was a pluralist; he used many arguments or styles to make his case, so that he might convert as many readers as possible.

It may also be that the ambiguities of the book reflect the ambiguities of Leopold's own life. I earlier alluded to the ideological split between conservation arguments. Leopold notes a dichotomy of his own and calls it "the A-B cleavage." He clearly yearned for us to listen to the B-side, to choose the arcadian view: "In all of these cleavages, we see repeated the same basic paradoxes: man the conqueror *versus* man the biotic citizen; science the sharpener of his sword *versus* science the searchlight on his universe; land the slave and servant *versus* land the collective organism."[23]

Leopold may have desired a kinder, gentler land ethic; nonetheless a maelstrom of killing swirls around *A Sand County Almanac's* beautiful evocations of sunrises and touching portraits of wildlife. Wilderness may be a locus for reflecting on the glories of biological diversity, but it is also a place for shooting many of those glories. During much of his career, Leopold ardently supported ridding wild lands of predatory "varmints"; he and his "sportsmen" friends clearly bore some responsibility for the events Leopold so lugubriously recounts. Despite his yearnings, Leopold simply could not leave nature alone. On a certain lake, he says: "You are seized with an impulse to land, to set foot on bearberry carpets, to pluck a balsam bed, to pilfer beach plums or blueberries, or perhaps to poach a partridge from out those bosky quietudes that lie behind the dunes."[24]

Leopold presents nature as a resource for people in part because to do otherwise would be hypocrisy. The land serves us by providing the basis for agriculture, meat to fill bodily cravings, beauty to satisfy our aesthetic cravings, experiences to fill our need for adventure, and moral lessons to help us live ethical lives. Biological diversity provides a diversity of functions for humans; even in protecting it and mourning its loss, we learn a lesson about what it means to be fully human.[25] Furthermore, as Nash suggests, it is possible that Leopold's ecological conceptions of the land organism led him to revere a more abstract system over its component creatures; the death of any one organism could certainly be sustained by the system.[26]

As other biologists would after him, Leopold perhaps desired some pure ideological state in which the extension of ethics to the land community was the inevitable, natural course of social evolution. Yet he realized that even he had been unable to live the gentle life he advocated. Furthermore, as with some of the biologists who would follow him, Leopold's feet were planted in separate camps. He attempted to reconcile the conflicting demands of objectivity and subjectivity, analytical thinking and deep emotion, lifeless scientific prose and the poetic musings of the naturalist, academic distance and engagement in conservation. He had few role models available to him. His great predecessor John Muir had eschewed the label *scientist*, scientific though his ecological observations were.[27] Leopold charted new ground for science. His arguments—both the facts and the feelings—come from his science, broadly interpreted. Leopold's aesthetics, too, are informed by his understanding of ecology and evolution. He asks:

> What value has wildlife from the standpoint of morals and religion? I heard of a boy once who was brought up an atheist. He changed his mind when he saw that there were a hundred-odd species of warblers, each bedecked like to the rainbow, and each performing yearly sundry thousands of miles of migration about which scientists wrote wisely but did not understand. No "fortuitous concourse of elements" working blindly through any number of millions of years could quite account for why warblers are so beautiful. No mechanistic theory, even bolstered by mutations, has ever quite answered for the colors of the cerulean warbler, or the vespers of the woodthrush, or the swansong, or—goose music. I dare say this boy's convictions would be harder to shake than those of many inductive theologians.[28]

Note here that it is an anonymous boy, not necessarily Leopold, who gets religion from observing and understanding wildlife. Leopold clearly identifies with the boy, but he is not letting on whether *he* has got religion from nature, only that he knows it to be a reasonable, even desirable, response, one he wishes to help evoke.[29] Leopold knows one can find religion in nature, and one can find it more fully with science abetting the search, even if scientists are sometimes mistaken in their interpretations. Understanding ecological systems even helps us understand ethical systems:

> The extension of ethics, so far studied only by philosophers, is actually a process in ecological revolution. Its sequences may be described in ecological as well as in philosophical terms. An ethic, ecologically, is a limitation on freedom of action in the struggle for existence. An ethic, philosophically, is a differentiation of social from anti-social conduct. These are two definitions of one thing. The thing has its origin in the tendency of interdependent individuals or groups to evolve modes of co-operation. The ecologist calls these symbioses. Politics and economics are advanced symbioses in which the original free-for-all competition has been replaced, in part, by co-operative mechanisms with an ethical content.[30]

Leopold the ecologist possessed an ecological understanding of ethics, and he evolved an ethic grounded in ecological laws. Ecology led him to realize that we needed a new ethic. Ecology helped him understand how an ethic works in society. And ecology provided him with the specific planks from which to construct that ethic. From ecological science comes ecological conscience.

Leopold extended ecological frontiers, both where ecologists could go—what paths were proper for them to pursue—and how far the principles and values taught by ecology would spread into the hearts, minds, and ethical codes of his fellow citizens. His ultimate goal was to translate what scientists knew and felt into words that would encourage everyone to get to know and feel the same things. He was not the first—nor, as we shall see, the last—to be well aware that how we think of and feel about nature determines how we treat it. He knew that individuals could shape the thoughts and values of others, as such sportsmen as Theodore Roosevelt, through their words, "created cultural value by being aware of it, and by creating a pattern for its growth." Leopold tried to create cultural value, to plant the seed around which a new ethic would crystallize: "What conservation education must build is an ethical underpinning for land economics and a universal curiosity to understand the land mechanism. Conservation may then follow." Stimulate universal curiosity and get people out into nature—especially as amateur scientists—and what had been revealed to Leopold and his fellow ecologists might be revealed to everyone.[31] People would see the true value of the land, and their conservation conscience would spring forth.

Leopold realized that ecological facts and economic valuations alone do not a conservation ethic make, since "no important change in ethics was ever accomplished without an internal change in our intellectual emphasis, loyalties, affections, and convictions. The proof that conservation has not yet touched these foundations of conduct lies in the fact that philosophy and religion have not yet heard of it. In our attempt to make conservation easy, we have made it trivial."[32] Leopold sought to change this state of affairs—to incorporate the deep value of the biota into philosophy and religion, as modern conservation biologists are trying to do in new (and in Leopold's day, impossible) ways.

"It is inconceivable to me that an ethical relation to land can exist without love, respect, and admiration for land, and a high regard for its value. By value, I of course mean something far broader than mere economic value; I mean value in the philosophical sense," Leopold writes.[33] Diving woodcocks in a Wisconsin woodlot or chattering parrots on a Mexican mountain were charismatic stand-ins for the more prosaic, more difficult to understand, organisms and processes that we must come to value. Through stirring evocations of beautiful vertebrates and through his reflexive descriptions of his own rapt reactions, he laid the groundwork for inculcation of the values we must come to understand:

> I have purposely presented the land ethic as a product of social evolution because nothing so important as an ethic is ever "written." Only the most superficial student of history supposes that Moses "wrote" the Decalogue; it evolved in the minds of a thinking community, and Moses wrote a tentative summary of it for a "seminar." I say tentative because evolution never stops.

The evolution of a land ethic is an intellectual as well as emotional process.[34]

Leopold fostered cultural evolution toward a land ethic by attempting to convert as many individuals as possible. He aimed to instigate a process of social evolution in which everyone would choose to agree with him. Curt Meine portrays Leopold as a conservative man, not prone to demagoguery; so conversion had to be a social choice, abetted by facts learned about nature and beauty seen in nature, rather than a process mandated by law. Nor was conversion a cosmic transformation that occurred when people got into nature and discovered the values or the ethic toward which we are "naturally" evolving. That is to say, the expanded circle of moral concern Leopold desired was not teleological or inevitable, but merely desirable as a freely chosen option, albeit one required for survival.

Like later conservation biologists, Leopold urges us to share the values he as a biologist understands and we as individuals and as a species need. He appeals to our intellect with ecological facts and to our emotions with beautiful pictures colored, too, by ecological science. He does not lay out all the elements of the ethic in itself; rather, he paves the ground for what we should think when we get into nature, so that we may replicate for ourselves the process he went through to devise his own deeply held ethic.

The way of thinking he would have us replicate requires ecological understanding; contrary to his own words, he had, in fact, to write some of the concrete tenets. At the core of the land ethic he proposed is a now-famous dictum: "A thing is right when it tends to preserve the integrity, stability, and beauty of the biotic community. It is wrong when it tends otherwise."[35] Things are to be judged as "right," not because they mesh with cosmic precepts Leopold discovered, but rather because they contribute to the continued ecological functioning and evolutionary unfurling of the land, its creatures, and the humans that are interdependent with the land and its creatures. Leopold's ethic is imbued with biological wisdom; it is an ecological and evolutionary ethic in two senses. First, one must understand ecological and evolutionary biology to appreciate nature properly, and thus understand the ethic; and understanding it helps ensure ecological and evolutionary permanence for other organisms and for ourselves.

Leopold uses all the tools in his biologist's kit—the collection of values, images, and facts a biologist is privileged to witness, shape, and feel by virtue of his prolonged contact with nature—to get us into the natural world. He provides us with guidelines to help us see nature as he does, for ecological literacy (which encompasses both facts and values stemming from the science) is a necessary precursor to feeling the land ethic. He would shape our very perception, for "that thing called 'nature study,' despite the shiver it brings to the spines of the elect, constitutes the first embryonic groping of the mass-mind towards perception. . . . To promote perception is the only truly creative part of recreational engineering."[36] With our expectations shaped by the power of Leopold's evocative images, the power of nature may guide us to derive a personal ethic like his, which we shall not merely understand with our intellects but feel with our souls.

"Recreational development is a job not of building roads into lovely country, but of building receptivity into the still unlovely human mind."[37] These words, which conclude my edition of *A Sand County Almanac*, sum up the book's intent. As others would after him, Leopold sought to engineer our perceptions. He did so as a scientist. His unique view derived from a professional perspective that simultaneously scientized and sacralized the natural world. He deployed the amalgam of natural values his career as a scientist and his avocation as a sportsman-naturalist had revealed to him to prompt people to discover for themselves what he exalted. With our perceptions so prepared, we may be profoundly transformed; our "unlovely minds" will be made more beautiful; and we shall not merely understand but *feel* the land ethic that Leopold evokes in *A Sand County Almanac*.

CHARLES S. ELTON

In tracing the arcadian and imperial strands of ecology's history, Donald Worster casts British ecologist Charles S. Elton (1900–1992) as something of a villain: Elton led twentieth-century ecology down the less desired "imperial" path, paving the way for a public conception of nature calculated in unappealing economic figures and measured by energy meters.[38] Categories Worster presents as dichotomous may in fact be interpenetrating: Leopold took Elton's ecological systems and biotic pyramids and made them fundamentals in his land ethic. And when Elton began to tackle conservation, he had clearly been influenced by *A Sand County Almanac*.[39]

In *The Ecology of Invasions by Animals and Plants* (hereafter *Invasions*), which appeared in 1958, Elton attempts to derive conservation lessons from decades of ecological research. He describes Earth's six biogeographic realms, where magnificent, unique flora and fauna developed in splendid isolation over the course of millennia. The peripatetic species *Homo sapiens* has eliminated the boundaries that separated them, Elton laments, and as a result the natural world has become more homogeneous, less attractive, and more dangerous: "It is not just nuclear bombs and wars that threaten us, though these rank very high on the list at the moment: there are other sorts of explosions, and this book is about ecological explosions."[40]

Elton devotes a large part of *Invasions* to graphic illustrations of organisms running, swimming, flying, and crawling amok when transplanted by humans to places where they had previously been absent. The European starling took North America by storm after a bevy was released in New York's Central Park; from five individuals imported to Czechoslovakia, the North American muskrat exploded through Europe. Freed from biological controls, imported plants, animals, and microorganisms frequently outcompete native organisms, parasitize native hosts, change the shape of native landscapes, and otherwise wreak havoc on native ecosystems.

Remedying this situation "depends very much on our attitude to wild life and to nature in general."[41] Like Leopold, Elton puts ecology to work to make that attitude more attuned to ecological facts and values. Elton, too, sought a reasonable ecological ethic to which many would assent, so that people would opt to conserve what he had dedicated his life to studying. An environmental ethic, according to Elton, can be justified three ways:

> The first, which is not usually put first, is really religious. There are some millions of people in the world who think that animals have a

right to exist and be left alone, or at any rate that they should not be persecuted or made extinct as species. Some people will believe this even when it is quite dangerous to themselves. Efforts to control plague rats in some Indian warehouses have sometimes been frustrated because the men in charge put out water for the rats to drink. Ideas of this sort will seem folly to the practical Western man, or sentimental.[42]

One need not read too far between the lines to suggest that Elton did not see this as sentimental at all. We may assume it to be his own belief when he says, "You may think the astonishingly diverse life of the globe was not evolved just to be used or abused, and perhaps largely swept away."[43] Like many conservation biologists who would come after him, he most likely accepted that wild species had intrinsic value, a right to continued existence.

Like those conservation biologists, Elton was also a pragmatist, however, and realized that effective environmental ethics should be rooted in anthropocentric appeals. So his second rationale for conservation is "aesthetic and intellectual": "nature—wild life of all kinds and its surroundings—is interesting, and usually exciting and beautiful as well. It is a source of experience for poets and artists, of materials and pleasure for the naturalist and scientist. And of recreation. In all this the interest of human beings is decidedly put first."[44]

Yet even these reasons may not convince the hardheaded pragmatist. So Elton matches that pragmatism: "The third question is the practical one: land, crops, forests, water, sea fisheries, disease, and the like. This third question seems to hang over the whole world so threateningly as rather to take the light out of the other two. The reason behind this, the worm in the heart of the rose, is quite simply the human population problem." Accordingly, "it is no use pretending that conservation for pleasure or instruction, or the assigning of superior rights to animals will ever take precedence over human survival. Nor should it."[45] Elton does not propose banishing people from natural reserves; rather he would incorporate nature into everyday life, use it for human betterment.

Elton was also a forerunner in putting ecological science to work "to harmonize divergent attitudes," in "looking for some wise principle of co-existence between man and nature, even if it has to be a modified kind of man and a modified kind of nature. This is what I understand by *Conservation*."[46] Much of *Invasions* documents a global upset in the balance of nature. In the book's final two chapters, Elton uses ecological theory to highlight the positive correlation between global diversity and global stability. By harnessing his readers' interest in the latter, he may be able to get them interested in the former.

Elton's final chapter, "The Conservation of Variety," reveals his ecologist's slant on the world. He is not concerned with individual species, so much the focus of conservation efforts then and later. Rather, he is interested in a forerunner of biodiversity, "ecological variety," which he treats as a concrete entity requiring our concern and protection. What counts is the dynamics of the ecological community, the innumerable complex interactions that keep systems stable. The ideal state of nature is diverse wilderness, epitomized by the tropical rain forest. But Elton recognizes that exploding populations over the globe are banishing the wilderness, and so it comes down to this: "If the wilderness is in retreat, we ought to learn how to introduce some of its stability and richness into the landscapes from which we grow our natural resources."[47]

Balance, stability, richness, diversity, and complexity: these qualities good in human communities are said to be good in ecological communities, too. Ecology has a history of its practitioners reading their desires into nature and then reading those desires back out to form the basis for a conservation ethic. It would be several decades after Elton before ecologists would suggest that it's a jungle out there: balance has been superseded by chaos. With the advent of conservation biology, new traits would be read in nature, and the conservation ethic would be adjusted accordingly.

Elton's conservation ethic in *Invasions* aims at practical ends that are both ambitious and modest. He was a political scientist: he helped form the British Nature Conservancy and then used it to lobby for conservation goals. His broad goal was "the keeping or putting in the landscape of the greatest possible ecological variety — in the world, in every continent or island, and so far as practicable in every district." He supported a form of restoration ecology too: ultimately, "there is a prospect of being able to handle our biological affairs by the better planning of habitat interspersion and the building up of fairly complex plant and animal communities."[48] Of course, this ambitious project would require the ecologist's steady guiding hand.

As an example of practicable measures satisfying all rationales for conservation and requiring ecological expertise and guidance, Elton argued for the conservation of the natural variety in Britain's roadside and farm hedges, which "form, as it were, a connective tissue binding together the separate organs of the landscape." Noting that these corridors preserve a large expanse of beautiful and useful plants and animals in what seem to be stable communities, Elton writes:

> From now on, it is vital that everyone who feels inclined to change or cut away or drain or spray or plant any strip or corner of the land should ask themselves three questions: what animals and plants live in it, what beauty and interest may be lost, and what extra risk changing it will add to the accumulating instability of communities. That is: refuge, beauty and interest, and security. This outlook may enable us to put into the altered landscape some of the ecological features of wilderness.[49]

The Ecology of Invasions by Animals and Plants attempts to change established attitudes, to forge new ethics. As a pragmatist, Elton consigned wilderness to the patron saint of lost causes. He was willing, instead, to settle for the preservation or re-creation of remnant scraps of ecological variety in the managed landscape. Ecology provides the scientific facts and the scientific fears that will prompt others to see the desirability of his prescriptions, along with an appealing view of nature untrammeled: a realm where beauty and stability would reign if only we would allow it. Elton the ecologist also warns of the perils that face us if we ignore his theories and prescriptions. He sought to put his decades of scientific research to work to change the way people viewed the landscape, so that they would change the landscape that they viewed.

RACHEL CARSON

Wildlife biologist and best-selling author Rachel Carson (1907–64) also used ecological science to change the way we see and feel.

Her early works focused on the wonders of nature in general and the marine environment in particular. By her indelible descriptions, she wished her readers to see what she saw and feel what she felt. In *Silent Spring* (1962), she channeled Elton's ecological ideas toward a more potent and political end than he did.[50] *Silent Spring* documents the terrifying threat uncontrolled use of pesticides posed to nature's beauty and human health—a world threat that could be overcome only by a thoroughly ecological worldview.

Silent Spring's phenomenal success transformed ecology from an obscure science to a cause célèbre. As with Leopold and Elton, Carson's attempts to make her readers see the world as she saw it pivoted on the tenets of scientific ecology: "For each of us, as for the robin in Michigan or the salmon in the Miramichi, this is a problem of ecology, of interrelationships, of interdependence." It was critical that we understand the expertise of ecologists, even though that expertise was not always impressive. Ecologists knew so little about the mechanisms by which chemicals make the spring silent and about how nature's elaborate web of organisms interact: "No one yet knows" what the effects of DDT will be; the effects of herbicidal destruction of the Western sage lands "are largely conjectural."[51]

Still, ecologists must be heeded, because they know more than anyone else—they introduced us to the complex interrelatedness that characterizes living systems—and they promise to know more in the future. Ecologists know enough to be afraid and to be humble: "Life is a miracle beyond our comprehension, and we should reverence it even where we have to struggle against it. . . . The resort to weapons such as insecticides to control it is a proof of insufficient knowledge and of an incapacity so to guide the process of nature that brute force becomes unnecessary. Humbleness is in order; there is no excuse for scientific conceit here."[52] This ecologically informed fear and humility underlies the book's persuasive power.

In *Silent Spring*, scientists are the enemies who have begotten the chemicals that beget death and destruction: "The concepts and practices of applied entomology for the most part date from that Stone Age of science. It is our alarming misfortune that so primitive a science has armed itself with the most modern and terrible weapons, and that in turning them against the insects it has also turned them against the earth."[53] But scientists are also the sources of salvation, creating alternatives to those chemicals and revealing hitherto obscure interrelationships within nature and human relationships with nature. For Carson, ecologists repay the debt of science to society by forging a worldview for us all to adopt so that we may, before it is too late, come to appreciate the services natural variety performs for us.

Like Leopold and Elton, Carson realized that citizens needed to view the world differently.[54] People took nature for granted; it was just there, omnipresent, unvanquishable. Carson further popularized the trend among ecologists of reifying nature as a concrete entity, in her case as "natural variety." Nature grew more tangible; it became a commodity that could be exchanged, valued, lost, gained, depleted, restored, quantified, scientized. This reification was essential if we were to hang a price tag, or a broader value tag, on nature. Only through its loss do we become aware of nature's existence; only thus are we able to see it as some *thing* that could be lost in the first place.

Carson's brief, arresting first chapter, "A Fable for Tomorrow," sets the stage for her dramatic argument and for this reification that has been pivotal in arguments on behalf of biodiversity conservation. We visit a bucolic setting: "There was once a town in the heart of America where all life seemed to live in harmony with its surroundings." This American paradise is paradisiacal in large part because of the splendor of its natural variety. maples blaze, birds soar, wildflowers bloom, foxes bark, and trout spawn. Paradise is lost, however, because "some evil spell had settled on the community."[55] The evil spell—the rain of death brought by pesticide spraying— sickens the residents, silences the songs of birds, and destroys the natural variety that previously would have been thought indestructible—and thus would not have been thought of at all.

Once commodified as a thing apart, we can think about natural variety's worth to us and talk about preserving it. Thus, for example, Carson discusses "a matter of great scientific importance—the need to preserve some natural plant communities." This is desirable "as a standard against which we can measure the changes our own activities bring about. We need them as wild habitats in which original populations of insects and other organisms can be maintained, for . . . the development of resistance to insecticides is changing the genetic factors of insects and perhaps other organisms. One scientist has even suggested that some sort of 'zoo' should be established to preserve insects, mites, and the like, before their genetic composition is further changed."[56] With a nod to Elton,[57] she intertwines two novel concepts: genetic diversity is essential to ecological health, and it is a commodity that humans can value and exploit. Therefore it must be preserved.

Carson derides those who see wildflowers only as "weeds" and insects solely as "pests."[58] In her worldview, even the most humble organisms merit ethical consideration, as do the "threads that bind life to life." Carson would have us join her in adherence to an ethic in which each species, all species, and the largely unexplicated forces that control the interactions between species deserve moral consideration and enduring conservation. The right to exist belongs to our fellow species, and the right to enjoy these species belongs to humans: "To the bird watcher, the suburbanite who derives joy from birds in his garden, the hunter, the fisherman or the explorer of wild regions, anything that destroys the wildlife of an area for even a single year has deprived him of pleasure to which he has a legitimate right." She rails against senseless spraying that reduced a favorite stretch of road to something "to be endured with one's mind closed to thoughts of the sterile and hideous world we are letting our technicians make. But here and there authority had somehow faltered and by an unaccountable oversight there were oases of beauty in the midst of austere and regimented control—oases that made the desecration of the greater part of the road the more unbearable. In such places my spirit lifted to the sight of the drifts of white clover or the clouds of purple vetch with here and there the flaming cup of a wood lily."[59]

Carson's reverence for nature is as clear as her moral stance: "The question is whether any civilization can wage relentless war on life without destroying itself, and without losing the right to be called civilized." As she crusaded against toxins, she also crusaded for a way of looking at the world that would foster conservation and so enrich our lives. For we *need* nature, as both ecological support system and provider of beauty and wonder: "Who has decided—who has the *right* to decide—for the countless legions of people who were not consulted that the supreme value is a world without insects, even though it be also a sterile world ungraced by the curving wing of a bird in flight? The decision is that of the authoritarian temporarily entrusted with power; he has made it during a moment of inattention by millions to whom beauty and the ordered world of nature still have a meaning that is deep and imperative."[60]

To possess a worldview where this kind of conservation is, literally, conceivable requires a moderate degree of ecological literacy. This is part of Carson's mission: to scientize the way we view our surroundings. Like Leopold and Elton, she must teach us ecological tenets so that we more fully understand our dependence on nature and more fully appreciate the beauty of nature. To share her ecosystems view, we are shown some of the intricacies of interrelatedness in air, water, and soil: like "variety," these relationships must become tangible, fragile, and important to us. Even humans are a bundle of complex ecological relations, since "there is also an ecology of the world within our bodies. In this unseen world minute causes produce mighty effects; the effect, moreover, is often seemingly unrelated to the cause, appearing in a part of the body remote from the area where the original injury was sustained."[61]

This new worldview requires an understanding of the delicate ecological balance created by eons of evolution: "It took hundreds of millions of years to produce the life that now inhabits the earth—eons of time in which that developing and evolving and diversifying life reached a state of adjustment and balance with its surroundings." As we saw earlier, along with this propitious state of "balance" comes the similarly desired "stability." Both accrue from variety: "Nature has introduced great variety into the landscape, but man has displayed a passion for simplifying it. Thus he undoes the built-in checks and balances by which nature holds the species within bounds."[62] For Carson, nature left alone serves up ever more ecological variety, which in its unfathomed intricacy provides our world with ever more ecological stability. Humans diminish variety

(a commodity to be cherished) and thus remove the checks and balances, yielding a world that is less stable, and therefore less pleasing
and more dangerous.[63]

All science is, in part, a social construction, and the science
Carson, Leopold, and Elton (and those who came after them) used to
cajole their readers reflected the values of the cajolers. The vast and
poorly understood relationships ecologists and conservation biologists study leave much room for interpretation, and the relationship
between the biologist and dwindling nature is in any case a complicated and emotional one. In hindsight, we can see that a healthy
dose of Carson's conservation values modified the scientific "facts"
that changed the way millions of her readers viewed the world. The
uses of the term *biodiversity* likewise cannily reflect the inextricability of biology facts and biologists' values; the term's inventors
make their arguments in such a way that if the ecological tenets
they advance should prove untenable, firm foundations for their
conservation arguments will nonetheless have been laid.

DAVID EHRENFELD

During the 1960s and 1970s, the growing reification of diversity
kept pace with perceived threats to the natural reality to which the
term referred. Adherents flocked to the ethic of Leopold, Elton, and
Carson. The counterculture adopted *ecology* as a protest-movement
buzzword. Widened awareness of species extinctions led to protective legislation in the United States in 1966, 1969, and, especially,
1973, when a law with teeth—the Endangered Species Act—was promulgated with scant congressional opposition.[64]

Biologists sustained their push to promote broad, deep awareness of the consequences of diversity's loss. During the 1970s, David
Ehrenfeld promoted the role of the biologist as "advocate for the
natural world" in two books, *Conserving Life on Earth* (1972) and
The Arrogance of Humanism (1978).[65] Ehrenfeld (who has a medical
degree and earned a doctorate studying sea turtles) sought simultaneously to bring the ecological perspective to a broad audience,
to promote nonanthropocentric reasons for diversity conservation,
and to get as many people as possible to experience nature for them-

selves, so that they would not need a host of convoluted rationalizations to become convinced of the values of biological diversity.

In *Conserving Life on Earth*, Ehrenfeld manifested the growing politicization of biologists, deriding the failures of legislators and federal regulators and balancing criticisms of overpopulation with admonitions against the wasteful ways of Westerners. As a unifying theme, he chose the grim specter of diminishing diversity, which he saw both as a potent symbol of humanity's arrogant loss of perspective and as a frightening crisis in its own right: "Like the Beast in the Jungle, in Henry James' great short story of that name, the central problem we must recognize and confront is not a faceless terror of the future; it is here among us, unseen but scarcely hidden to those who look, and it has been with us quite some time. For now, we must be content to call the Beast the loss of irreplaceable diversity." Ironically, this author dedicated to moving away from viewing components of nature as usable resources for humankind quite starkly portrays diversity as a usable commodity. The growing conception of diversity as a commodity, and an important one at that, is evident: "The great tragedy of the Green Revolution as currently pursued is that it tends to destroy the very diversity that it and the world need to survive and prosper."[66]

Like many other biologists, Ehrenfeld felt it imperative that people see and feel for themselves what ecology (in both the scientific and popular senses) was all about. *Conserving Life on Earth* was based on his 1970 college textbook *Biological Conservation*, but it was aimed at a more general audience. Ehrenfeld realized the paradox of what he was attempting, as "writing about biological conservation in the 1970's is a little like advertising color television on black-and-white screens: one can assert, persuasively, how beautiful and rich the colors are, but acceptance of the idea is still an act of faith on the part of the inexperienced audience. Here I can only hope that this book provokes some people into seeing for themselves what all the shouting is about."[67]

Like others, he wants his readers to get out in nature: the deeper, and to Ehrenfeld, the more important, reasons for conservation are only divined that way: "Who knows the world so well that he can say that the scientific, 'objective' reasons for saving alligators are ultimately more important than the emotional, 'subjective' ones? The former are heavily emphasized in this book, but only because the latter are gained by personal acquaintance, not by reading intellectual expositions." This "other kind of conservation is hard to explain or justify in writing, at least to those who haven't already experienced and felt the meaning of diversity."[68]

Ehrenfeld wants people to move from the "resource" school to the "holistic" school of conservation, to adopt a perspective corresponding to the "B" side of Leopold's "A-B cleavage."[69] He wants us to stop viewing the contents of nature as a stockpile of goods and services that, if conserved at all, await exploitation at some future date. By *holistic*, Ehrenfeld means "the complexity of ecological relationships and the high degree of connectedness binding together the biological world, the atmosphere, the surface of the earth, the fresh and salt waters, and the artifacts of human civilization," which, when properly grasped, lead us to "a humility and caution in the face of great forces dimly understood." As did Leopold, Elton, and Carson, and as would future conservation biologists, Ehrenfeld deduces from the known tenets of ecology and from what we don't know—our "ecological ignorance"—the admonishment to fear.[70]

Despite Ehrenfeld's personal, biocentric reasons for valuing diversity, he devotes most of *Conserving Life on Earth* to more pragmatic arguments for diversity's value, understanding that such a compromise cannot be avoided in modernist society.[71] While ecology is part of the supremely humanistic institution of science, and the corresponding "scientism," which got us into our environmental mess, it is nonetheless essential for viewing the world holistically and thus supporting holistic conservation.[72] Only through understanding ecological relations can we fully appreciate the value of biological diversity.

Ecology got Ehrenfeld into trouble, although considering his views on the fragility of our humanistic mastery of nature, that probably did not surprise him. Like Leopold, Elton, and Carson, he relied heavily on the tight interlocking between biological diversity and community stability. This theme pervades his book, although he would later describe it as "vexing and embarrassing," noting that the diversity-stability hypothesis "turned out to be a rallying point for conservationists who wished to justify with scientific reasons their original emotional desire to protect the full richness of Nature, including the apparently useless majority of species."[73]

Even anti-humanists make humanist mistakes. What Ehrenfeld only touched on briefly in *Conserving Life on Earth*—"the vague but growing feeling that the wholly anthropocentric perspective is not always the wisest and best"[74]—he delves into with great gusto in *The Arrogance of Humanism*. Here Ehrenfeld launches a diatribe against our slavish devotion to humanism, "a supreme faith in human reason—its ability to confront and solve the many problems that humans face, its ability to rearrange both the world of Nature and the affairs of men and women so that human life will prosper."[75] The twin gods of science and technology reign as the apotheoses of humanism. Ehrenfeld commits apparent apostasy in condemning our overweening adherence to the very system of knowing the world in which he has been so extensively trained.

In his chapter "The Conservation Dilemma," Ehrenfeld rails against the predominant tendency to derive ever more obscure monetary values for species. Like Leopold, Ehrenfeld understands that most species possess no conventional value; at least in our present state of knowledge, we lack the ability to dice, grind, or splice most organisms into some product that can be bought or sold. This does not stop biologists and others interested in conservation from inventing resource values for species. Ehrenfeld would terminate this practice, as "rationalizations being what they are, they are usually readily detected by nearly everyone and tend not to be very convincing, regardless of their truthfulness." Rather than support more research into the true resource potential of diversity, Ehrenfeld would scrap the whole system that considers worthy of conservation only the economically gainful; he rejects the mind-set where only what is *"logical, practical"* merits our conservation concern.[76]

Even arguments that we should preserve nature because of its beauty suffer from humanistic arrogance; if something has value because it is beautiful, that beauty is nevertheless in the eye of the beholder. We must look beyond anthropocentrism. Citing Elton's religious basis for conservation, Ehrenfeld would have us adhere to the most unscientific view that some value just *is*. He would have us adopt what he calls "The Noah Principle," named for the sage who did not consider how "useful" a species was when he gathered its members onto the ark: "This non-humanistic value of communities and species is the simplest of all to state: they should be conserved because they exist and because this existence is itself but the present expression of a continuing historical process of immense antiquity and majesty. Long-standing existence in Nature is deemed to carry with it the unimpeachable right to continued existence."[77]

This biologist opposed to humanism, a worldview that underwrites much of the scientific enterprise, does not abandon science altogether. In 1987, Ehrenfeld was pivotal in launching conservation biology, a bold endeavor to marry science with nonhumanistic values that have traditionally eluded most scientists but have been dear to many conservationists. As founding editor of the discipline's flagship journal, *Conservation Biology*, Ehrenfeld would integrate a variety of value perspectives as part and parcel of a normative, political science. He saw to it that alongside the charts and graphs readers might expect to encounter in a scientific publication, they found discussions of conservation education and values. When Ehrenfeld and his peers reveal both the pragmatic and the value-laden beliefs underlying their conservation yearnings, "whatever the outcome, we will have had the small, private satisfaction of having been honest for a while."[78]

INVENTING BIODIVERSITY

In the 1970s and 1980s, ecologists developed detailed arguments for the economic and ecological value of biological diversity. Norman Myers's *The Sinking Ark: A New Look at the Problem of Disappearing Species* and Paul and Anne Ehrlich's *Extinction: The Causes and Consequences of the Disappearance of Species* graphically depict the severity of the biological diversity crisis.[79] Each outlines the gamut of reasons, from economic and ecological to aesthetic and ethical, why biological diversity has supreme value for all of us.

While they present many pragmatic arguments for biological diversity's importance, both books operate from their authors' deeply felt belief that humans have no right to eliminate other species.

Myers has revealed to me his personal commitment to the intrinsic value of other forms of life.[80] Yet in *A Wealth of Wild Species: Storehouse for Human Welfare* (1983), he delivers a comprehensive explication of the economic rationale for biological diversity conservation. He ruefully notes, "However much I may agree that every species has its own right to continued existence on our shared planet, I do not believe that the world yet works that way."[81]

The Ehrlichs, too, are driven by deeper motives: "This is fundamentally a religious argument," they write. "There is no scientific way to 'prove' that nonhuman organisms (or for that matter, human organisms) have a right to exist—it is rather an ethical view held by a portion of humankind that includes Ehrenfeld, the great English ecologist Charles Elton, and many others who have been concerned about conservation. . . . Along with many other ecologists, we feel that the extension of the notion of 'rights' to other creatures—indeed, even to such inanimate components of ecosystems as rocks and land forms—is a natural and necessary extension of the cultural evolution of *Homo sapiens*."[82] Myers and the Ehrlichs extend the reach of science further into the realms of politics, economics, ethics, and religion. They borrowed from other teachers of how to put biology into service for conservation; they, in turn, would become teachers for others.

In the 1980s, a new discipline, conservation biology, was formalized. Its founders aimed to unite ecology and evolutionary biology with praxis to conserve biological diversity. Conservation biologists describe their discipline as "mission-oriented." Their mission is not merely to document the deterioration of Earth's diversity but to develop and promote the tools that would reverse that deterioration.[83]

The National Forum on BioDiversity, sponsored by the Smithsonian Institution and the National Academy of Sciences, proved a critical juncture in conservation biologists' mission. Founded in 1863, the NAS endures as a much-respected nongovernmental nonprofit institution. Its mandate is to monitor and promote the scientific enterprise. Current members elect new members, chosen from the upper echelons of scientific achievement. Its reputation is for extreme conservatism, particularly in matters of scientific authority and objectivity: science must be kept above the fray of politics and squabbles if its word is to carry the considerable weight of objectivity, truth, and value-neutrality.[84]

It is ironic, then, that the term *biodiversity* and the politics it has engendered sprang from this august and cloistered institution.

In the mid 1980s, we meet Walter G. Rosen as a senior program officer for the Board of Basic Biology, a section of the Commission on Life Sciences, which itself is a subdivision of the National Research Council (NRC). The NRC is the advice-giving arm of the NAS, dispensing wisdom on scientific matters of national importance. Rosen had arrived at the National Academy via a circuitous route. Branching out from the field of plant physiology, he had taught courses on subjects such as "Technology and the Biosphere" during the heady days of the counterculture 1960s at the State University of New York at Buffalo. Later, as a dean at the University of Massachusetts at Boston, Rosen attempted to create a college focused on recognizing human dependence on the natural environment. This, he says, has been the centering vision of the disparate strands of his career. During a sabbatic leave, he worked for the Office of Toxic Substances at the Environmental Protection Agency, using his knowledge of plant physiology to trace the environmental impacts of toxic chemicals on plants. His sabbatical became permanent, and eventually he moved to the NAS, where he remained until the late 1980s.

The Board on Basic Biology advises the NAS on biological issues that deserve priority research focus. According to Rosen, the issue "that kept rising to the surface was the loss of biological diversity, the increasing frequency of extinctions." Board members also wished to do more to inform the public at large on this issue. Rosen "took these two ideas and put them together and said, 'Why don't we have a forum on biodiv—on biological diversity?' . . . they bought that because it was directly responsive to what they said they felt there was a need for."

The idea of a national forum on the issue crept up the review system at the NAS, a process Rosen calls "ponderous and cautious." Along the way, it was suggested to Rosen that he invite the Smithsonian Institution to cosponsor the forum. In response, the Smithsonian's secretary, Robert McCormick Adams, enthusiastically committed his organization to the project. The Smithsonian, as Rosen puts it, "is massive, and it does things in different ways. And it does some things on a very large scale. And that's what happened here." The forum's dimensions expanded: press kits were dispatched, press briefing sessions were held, an eye-catching event poster (which eventually became the cover of the resulting volume of conference proceedings) was created, and the Smithsonian put together a traveling poster exhibition reflecting the theme "Diversity in Danger." Prominent biologists such as E. O. Wilson and Peter Raven were involved early on.[85]

Perhaps the most potent recruiting tool to emerge from the conference was the term *biodiversity*. Rosen first penned the neologism as convenient shorthand. He recalls, "It was easy to do: all you do is take the 'logical' out of 'biological.'" He means this ironically: "To take the logical out of something that's supposed to be science is a bit of a contradiction in terms, right? And yet, of course, maybe that's why I get impatient with the Academy, because they're always so logical that there seems to be no room for emotion in there, no room for spirit." Wilson at first opposed the term as "too glitzy," but he now admits that Rosen "turned out to be completely right." The glitziness of the word contributed to the speed of its spread, and Wilson told me that he is pleased *biodiversity* is now so "thoroughly ensconced."

I asked Dan Janzen, who was invited by Wilson to speak at the forum, why *biodiversity* has become such a potent buzzword in the conservation movement. "Several reasons," he replied: "One is deliberate. The Washington Conference? That was an explicit political event, explicitly designed to make Congress aware of this complexity of species that we're losing. And the word was coined—well different people get credit for coining the word—but the point was the word was punched into that system at that point deliberately. A lot of us went to that talk on a political mission. We were asked, will we come and do this thing? So we did."

The National Forum on BioDiversity almost failed to come off. Rosen says he "walked on eggshells in seeing to it that the idea was kept alive. . . . While I was all for it, I was also looking over my shoulder all the time and thinking that the Academy might shut us down." According to Rosen, the NAS was "very, very upset about" the possibility that the forum might "turn into an exercise in advocacy. . . . From their point of view, the Academy owes its reputation for objectivity and what have you to the fact that they're *not* advocates." One portion of the forum was to be about the value of biodiversity, by which they meant the economic value: "But when they saw the word *value*, they got very concerned that that implied advocacy and, you know, love of biodiversity."

Eventually it came to imply just that. The forum did go on, from 21 to 24 September 1986, and all that the guardians of objectivity at the Academy might have feared came to pass. It was here, as Laura Tangley notes, that biologists who loved biological diversity came out of the closet. Development experts, economists, and even ethicists and theologians joined the impressive array of biologists to discuss the biodiversity crisis.

They also came to *create* the biodiversity crisis, at least in the minds of the press, the politicos, and the public. The forum, Rosen said, was "an exercise in consciousness raising,"[86] and a publicity juggernaut was launched to achieve that goal. A group of eminent biologists in attendance (Jared Diamond, Paul Ehrlich, Thomas Eisner, G. Evelyn Hutchinson, Ernst Mayr, Charles D. Michener, Harold A. Mooney, Peter Raven, and E. O. Wilson) christened themselves the Club of Earth and held a press conference to broadcast the importance of biodiversity. "The species extinction crisis," they proclaimed, "is a threat to civilization second only to the threat of thermonuclear war."

About 14,000 people attended the forum. Stories about the event appeared on the front pages of major newspapers "above the fold," Rosen recalls. A teleconference was beamed to over 100 venues, and as many as 10,000 people viewed the live event from sites around the world.[87] The volume of papers from the conference edited by Wilson has become a best-selling title of the NAS Press. A videotape of the teleconference, interspersed with stirring images of the natural world, sold out. A review in *Ecology* predicted that "*BioDiversity* is bound to enlist new believers. . . . Its companion, *BioDiversity: The Videotape*, provides further enlistment of future crusaders."[88]

For Rosen and the Board on Biology, the forum "accomplished exactly what was hoped for, and did so to a greater degree than was ever expected." As a consciousness-raising event and media spectacle, it succeeded beyond the organizers' dreams. At one stroke, "the biology and the focus of biodiversity was recognized as a concern of a large array of disciplines," Wilson remarked to me. Rosen received scant recognition from his superiors at the NAS, but his efforts may have signaled new directions for the role of its scientists in public advocacy. For example, since the Forum on BioDiversity, a panel of scientists with the NAS's imprimatur has issued strong evidence and concomitant warnings on the impending peril associated with global climate change.[89] The boundaries between "science" and "society" may never be drawn the same again, at least by the scientists who have steadfastly demarcated the terrain.

The Forum also signaled a new direction in conservation thinking and action. According to Wilson:

> "In 1986, there wasn't any word or simple phrase that could capture the broadened sweep of concerns represented at the Forum, and which were soon thereafter to coalesce into a new direction in the international conservation movement, and even as a discipline. So that biodiversity studies, or biodiversity issues, however you want to phrase it, so that the forum came to be not just about the biology of the origination of diversity and extinction, but also all of the other concerns, through ecology, population biology, and in the most novel development, economics, sociology, and even the humanities. So in one stroke, the biology and the focus of biodiversity was recognized as a concern of a large array of disciplines."

In the few short years since the National Forum on BioDiversity, the term has been promoted vigorously; it has been transformed from a bit of scientific esoterica into a buzzword of popular culture. In 1988, *biodiversity* did not appear as a keyword in *Biological Abstracts*, and *biological diversity* appeared once.[90] In 1993, *biodiversity* appeared seventy-two times, and *biological diversity* nineteen times.

The new term pops up in newspapers and magazines on a regular basis. Conferences on the topic of biodiversity have become commonplace. Environmental groups use it as a focal point in lobbying and fund-raising efforts (e.g., "For as you know, understanding and protecting biodiversity is what the Nature Conservancy is all about").[91] Biodiversity was one of the major items on the agenda at the Rio Earth Summit, where the United States was lambasted for refusing to sign the treaty dedicated to biodiversity's protection. By giving money and land and exerting pressure to protect biodiversity, the public has joined biologists in attempts to sculpt the physical, political, and normative landscape to diversity's needs.

Many biologists had always loved the Earth but had repressed their desire to translate that love into action. Rosen provided a forum, a vent for the forces that rumbled under the surface, and a rallying flag that has since been hoisted innumerable times in defense of the Earth. *Biodiversity* burst forth as the right word, symbolizing the right idea at the right time for biologists who were certain they were right to cross the safe boundaries of their hermetic disciplines to defend Earth's biotic bounty. With biodiversity, the efforts of Leopold, Elton, Carson, and others found fruition: biology, nature, and the way we conceive of the natural world would never be the same again.

WHY & WHENCE

THE TERM

BIODIVERSITY?

 3

Since the 1986 National Forum on BioDiversity, many conservationists have switched from acting on behalf of *endangered species* (which itself was a switch from other terms, such as *nature* and *wilderness*) to promoting *biodiversity*. Shifts in rhetoric, particularly in the fractious world of conservation, are seldom accidental. We must ask: why *biodiversity*? Why has this neologism proven so successful in attracting concern, financing, and action for conservation? To whom does it speak, and what does it say to them?

BIODIVERSITY AND WILDERNESS

The term *biodiversity* serves biologists' purposes better than *wilderness*. Like the idea of nature, Max Oelschlaeger points out, "the idea of wilderness is whatever anyone or group cares to think."[1] Such nonspecificity may ill serve conservation advocates, even though, as we shall see, the biodiversity concept faces the same charges.

Furthermore, the value—even the existence—of wilderness is far from universally recognized. In *Wilderness and the American Mind*, Roderick Nash warned: "Friends of wilderness should remember that in terms of the entire history of man's relationship to nature, they are riding the crest of a very, very recent wave."[2] As such, *wilderness* lacks roots deep enough to support a broad, flourishing conservation movement. Although posited as a universally important environmental paradigm, *wilderness* in reality fixes a particular time in a particular culture. Ramachandra Guha depicts wilderness as an obsession of rich Northerners with too much time, and power, on their hands.[3] Environmentalists who foist wilderness conservation on poorer nations, Guha suggests, enforce a kind of siege mentality in local people and divert attention from more pressing environmental problems.

And wilderness, that vast unspoilt territory beyond the tainting reach of humans, may not really exist. More and more, historians and scientists suggest that what we have traditionally viewed as wilderness has, rather, resulted from long periods of human management. William Cronon and William Denevan have shown how the "wilderness" supposedly encountered by early European settlers had actually resulted from Native American land management. Susanna Hecht and Alexander Cockburn reveal that the Amazon, so often thought of as the vastest, most pristine wilderness on earth, has in reality been intensively managed for generations. Carol Kaesuk Yoon reports that at Costa Rica's La Selva Biological Station, researchers are turning up pottery shards and crop residues that point to past civilizations where until recently we had imagined only wilderness. Bill McKibben argues in *The End of Nature* that even if places beyond the sullying hand of humans have endured until recent times, they are now gone: global warming and toxic spew render quixotic the search for such pristine realms.[4]

A number of scholars point out that plotting conservation around wilderness is a dubious strategy. When we prize only the pristine, we establish a dichotomy in which we preserve a small amount of undefiled nature while leaving the rest open for any and all to despoil.[5] Still, the idea of wilderness has powerful appeal in some quarters and is cited even by those who argue for biodiversity. Thomas Eisner "can get emotional over the prospect of wilderness and its primordial state forever disappearing, being one of the worst possible consequences of what we're doing to the planet." Reed Noss sees "amateur conservationists, using the term *biodiversity* in very, very loose ways. And what they're really interested in is wildness— wild areas, natural areas."

I am not arguing against wilderness preservation. But as a conservation strategy, particularly one biologists can control, it is redolent of class privilege, culturally rooted, and ontologically precarious. A strategy based on biodiversity may eventually be judged similarly; but, as we shall see below, it has many strengths and co-opts much of what makes the idea of wilderness so attractive. Desire for habitat and ecosystem protection may be a proxy for wilderness preservation, the desire for untrammeled vast reaches; these can be smuggled in under the biodiversity concept.

And wilderness may figure in biodiversity in another sense. Wilson argues that "humanity needs an unending frontier, an ability of unlimited promise." He declares that "biodiversity is the frontier of the future."[6] For Wilson, biodiversity is a vast wilderness, an uncharted, wondrous realm in which to indulge the human senses. We are urged by many biologists to preserve wilderness in the traditional sense, so that they, and we, may explore the wilderness of biodiversity. Perhaps Noss is right: "Wilderness and biodiversity need each other"[7]—both as concrete phenomena and rhetorical strategies.

BIODIVERSITY AND CULTURAL DIVERSITY

Within the academy and out in society, canons and prototypes have given way to pluralities and multiplicities: we laud diversity everywhere. Some biologists who boldly assert that biodiversity is a normative good associate the claim with the more widely familiar one that cultural diversity is a normative good. As biologists link themselves with the forces promoting the multicultural ethic that has made normative and political headway in our society, different kinds of diversity thus become symbiotically and metaphysically linked in inherent "goodness."[8]

Deep ecologist Bill Devall states that "one of the ultimate norms based on a deep ecology insight is that diversity has worth. . . . Divisiveness between feminists and male philosophers, or prejudice against homosexuals, non-male, or non-female individuals distracts us from the real work."[9] For biologist Peter Raven, a true understanding of ecological interdependence makes us appreciate social interdependence:

> "If the world is one unity, one place that we all live, one place that supports us, in which biodiversity is the leading force, then I don't think you could even dream of inhumanity to other people. I don't think you

could dream of the kind of runaway greed that characterizes our society in the United States. I don't think you could turn your backs on the cities. I don't think you could even fantasize not giving blacks the kinds of—not allowing blacks the kinds of advantages that everybody else has. I don't think you could really be inhuman to women. I don't think you could be inhuman to children. I don't think you could really have the elderly grabbing all the money because they have all the power. I don't think any of those things would be possible if you really once and for all understood that you're all living on one single world and that's all there is and that parts of that have to fit together."[10]

Of course, strategies for preserving both cultural and biological diversity are linked, as are the forces making both disappear.[11] Environmentalists can tap into political ferment on behalf of indigenous land rights; indigenous peoples can hop on the biodiversity bandwagon. All can seek the same ideal, harmony through variety: harmony within and between ecosystems, within and between cultural systems, and harmony of humans coexisting peaceably within ecosystems. For example, *Global Biodiversity Strategy: Guidelines for Action to Save, Study, and Use Earth's Biotic Wealth Sustainably and Equitably*, a document prepared in conjunction with the Rio Earth Summit, states that "cultural diversity is closely linked to biodiversity. Humanity's collective knowledge of biodiversity and its use and management rests in cultural diversity; conversely, conserving biodiversity often helps strengthen cultural integrity and values."[12]

Some biologists express the value of cultural diversity, or at least the value of indigenous groups said to be the keepers of knowledge about biodiversity. These people are perceived as resources whose allegiance must be won if they are to serve as guardians of biodiversity, as biologists wish them to do. Or they are posited as "naturally" serving as protectors of biodiversity and thus must be valued and protected so that they may fulfill this role. Some advocates of cultural diversity deride biodiversity advocates as imperialist ecofascists, insensitive to the needs of indigenous groups.[13] The appeal to cultural diversity by biodiversity advocates may be a response to these critiques, showing how the goals of the two groups are mutually reinforcing, not mutually exclusive.

The linkage of cultural and biological diversity symbolizes a shift in conservation thinking and strategy. As *Global Biodiversity Strategy* suggests, conserving biodiversity "entails a shift from a defensive posture—protecting nature from the impacts of development—to an offensive effort seeking to meet people's needs from biological resources while ensuring the long-term sustainability of Earth's biotic wealth."[14] We need no longer think of "conservation and development" as antithetical; rather, we proclaim their indissoluble linkage.

Note that love of diversity in all its forms is sometimes taken as a sine qua non of human experience. Philosopher Bryan Norton alludes to the human love for all kinds of diversity. Hugh Iltis echoes him, but with a twist: this love of diversity is biologically based and expressed in broad, deep biophilia.[15] This thing called "diversity" has thus been reified: a previously intangible or abstract concept has been made into a definable, graspable entity. One can promote it as a concrete good and can work with it, mold it, speak for it; a reified diversity can be encompassed within the broad realm of biologists' expertise.

Other biologists, including Thomas Lovejoy and David Ehrenfeld, suggest that we do not yet value diversity of all kinds enough, but must learn to do so.[16] I find Ehrenfeld's position paradoxical. While pleading with us to come to value diversity, he argues that we unfortunately live in an age of generality. This infects ecological science, which Ehrenfeld claims is now governed by the search for patterns and overriding theories of systems at the expense of particular knowledge of specific places or organisms: "Generality is power," he proclaims.[17] Knowledge and love of bio*diversity*, by Ehrenfeld's reckoning, must result from a branching, an amalgam, of individual expertise and appreciation. We need to appreciate and study specificity to come to know and love and save diversity. As I see it, the biodiversity idea bridges this tension between focus on the specific and on the general. It builds on a base of specifics of knowledge about individual species and places—and conservation efforts to save those species and places—yet also includes knowledge (and attempts to save) ecological pattern and process.

In this book, I often use the word *political* the way sociologist of science Bruno Latour uses it: "By politics you mean to be the spokesman of the forces you mould society with and of which you are the only credible and legitimate authority."[18] Yet in at least one case, someone has reaped more prosaic political benefits by articulating biological to cultural diversity. Seeking office as a people's deputy in the first elections under *perestroika* in the former Soviet Union, Professor Nikolay Vorontsov seized upon the Russian-American geneticist Theodosius Dobzhansky's emphasis on diversity in natural populations and extrapolated that polymorphous diversity had similar importance in economic, cultural, and political life. If Russians wanted to save their country, Vorontsov asserted, they had to focus on diversity in all sectors. This played well in a society where uniformity and monopoly had harmed public and private life. Vorontsov won, and he later became minister of nature conservation.[19]

DEFINITIONS OF BIODIVERSITY

I began nearly all my interviews with conservation biologists by asking for a definition of *biodiversity*. Here are some of their responses:

PETER BRUSSARD: "The standard definition is species diversity, and then diversity of communities or habitats that the species combine into, and then, on the other side of the scale, the genetic diversity that the species are comprised of." Other comments: "I think buzzwords have a way of going through here. I [look at] *biodiversity* as a more inclusive term than even *habitat* or *ecosystem*."

DAVID EHRENFELD: "I don't have a definition of *biodiversity*. I've tried very hard to stay away from formal definitions. When I deal with it in the journal [*Conservation Biology*, of which he was the founding editor]—and it's one reason I don't much like the word—it obviously means to some people species diversity; other people expand that to include populations. To other people it means really genetic diversity, heterozygosity, allelic diversity, often within populations. To many people, it means variety of ecotypes or ecosystem types, landscape types. Obviously, it's all of those things. But mostly when I think of biodiversity, I think of plain, ordinary species diversity. And by the way, I don't really value it, value the term, as highly as some people do. I think it's one of those wonderful catchwords like *sus-*

tainable development, that, because it's vaguely defined, has a broad appeal, like motherhood."

PAUL EHRLICH (from a 1992 paper with Anne Ehrlich): "The variety of genetically distinct populations and species of plants, animals, and microorganisms with which *Homo sapiens* shares earth, and the variety of ecosystems of which they are functioning parts."[20] In interview: "To me, biodiversity is the living resources of the planet."

THOMAS EISNER: "Biodiversity is the total number of genetic lineages on earth. I just made that up; if I think about it, chances are I'll change my definition rapidly." "I usually don't use it globally, I use it regionally. It's a convenient term to basically encompass anything that's living in a given area, of which most things are unknown, so you're including the known and the unknown . . . not only all that's described, but everything that's there."

TERRY ERWIN (in a 1991 paper): biological diversity (which he uses interchangeably with *biodiversity*) "is in fact the product of organic evolution, that is, the diversity of life in all its manifestations. Biological diversity is thus holistic, and this is indicated in the very nature of the word's root meanings; it is the sum of earth species including all their interactions and variations within their biotic and abiotic environment in both space and time."[21]

DONALD FALK: "I think of it as fundamentally a measure of difference. And the most important aspect of the definition for me is that it exists at many different levels of biological organization, even though we tend as a mental habit to focus most on species diversity. That's only one, and in many respects not even the most interesting level of diversity. So I guess I would describe it as the dimension of difference at multiple levels of organization." "And so *biodiversity* might not be the best term in an absolute sense. But it is the best way that we have of explaining that we're talking about all these different entities and all the different processes. And process is as integral to the concept of diversity as the object we see at any given moment."

JERRY FRANKLIN: Biological diversity (for which *biodiversity* is "just a shorthand") is "the complete array of organisms, biologically mediated processes, and organically derived structures out there on the globe." "And it very definitely, as far as I'm concerned, goes way beyond individual species. It goes beyond genetic variance."

VICKIE FUNK: She laughed when I asked for her definition. When I said it was the hardest question of the whole interview, she replied, "It really is, and I'm sure you get long, rambling answers and explanations. Because this is something we've gone round and round about." "I guess . . . *biodiversity* to me means importance: which areas do we have to concentrate on, or which groups are more 'important' in terms of preserving than others." "I tried to find a definition for *biodiversity* years ago and just couldn't come up with one. And so I don't have one. And I've asked, I don't know how many people, you know, Lovejoy and other people here, what their definitions are, and they don't have one either." She asserts that different disciplinary perspectives give different definitions and different considerations of conservation priorities.

HUGH ILTIS: "Well, it's just the diversity of living things on the face of the earth." "It's the number of species and the uniqueness of species."

DANIEL JANZEN: "The whole package of genes, populations, species, and the cluster of interactions that they manifest."

K. C. KIM: It is "the variation or the variability or the variety of living organisms. . . . , which includes intraspecific variation. . . . You're looking at the community level, you're looking at ecosystem level, at landscape level, and so on."

THOMAS LOVEJOY: "The term is really supposed to mean diversity at all levels of organization. But the way it's most often used is basically relating to species diversity. The way it relates to species diversity used to mean just the number of species and their relative abundance and various measures of it." "I think for short operational purposes, that *species diversity* is good shorthand. It's not the whole thing, but as you're rushing around trying to do some things, it's the most easily measured, and it's the one at which the measures are the least controversial. But you're really talking about more than that. You're talking about the way species are put together into larger entities and you're talking about genetic diversity within a species."

JANE LUBCHENCO: "Biodiversity encompasses the number, the variability, and the variety of life on Earth. And it's usually considered at three different levels of biological organization: genetic, species, and ecosystem."

S. J. MCNAUGHTON: "The total biotic diversity as indexed by the number of species and genetic diversity they encompass."

REED NOSS: "Well, *biodiversity*, to me, is shorthand for all the richness of life. You know, when I give lectures on biodiversity, I define it simply as the variety of life and its processes. . . . By itself, it's vague enough as a concept that it can be misconstrued. It can be deliberately misinterpreted by people who say, 'Oh, we're increasing biodiversity by putting clear-cuts on the landscape.' And so I try to head that off right away by . . . describing some things that biodiversity is *not*. And giving some examples of how diversification at one scale, say by clear-cutting a pristine forest landscape, may in fact lead to impoverishment at a broader scale as you lose sensitive species from the landscape and add nothing but cosmopolitan weeds. So that's how I describe it. But again, when I think about biodiversity, the images I get are just *everything*. And you know, it includes to me related concepts like ecological integrity, which I think go along with it. So I think it's an appropriate organizing concept for what it is we're interested in protecting." Biodiversity "by definition forces us to consider many different species, not just vertebrates, not just plants, but things that we don't see or understand nearly as well, such as bacteria and fungi. And the processes that tie these all together, the interrelationships between them. I think it forces us, if we do justice to the concept, it forces us to think much more broadly than we have in the past."

GORDON ORIANS: "*Biodiversity* is a very comprehensive term which reflects the diversity of living organisms at all levels, from . . . populational—genetic and geographical—diversity to species, to lineages, and higher taxonomic categories, to ecological systems."

DAVID PIMENTEL: The term *biodiversity* "really focuses on the mix of species or diversity—I'll use the word *diversity*—of species that exist in a particular ecosystem—or world, for that matter, which takes the whole system."

PETER RAVEN: "The sum total of plants, animals, fungi, and microorganisms in the world including their genetic diversity and the way in which they fit together into communities and ecosystems."

G. CARLETON RAY: "[The biologist Otto Solbrig] starts out by saying that it's a fundamental property of life. And what we're looking at in biodiversity is the history of biology, the history of life in all of its forms over the entire time it's existed on our planet." "And one of the things I think it should *not* be seen as—that's the best way to approach it, I think—is just an accounting of the number of species. So you could talk about species richness, which is different than just 1-2-3, obviously. Or you could talk about [it] from molecules to ecosystems, as Otto Solbrig has put it."

WALTER ROSEN: "I'd rather not [try to define *biodiversity*]. I'm content with the definitions that are out there. . . . simply noting that biodiversity is something that occurs at a community level, at a species level, you know, that it's genetic as well as—whatever. That it manifests itself in more than one way."

MICHAEL SOULÉ: There are "only two ways to define *biodiversity*. . . . The short way is the best. And it's life in all of its dimensions and richness and manifestations, not only at the level of individuals and species, but at the level of aggregations, communities, or what have you." "You always have to explain to people what it is. And by the time you're done, their eyes have glazed over. Because they don't care, most people don't care about genetic variation within species and the difference between that and species diversity. And the difference between that and alpha, beta, and gamma diversity. . . . So that's why I take a more intuitive approach. I mean, the way I do it with a public group is basically to say, 'Say you're standing in (some place that's familiar to them). And then you walk over into the canyon or to the forest. And then you walk out into the meadow. And you're hearing these sounds and you're seeing these organisms. And that's biodiversity. And then if you went to another continent, it'd be a lot different, but you'd still recognize it as life.' "

E. O. WILSON: "Biodiversity is the variety of life across all levels of organization from genic diversity within populations, to species, which have to be regarded as the pivotal unit of classification, to ecosystems. Each of these levels can be treated, and are treated, independently, or together, to give a total picture. And each can be treated locally or globally."

DAVID WOODRUFF: "I take a very holistic view of biodiversity. It's made up of individuals that make up populations that make up species. And . . . most species cannot survive without other species around them. So I have a definition of biodiversity that runs all the way from individuals and species, the traditional elements, through to community-level relationships, which are essential for the survival of individual elements. And from there you go along towards the way those systems affect the atmosphere and the oceans of the planet. And although I don't believe in the goddess Gaia and the Gaia model, I think that there are elements of physics that make the planet habitable that are entirely due to biodiversity. That's what I mean by a holistic view."

Most of these definitions reflect biological totality. Only the definitions of Iltis and Pimentel restrict biodiversity to the realm of

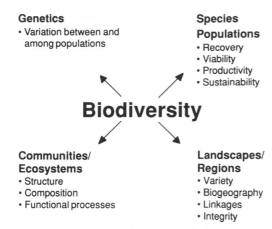

Genetics
• Variation between and
  among populations

Species
Populations
• Recovery
• Viability
• Productivity
• Sustainability

Biodiversity

Communities/
Ecosystems
• Structure
• Composition
• Functional processes

Landscapes/
Regions
• Variety
• Biogeography
• Linkages
• Integrity

FIG. 3.1 Salwasser's (1991) "focal elements for ecosystem management."
Used with permission of Island Press.

species diversity. Otherwise, these biologists use the term *biodiversity* to represent multiple levels of biological hierarchy—genes, populations, species, communities, ecosystems, as well as the interactions among levels, and the processes that have given rise to them. Biodiversity is "the whole package of genes, populations, species, and the cluster of interactions that they manifest" (Janzen). It is "the living resources of the planet" (Ehrlich). It "is shorthand for all the richness of life" (Noss), "the history of life in all of its forms over the entire time it's existed on our planet" (Ray). It is "the sum of earth species including all their interactions and variations within their biotic and abiotic environment in both space and time" (Erwin); or "life in all of its dimensions and richness and manifestations" (Soulé). "You're talking about a subject that is literally as large as the world itself," Falk reminds us.

Published definitions can be similarly broad; few restrict biodiversity's breadth. Hal Salwasser prescribes an "approach to conserving biodiversity through ecosystem management" (fig. 3.1).

Figure 3.1 suggests the confusion surrounding just what ecosystems represent to different biologists, and the confusion about what such a broad definition of *biodiversity* might mean. Gordon Kirkland, Ann F. Rhoads, and K. C. Kim say that biodiversity is "not only the total of all native species in a particular region (species richness), but also the ecological roles of these species in ecosystems, and the genetic diversity contained within species populations."[22] *Global Biodiversity Strategy* calls biodiversity "the totality of genes, species, and ecosystems in a region," and tells us that "human *cultural diversity* could also be considered part of biodiversity."[23] Raven's article "Defining Biodiversity" does so as "a shorthand way of referring to the world's living endowment."[24] It is a broad banner, this biodiversity.

ECOLOGY, EVOLUTION, AND CHANGING CONSERVATION GOALS

The semantic evolution from *endangered species* to *biodiversity* parallels an evolution in how biologists view the world, with concomitant evolution in what they want preserved. And biodiversity's breadth resolves (at least temporarily) a number of prickly problems that arise from narrow focus on endangered species.

First, biologists tell us that species extinction, like death and taxes, is an unfortunate inevitability. "Nor is it likely that very many wonderful and admirable species can be saved from extinction, if for no other reason than that they always have been accustomed to doing each other in," notes the historian Joseph Petulla. When we assert that all species have a right to life, we fall, according to John Tierney, into a "fundamental contradiction: The species being preserved are alive today only because others have died."[25] The layperson responds with the obvious question: Why should we be trying to stop processes that would proceed without us anyway? Can't the Earth do without these species?

Reed Noss has a simple answer: "99 percent of all species that ever lived are now extinct. But I think we have an obligation now, in our generation, and in foreseeable generations, to try to protect every species, try to maintain every species, because virtually every species that is going extinct now is going extinct due to human activity, not because of natural processes." Robert May illustrates this more graphically: "Contemporary rates of speciation are of order 1 million times slower than rates of extinction. Were speciation rates plotted on the y-axis on a graph 10 cm. high, then on the same scale extinction rates would require an x-axis extending 100 km."[26] While such responses convince some people, the "naturalness" of species extinction remains prima facie plausible to others. By shifting the emphasis away from species, biologists avert this problem.

Other naysayers, led by Julian Simon, an economist and arch-opponent of conservation, opine that biologists widely and wildly inflate estimated rates of species extinction. Let's see the hard evidence, they demand.[27] Some biologists join them in calls for greater caution in making these estimates.[28] Although some still play the numbers game (see below), the switch to a strategy based on biodiversity can make species counts superfluous; instead, conservation comes to focus on pattern, process, and multiple hierarchical levels.

This debate treats "species" as unproblematic, concrete entities. They are not—at least, not always. Rather, like biodiversity, the species concept is a construction that blends the abstract with the concrete.[29] In *Conservation Biology*, Martha Rojas writes that "there is no agreement on what species are, how they should be delimited, or what they represent."[30] Different definitions of what constitutes a species have different implications for conservation, and each definition poses problems for recalcitrant taxa such as plants and asexual organisms.[31] Current debates in U.S. conservation swirl around whether, for example, the Florida panther and the red wolf are distinct species or merely hybrids.[32]

Such distinctions need not eliminate these animals from conservation consideration, according to Thomas Dowling et al.: "As conservation biologists, however, we must continue to protect entities like the red wolf to preserve the evolutionary potential of each species."[33] Such entreaties beg the question, though: if the government is mandated to save endangered species, and no one is quite sure what a species is, or if a given organism is or is not a member of a distinct species, then how are we to proceed? Vickie Funk recalls the furbish lousewort controversy, where "somebody came along and said, 'Ah, it's not a species, it's a subspecies.' [Makes game show buzzer sound.] You know, there goes the whole justification for saving the area they wanted to dam, because now it's not even a species anymore."

It should be noted, however, that the Endangered Species Act (ESA) requires protection of subspecies as well. Subspecies are geographic subdivisions of species that some biologists feel exhibit sufficient physical or genetic differences to merit some kind of taxonomic distinction, even if those differences do not confer full species status. A widely overlooked subplot of the sprawling northern spotted owl saga reveals that this bird is not even a full species, but a mere subspecies.[34] If biologists find it difficult to sort out species, imagine the difficulties in deciding what merits subspecies status; some biologists have simply abandoned the concept.

The environmental ethicist Holmes Rolston III sums it up: "It is difficult enough to argue from the fact that a species exists to the value judgment that a species ought to exist—what philosophers call an argument from *is* to *ought*. Matters grow worse if the concept of species is rotten to begin with. . . . Perhaps species do not exist."[35] Targeting species as the exclusive level of conservation becomes dubious practice when species turn out not to be species, and when that is not necessarily what we want to save anyway. We attempt to avoid such problems by moving to the broad aegis of biodiversity, while still retaining some of the features that have made species an appealing focus for conservation.

Earlier, I cited Ehrenfeld's contention that we can get to an appreciation of generality, of diversity, through a better understanding of the specific, of the local. Alan Pound, a biologist at Costa Rica's Monteverde Cloud Forest Reserve, told me he thought the focus of conservation should be on "maximum uniqueness—preserving what makes that place special." Through conservation endemicity, we arrive at global variety, or maximum global biodiversity. Norton echoes this: by allowing ecosystems to undergo succession to maturity, diversity decreases within habitat, but specialized assemblages within the habitat develop. Such endemic uniqueness contributes to global biodiversity.[36] Thus *biodiversity* can be viewed as an overarching term for multiple local endemicities.

Biologists and allied environmentalists lure public support for aggressive conservation policy with the bait that species are a tapped and untapped lode of economic benefit. (This is discussed more fully in Chapters 5 and 6.) "And there are so many species. How many geese that lay golden eggs are there apt to be in that number?" philosopher Elliott Sober asks. Noss holds that "[the economic benefit] argument becomes very strained when you try to apply it to all species. It is not, just not a credible statement that every species out there may hold some curative to cancer." This view is shared by Ehrenfeld, Brussard, and Orians, among others; Aldo Leopold held it a half century ago.[37] But just as in the ecosystem-function argument discussed below (and allied with it—for economic value need not be direct chemicals or food sources; it may be the free ecosystem services species provide), if we know that some species confer economic value (and we do know this), then it makes sense to strive for protection of all species and the natural matrix that nurtures them—a.k.a. biodiversity.

For diversity begets diversity: the more species we have globally and locally, the more species we shall have in the future—and so the more opportunities for human gain. This works in ecological time through processes of invasion and colonization. Following a natural disturbance, colonizing species must come from a nearby stock; the more diverse the stock, the greater the chance of a complex, tightly integrated colonized community. It works in evolutionary time through competition, co-adaptation, and all the processes of natural selection as organisms hone their niches—become increasingly specialized—and more species cluster in a given area.[38] Thus diversity itself becomes an important commodity to preserve if we wish to harvest the maximum possible yields from nature.

If diversity engenders diversity, then loss of a single species diminishes the prospects for creation of additional species. Furthermore, we should be concerned with downward spiraling of species. Eisner points out that due to co-adaptations, when one species disappears from a given place or from all places, its symbionts and co-dependents disappear as well. Myers admits that some species loss is acceptable, but then asks, "When do species losses shift from being marginal to becoming significant, serious, critical, crucial, and catastrophic?"[39]

Paul and Anne Ehrlich metaphorically describe this as the "rivet-popping" problem. Imagine yourself on an airplane about to take off. You see someone crawling along the wing, slowly removing one rivet after another. The plane can certainly fly without some of the rivets, but eventually the wing will fall off. By analogy, the wing represents the biotic processes of Spaceship Earth (what many now think of as biodiversity), and each rivet stands for a species or other functional component. The Ehrlichs argue that to preserve the wing intact, we must fight to save each rivet. Similarly, Lovejoy refers to the "incrementalist problem": actions that seem perfectly reasonable on a case-by-case basis eventually lead to environmental catastrophe, the collapse of the system's ability to support life.[40]

" 'Ought species x to exist?' is a single increment in the collective question, 'Ought life on Earth to exist?' " Rolston notes. Sober warns us of the "slippery slope argument": we might all agree that biological diversity as a whole is good and necessary, but once we admit that any given species is dispensable, the exigencies of modern life will permit it to be lost, and then another, and then another.[41] Once we set foot on the slippery slope, we slide till we hit bottom. Ecologically speaking, allowing the loss of one species can lead to utter biotic impoverishment. Philosophically speaking, once we admit that it is morally acceptable to relinquish one species, then it is a short slide to admitting that any species is relinquishable. And politically speaking, if we allow one to go, we open the door for any or all to go. But human expansion and political compromise entail that occasionally we must relinquish a species. We avoid the slippery slopes that spell peril for conservationists' arguments when we focus our efforts not on species but on biodiversity—that is, the totality whose implicit value underlies all arguments for individual species.

From the viewpoint of an ecologist concerned about maintaining ecosystem integrity, it is difficult to make a species-by-species case for conservation, because of our rudimentary understanding of the functional roles of species in ecosystems. We do know that many species have ecological equivalents: if one species is removed, other species may provide the same ecological services. Critics of species conservation can point to numerous cases in which alleged keystone species have disappeared and the system has continued to function well. If eastern hardwood forests seem to do fine without chestnut trees, formerly a keystone species there, then it makes sense that forests could do without species that have always been much scarcer: what difference could a rare orchid possibly make?[42] Here, again, we encounter the problem of incrementalism, or rivet-popping, or the slippery slope: at some point, the biologists agree, ecosystem function will be impaired and the system will degenerate. *This* is what's important, and this is addressed more effectively by focus on biodiversity rather than focus on individual species that may or may not be important to sustaining overall biotic processes.

Many conservation biologists will agree that they have some general idea of what drives a system, what provides the most crucial ecosystem functions. These are the plants and bacteria, the insects and their humble, teeming kin whose protection on a species-by-species basis proves unfeasible. Most skulk unknown, unlabeled, and unloved. If identified, they are poorly studied: who knows what they do or whether they face extinction? The few whose conservation status can be documented simply lack the cachet or the charisma to attract public and political support. Virtually no insects are currently protected under the Endangered Species Act. What Wilson calls "the little things that run the world" produce oxygen, fix nitrogen essential to crops and forests, pollinate flowers, disperse seeds, provide nonchemical pest control, and, according to biologists, make the world livable, while saving humanity billions of dollars each year.[43] Under the aegis of biodiversity, they gain clout and protection.

Next, species-by-species conservation ignores the impending threat of global warming. With an endangered-species approach, when a species is deemed threatened, biologists identify its critical habitat; when all goes well, land is set aside to safeguard that habitat. If global-warming scenarios play out as cast, such strategies will prove virtually useless as the forces of selection act to make the species "migrate" to more suitable climes—usually north in latitude or higher in altitude. If unable to migrate fast enough, or if no suitable habitat exists, species will perish. Organisms that depend on a narrow range of ecological conditions might find those disappearing; coastal organisms could be swamped by rising seas. A shift to focus on biodiversity makes it more feasible for more species to survive global warming, as emphasis would be placed on regional or landscape management for maximum perpetuation of species and ecosystem processes.[44]

Single-minded dedication to species preservation also ignores other levels of the biological hierarchy that merit conservation. Rather, we might focus on maximum genic diversity. Geographically or genetically distinct populations maintain the variety necessary for evolution to continue in the context of environmental change.[45] Communities of tightly intertwined organisms should be targeted for preservation. Or perhaps it is ecosystems that are important for conservation efforts. Biodiversity covers all these levels of biological hierarchy, as well as the interactions between them.

When biologists manage for endangered species, they first attempt to ascertain a minimum viable population (MVP) size at which a species or a population has a solid chance of surviving in the face of stochastic processes like random genetic drift or catastrophic weather events. Not only are MVP's difficult to ascertain; the MVP concept also suggests that culling can be allowed over and above that MVP. This lessens the chances of a species in the face of chance.[46]

Furthermore, development may continue unabated when no endangered species are known to be present, or when it is deemed "safe" to eliminate certain individuals of a species. G. Carleton Ray points out that some of the more valuable species for humans or for ecosystems may be plentiful, rather than rare, and that no policy exists to protect those. And species that are currently plentiful (e.g., fur seals in the North Bering Sea) may actually be declining at a precipitous rate. North American passenger pigeons once flew in flocks so vast they allegedly blocked the sun, but protection efforts only began once the bird was recognized as rare, and by then it was too late to halt the pigeon's slide to extinction.

If protection of endangered species has proven difficult and contentious in the United States, it has nonetheless had its successes.[47] But the approach becomes nearly impossible in other contexts. In ocean conservation, exploration is rudimentary, new species turn up regularly, species ranges regularly include the territorial waters of several nations, and the presence, absence, or abundance of given species fluctuates wildly: basing ocean conservation policy on the endangered-species concept is impracticable.[48] In tropical nations, the situation can be worse. Most of these countries, rich in species, lack even rudimentary species checklists. (In fact, the United States has no thorough biological inventory, either.) Endangered species protection is politically challenging and frightfully expensive, and most Southern nations lack the will, the cash, and the knowledge to do so. "I find it fascinating," writes Jerry Franklin, "that conservation biologists who would not for a moment consider a species approach to the conservation of biological diversity in tropical rainforests will defend such a strategy for the temperate regions."[49] Managing for biodiversity allows for a broader range of conservation goals and takes into account the political and ecological uncertainties of oceanic or tropical terrestrial systems.

Battles over endangered species, at least as they have been fought in this country, tend to dissolve into what Lovejoy calls "idiotic dichotomies": the northern spotted owl versus timber jobs, or the snail darter versus the Tellico Dam. The snail darter—a dangerous symbol, which is "coming back to haunt us now," as Erwin observes—was only the tip of an iceberg of reasons why the dam should not be built, but it became the poster fish of those who questioned the wisdom, goals, and even the sanity of environmentalists. "Who cares about one species of little fish [enough] to hold up jobs and billions of dollars?" Erwin asks. Especially, as Eisner admits, when "it so happens that nobody really gives a damn about the snail darter." In the view of biologists promoting it, the concept of biodiversity gets away from simpleminded debates over symbols about which biologists and other concerned parties may not care that much in the first place.

AVOIDING TRIAGE

What everyone does care about is even trickier. Our answers will depend on who "we" are. Many biologists assert, albeit reluctantly, that given the sociopolitical situation and the number of endangered species, we must make decisions about which species to save and which to let go. As Peter Raven puts it, "If everything is being lost in the tropics, for example, at least we should be making choices about studying and keeping something."

In *Noah's Choice*, Charles Mann and Mark Plummer describe the tragic choices that confront a society that wants to preserve each endangered species, yet also does not want to give up any of its amenities, such as hospitals, highways, and golf courses. Mann and Plummer urge that we devise a logical system for making the necessary excruciating decisions.[50] Yet how to implement triage—the deliberate abandonment of ecologically or politically hopeless cases to concentrate limited resources on species that have a chance to make it—is itself nearly a hopeless case. This is another entry point for the concept of biodiversity.

Of course, some biologists reject triage outright. "I think we ought to save it all, as a basic principle," Wilson asserts. Noss is unequivocal: "I reject triage. . . . It's ethically pernicious to me. . . . I think that the argument that there isn't enough money, there aren't enough resources in general to protect all species and some of them we're just going to have to let go, is disingenuous." Falk is adamant: "You cannot persuade me that we are in a triage situation with respect to the resources to save life on earth. I do not accept it. I think it's a completely fallacious argument. Just as medicine tries to save every patient they can, that has absolutely got to be our mission. If we fail, we fail."

Soulé suggests "we should *attempt* to save each and every species, but we shouldn't put the same amount of work into saving each and every species. And in some cases, we probably shouldn't devote much work at all to saving some . . . where it's hopeless." But how to choose? Which species are more valuable than others? That a number of biologists assert we must have triage, then provide no guidelines for how we go about this, itself speaks to the problem. "Let us recognize the urgent necessity of making choices among threatened species," Norman Myers admonishes us. "All species are worth saving; we simply cannot save them all." Here and elsewhere, however, Myers offers no guidelines as to how we might choose.[51]

A comment of David Woodruff's during my interview with him is telling:

"Pragmatically, we cannot save each and every species. If you force me— so then I think you have to rank or prioritize your species. I'd rather not use the word *triage*, because that means that you consciously have everything set out. You go past each stretcher and you put one of three colors on it. We don't know where the stretchers are, let alone who's on the stretchers. So instead, my attitude is to prioritize from what you do know. And then pick key species through—by working, by concentrating on those key species, you will be able to save ecosystems and larger units."

You can feel Woodruff's ambivalence about making decisions he knows must be made, while knowing that the criteria for making those decisions are not available, and that even if they were available, they would not be clear-cut. Switching rhetoric to biodiversity can help us avoid having to make these decisions, at least on a species-by-species basis.

Some, however, have begun the process of laying out how we might make choices. "It's self-evident," Colin Tudge declares. "The justification is gut feeling and common sense; and neither should be lightly put aside." Gordon Orians says that "environmental assessment and management are best served when people explicitly choose a limited number of 'valued ecosystem components' (VEC's). . . . A VEC can be a single species of economic (deer) or aesthetic (California condor) value, systems of interacting species (bees with the plants they pollinate), or an entire ecosystem (a wetland or a rain forest)."[52] This still returns us to Tudge's realm of "common sense," and my common sense may well dictate different solutions than yours; I may not even have any common sense. With the bewildering array of possibilities, how can we ever set priorities? And who are "we," anyway?

To begin to delineate priorities, Robert May and others have begun what he calls "the calculus of biodiversity."[53] May engages in intricate mathematical calculations of what to preserve based on genetic distinctions among lineages: he would have us focus conservation efforts on the "most unique" lines. Such belabored taxonomic calculations, while well-intentioned, are easy to criticize (see Ehrlich's comments in the next paragraph) and do not really seem to get at what many biologists and most environmentalists and lay folk find important either. We can avoid tedious mathematical calculations of relative species value by switching to biodiversity.

Others admit outright, or at least hint broadly at, the difficulty (if not impossibility) of making triage choices among species. Lovejoy told me he believes that something like triage may be necessary, but still finds it "inadequate to the situation. . . . Conceptually, it's

an easy way out. To actually do it intelligently would be a bitch."
Daily and Ehrlich deride such attempts as " 'crackpot rigor' (detailed
mathematical analyses of an intractable problem) or 'suboptimiza-
tion' (doing in the very best way something that should not be done
at all)."[54] Ehrlich also calls it "mental masturbation." Funk told me
she sees an intractable difficulty with triage of species and places, in
that "the problem is *be sure*. I mean, who do you trust? Who do you
believe? I mean, some government can pass a—say, 'Okay, we'll stop
putting pressure on you for this area if you'll give us that area.' But
then ten years down the road, they say, 'Oops, never mind.' What
do you consider an adequate guarantee . . . —I mean, how many
treaties did we sign with the Indians? [laughs]."

Facing the difficulty of triage, Norton concludes that "once the
problem of endangered species is looked at more broadly, as a problem of protecting overall biological diversity rather than as a problem of saving individual species, it can be formulated in a different
way: How ought insufficient funds and efforts best be spent to meet
threats to biological diversity?" Precisely. Or, as the economist Gardner M. Brown, Jr., puts it, "If species truly cannot be ranked, yet are
thought to be important, a rational prescription is to maximize the
number of species saved per period for a specific outlay."[55]

But, according to many biologists, few of us are that rational.
"It seems doubtful that humans can make unemotional decisions
on what to save," opines David Challinor, assistant secretary of the
Smithsonian Institution. Evolutionary biologists, theoretically, do
not discern higher from lower. Brussard asserts that for a conservation biologist, a butterfly is worth as much as an elephant. But few
biologists (never mind the rest of us) seem to be able to apply this
dictum of their science to the desires of their consciences. Soulé suggests a kind of free market for conservation decisions, where people
who like certain species fight to preserve them, and the diversity in
human preferences would end up saving the diversity of nonhuman
species. But here we end up with what K. C. Kim calls a "dueling
species" approach, as various groups battle for their species' priority,
rather than working together to save as much as possible.[56]

Not surprisingly, many biologists urge us to focus on preservation of keystone species, the ones crucial for sustaining functional

ecosystems (while acknowledging how little we know about what constitutes a keystone species in a given ecosystem).[57] Stephen Kellert's research might make us think twice about this approach. He has found that many people are willing to sacrifice a great deal to preserve certain species that are "large, aesthetically attractive, phylogenetically similar to human beings, and regarded as possessing capacities for feeling, thought, and pain." He calls these the "cognitively meaningful" organisms. Biologists Andrew Dobson and Robert May note that "no conservation group mourned the passing of the wild smallpox virus, and no organization speaks for the conservation of gut nematode species. This juxtaposition is an extreme example of a more general aspect of the way the conservation ethic is applied differently to different taxonomic groups. . . . Intensity of public opinion tends to rise along a continuum that matches the deliberately oversimplified characterization of 'r-selection' to 'k-selection' in life history."[58]

Some biologists decry this trend—David Western calls it a "knee-jerk response"—and, as noted above, urge us to preserve the "little things," but many are going with the flow of human nature. The arch title of a recent article—"All Animals Are Equal but Some Are Cetaceans"—speaks volumes.[59] And so some biologists support making conservation decisions to save those cute species that have hair, feathers, and/or big eyes, the ones that appeal to us emotionally, nicknamed the "charismatic megavertebrates." A few scientists believe this, not as a matter of strategy, but as a matter of principle. "Millions of new beetles do not compensate for the loss of lions, tigers, and elephants," according to the editor of *Science*.[60] Soulé looks "at it in terms of how difficult it would be to reevolve a species. And it would be awfully hard to reevolve a rhinoceros if we didn't have any rhinoceroses left. It's less difficult to reevolve some beetles if some beetles go extinct. I don't think all species are equal, by any means. I think we should work hard to protect the larger organisms because they're more extinction-prone by far. They would be very difficult to reconstruct, even if you believe in *Jurassic Park*."

The "big" organisms can also protect what biologists really want —namely, everything else, which is to say, biodiversity—in two ways. First, they act as "flagships" to rally emotional and financial support for broader conservation efforts. As Linda Graber describes this, "elite leaders attempt to mobilize mass public opinion by presenting an event or issue in terms of a cherished image; therefore, the event or issue must be simplified for public consumption. Dramatic outline virtually requires emptiness of detail." So, for example, the National Zoo chose the golden lion tamarin for captive breeding programs, not because it was the only endangered marmoset, but because they are "unbelievably beautiful." According to Russ Mittermeier of the World Wildlife Fund, the tamarin and the muriqui "have really become the flagship species for the entire region [the Atlantic forest of Brazil], and the campaigns using them as symbols are excellent examples of the way in which key groups of animals can be used to sell the whole issue of conservation, both in the tropical countries and in the developed world."[61]

Furthermore, larger organisms tend to live at the apex of the trophic pyramid and therefore require larger home ranges; so the emotionally appealing organisms are ecologically appealing too. They serve as "umbrellas": when we preserve their habitats, we also incidentally provide habitat for a host of other species.[62]

Under the aegis of biodiversity, biologists can have it both ways. Biodiversity includes the "cherished images" that the public finds "cognitively meaningful"; yet it also embraces the underappreciated organisms—the ones that often drive the system—and the system itself. Biodiversity helps raise consciousness and funds, while still remaining true to biologists' visions and priorities. It avoids the tough triage decisions biologists are not prepared to make. Setting triage priorities is the biologist's equivalent of "Sophie's Choice": recall the movie scene where a Nazi guard tells Meryl Streep that she can save either her son or her daughter, but not both. How could a biologist ever choose? She need not when she chooses to argue for biodiversity.

GETTING PAST THE ENDANGERED SPECIES ACT

Much of the conversion to biodiversity in the United States has been a countermeasure to difficulties with the Endangered Species Act. The ESA of 1973 built upon nearly a century of wildlife protection legislation, including the largely symbolic ESA's of 1966 and

1969, which regulated international trade in endangered species and authorized the secretary of the interior to focus efforts to preserve endangered species, broadly defined.[63] The 1973 act mandated that species threatened with extinction in the United States be listed and recovery plans drawn up and promulgated. It brooked no exceptions and drew no major opposition. From a legislative body that often seems gridlocked and gutless, passage of a conservation law with such bite may seem like a miracle. But the conservation movement had gained considerable momentum by the early 1970s, and certainly few people realized the major confrontations the law's implementation would incite.

The very sharpness of the ESA's teeth further abets the switch from endangered species to biodiversity. A 1992 cover of the *Atlantic* blares: "Playing God: Why we shouldn't try to save every endangered species." Why, the authors ask, do we have to save *every* one? Only because this "bureaucratic horror" demands it. They chronicle the battle faced by a man seeking to put a golf course in the last habitat of the Oregon silverspot butterfly, lamenting that "society had chosen an insect over the dream of a human being, and for the life of him Schroeder couldn't see the logic in it, or how anyone was better off for it."[64]

The Ehrlichs note that people have refrained from filing lawsuits on behalf of endangered species for fear of public backlash that might lead to congressional abrogation of the ESA. Such fears may yet be justified: by the time you read this, the ESA may either have been repealed or significantly weakened.[65] Furthermore, some groups take advantage of the act's status as the only nearly unimpeachable environmental law, looking for endangered species to prevent development they oppose for other reasons. Although a plethora of economic and ecological reasons supported stopping construction of Tennessee's Tellico Dam, only a few nondescript fish held any legal clout. Opponents of the ESA ignore the fact that nearly all endangered species disputes are settled out of court, and that the cost of administering the ESA equals the cost of one mile of urban interstate highway construction.[66] A comment by Peter Brussard may be telling here: if so many people think the law is too strict and too oppressive, "that means probably precisely that it *is* working exactly as it should." Still, Lovejoy and Lubchenco noted to me that the ESA may simply be too expensive politically, and that this helps explain why the focus has shifted to biodiversity.

Others hold more fundamental objections to the ESA. Many biologists see it as an expensive and nearly futile attempt at damage control. Orians notes that a "species-by-species basis is *not* going to work. . . . Waiting till something is in deep trouble and then declaring it endangered and then trying to do something is one hell of a way to manage the biosphere." If biologists' estimates of the sheer number of species going extinct globally every year are accurate, it is folly to attempt to save them on a species-by-species basis. Others note the expense of "emergency room conservation," even going so far as to call the "doomed" attempts to save the California condor "lunacy." Malcolm Hunter calls the ESA a "pathetic actor" that serves only to "salve our conscience." Ehrlich notes that by the time a species makes its way onto the endangered species list, it probably no longer plays much of a role in its ecosystem—in functional terms, it is already gone.[67]

For so many biologists, last-minute damage control draws financial, political, and scientific resources away from places where conservation can still make a difference.[68] David Blockstein calls the ESA a safety net "sagging under the strain of the sheer magnitude of species needing attention. . . . Its focus must be broadened from a species-by-species program to conservation of entire habitats." Hal Salwasser submits that crisis management for a small number of species is not enough; instead, we need "ecosystem management."[69]

Although few of them want to see any species let go, what biologists *really* want is preservation of habitat, of ecosystems, of relatively unspoiled territory where the pageant of evolution can continue to unfurl. "Evolution is good" is one of the normative precepts by which Soulé defines conservation biology. Ehrenfeld's "Noah Principle" attests to the glory of the evolutionary process. A number of years ago, Iltis attempted to warn us that we were bringing about "the crucifixion of the evolutionary dream."[70]

Endangered species are too easily seen as static types, rather than as mutable evolutionary units. The latter view places more emphasis on intraspecific variation that facilitates continued evolution; the biodiversity concept incorporates this emphasis. By shifting toward conserving dynamic processes rather than unchanging species, we retain the potential for "biological communities to play out their evolutionary drama."[71] Frankel and Soulé urge a switch away from "preservation," the maintenance of static entities, toward "conservation," which allows for continued evolutionary change. Bill Willers says that given how much we don't know, biologists must develop schemes for standing back and letting evolution do its thing: "Do this: Take a powerful, rock-steady stand for pure process. That failing, stand aside and refrain from neutralizing those who will." David Western also recommends managing for continued change, and Daniel Botkin would like to incorporate the view of nature as constant flux into our way of thinking; our way of conserving nature should run in parallel flux.[72]

"It is not *form* (species) as mere morphology, but the *formative* (speciating) process that humans ought to preserve, although the process cannot be preserved without its products," Rolston writes. "Perpetuating a substantial part of the evolutionary process is what future generations will expect of us," predict di Castri and Younes; biodiversity is the key to this, inasmuch as it "embraces all levels of organization from the molecular unit (and also the chemical and physical ones), to the individual organisms up to the population, community, ecosystem, landscape, and biospheric levels." Protecting evolution must be the foremost goal of conservation, Erwin believes, and a "transcultural" goal for conservation should thus be to guard the recently evolved hotbeds of speciation in any cladogram, rather than concentrating efforts on interesting endemic relics (we shall return to this in Chapter 4). Soulé holds the "spiritual" view "that each species, each entity should be allowed to continue its evolution." Orians similarly expressed "concern about, and respect for, the processes of evolution, and the richness they've produced," saying that we should not "participate deliberately in activities that unnecessarily terminate lineages, that reduce the opportunity for evolution." Brussard has worried about species but now speculates that "maybe what we really need to do is make sure we preserve genetic diversity, a big piece of it, and preserve the processes that enable this to continue rather than worrying about individual species."[73]

Biologists place eminent value on the evolutionary process. By conserving biodiversity, we conserve variation at all the levels—genetic, population, species, community, ecosystem—that allow the evolutionary process to continue, that provide the capacity for nature to change, to rebound, to evolve. Protecting endangered species may connote a commitment to keeping things the way they are; biodiversity conservation is a broad mandate for change.

HABITAT, ECOSYSTEM, AND BIODIVERSITY

The 1973 Endangered Species Act mandates provision of "means whereby the ecosystems upon which endangered species and threatened species depend may be conserved" and "a program for the conservation of such . . . species."[74] This suggests we are preserving the ecosystem to save species, rather than saving species to keep the ecosystem functioning. But most biologists attribute greater value to the ecosystem and the services it provides.

Reid and Miller suggest that "species conservation is best achieved through conserving habitats and ecosystems—not through heroic rescues of individual species." "We must use another tool, one based on a higher level of biological organization, such as natural communities, to consider species in manageable groups," Hunter declares. Scott and his collaborators say that "if biodiversity is to be saved, our focus must be on saving functioning ecosystems."[75]

But Reed Noss and Larry Harris note that the focus on individual species "neither requires (in the short term) nor guarantees (in the long term) preservation of ecosystem processes." Elsewhere, Noss notes that species are components of ecosystems that provide us with essential services. This argument's focus is blurry: does the ecosystem maintain the species that provide us with these services? Or are the species functional cogs in the ecosystem wheel that rolls out these services? Dennis Murphy is unequivocal in his support of the latter view: "The single-species focus of the ESA has not been especially successful in protecting functioning ecosystems." His comment on the "use" of invertebrate species is telling: "Given historical rates of invertebrate [endangered-species] listings, the ability of invertebrates to confer protection on ecosystems at risk seems small."[76] No matter how crucial they are to functioning ecosystems, insects and their kin do not pack much in the way of political clout.

If you doubt that biologists' debates have had an impact in political circles, consider the graphic introductory comments of Congressman James D. Scheuer (D, New York) in a hearing on the proposed National Biological Diversity Conservation and Environmental Research Act: "It is a futile exercise to spend time, energy, and money declaring a species endangered when, almost inexorably, by the time you pronounce it endangered, it's gone. It's history. The only way we can do that is by looking at an ecosystem, and then the spotted owl isn't quite so important, and then the snail darter isn't quite so important. It's the ecosystem that we're trying to save, which has a lot of other things in it besides spotted owls and snail darters."[77]

ON THE GOALS OF HABITAT AND ECOSYSTEM PRESERVATION

Most biologists I interviewed shared the view that we must shift our conservation focus from species to processes, ecosystems, and habitat:[78]

PETER BRUSSARD: "Well, the battle over the spotted owl is not really whether we need a spotted owl. It's whether we're going to preserve some tiny remnant of old-growth forest. And I think there are lots and lots of sound arguments for preserving old-growth forests. But unfortunately we don't have an Endangered Old Growth Act; we have an Endangered Species Act. So that's why we're fighting over the spotted owl. Frankly, if the spotted owl were to vanish the day after tomorrow, I doubt there would be the slightest ripple in the ecology of the forest in the big sense. The flying squirrels and wood rats might get slightly more abundant until something else came along to eat them."

DAVID EHRENFELD: "I think that the highest priority for conservation ought to be relatively intact ecosystems."

PAUL EHRLICH: "I mean, obviously, I like the spotted owl. But there's the southern spotted owl, too, and so on. If it was only the spotted owl, we'd be in a lot deeper trouble than if we were talking about all the populations and ecosystems of the Northwest that the spotted owl is now a fragile shelter for. And I guess I believe in gradually bringing the public around to an honesty in things."

THOMAS EISNER (before testifying to a congressional committee investigating reauthorization of the ESA): "I find I've got some new arguments that are holding water, that put value on the unchanged natural enclave as opposed to the spotted owl. I think it would be fairly easy to identify biodiversity with habitat. And by the way, that's something I set myself specifically for this testimony—I have five minutes to do it in—and I want to still have some real concrete examples, where—and that's not going to be easy to do—where you shift the concern to the habitat. And by so doing, to the notion that that habitat represents."

TERRY ERWIN: "I'm not interested in species at all. I'm interested in ecosystems, and habitats, and microhabitats. My recent work in the Amazon indicates that the microhabitat is the unit of home for a complex little community of critters."

JERRY FRANKLIN: "[We should be] beginning to swing more towards putting energy into ecosystems than enlarging the continuing list of endangered species." "The spotted owl is a fantastic indicator species. And functionally, I don't think it's essential to the forests at all. But what it did was run a flag up and say, 'Hey, you're running out to the end of your old-growth forests. You're getting down to the bottom of the barrel.'. . . Isn't anybody in Congress thinks this is about spotted owls anymore. It isn't. It's about forest ecosystems, anadromous fish, a list of critters about as long as your arm."

VICKIE FUNK: "I think it's really not switching to biodiversity so much as it's switching to the concept, and then *biodiversity* is a handy word for it. But the concept [is] that you can't preserve the species without preserving the habitat. I mean, look at the condor. They think they're going to reintroduce the California condor. Bullshit. There's no way the California condor is going to exist without a habitat where it can exist."

DANIEL JANZEN: "People get caught up in missions, and they look for vehicles to carry the mission forward. And species are a lot easier to hang things on than habitats are. And it's a lot easier to talk about extinction of species than extinction of habitat."

K. C. KIM: "The ultimate goal in my mind is basically the biodiversity or preservation of ecological systems."

S. J. MCNAUGHTON: "I think that the best strategy for conserving biodiversity is to recognize ecosystems that are important reservoirs of bio-diversity and conserve those in their functional totality."

REED NOSS: "[Ecosystem management] forces us . . . to think much more broadly than we have in the past. . . . We can look at all that within this broader context of how is this going to affect diversity as a whole, over a region or a continent or the globe."

PETER RAVEN: "It hasn't really been biologists who have concentrated on endangered species, I think. It's been the general public—and the politicians—and I think they're still as focused on endangered species as they ever were. And I think that's a big problem. I think they should be focused on systems and bioregionalism. . . . They have a bioregional plan now in California, which I think has some hope of success, where you look at whole systems and try to think about their preservation rather than take it all down to individual species and fighting it out."

G. CARLETON RAY: "By looking at the function of ecosystems, by trying to understand how systems work, if we lose that we've lost the whole ballgame. So this seemed to be a paramount concern as well as the most challenging scientifically."

DAVID WOODRUFF: "Biodiversity conservation to me means conserving natural systems that make the biosphere habitable, that make it what it is. And species are just one element in that."

Some (Funk, perhaps McNaughton) suggest that only by focusing on the ecosystem or habitat can we save the species, while others (Brussard, Ehrenfeld, Ehrlich, Eisner, Erwin, Janzen, Kim, Noss, Raven, Ray, Woodruff, perhaps Franklin) suggest that focusing on species misses the main point of conservation, which is to save the ecosystem. Even Funk and McNaughton elsewhere suggest that it is habitat or ecosystem that is important to them.

Of course, many want both functioning ecosystems and the species that comprise them; after all, we cannot have one without the other. Bryan Norton and his colleagues mulled over these issues, asking, "Is concern for ecosystems a mere byproduct of concern for species? Or is the preservation of species valuable mainly because it contributes to the protection of ecosystems? Indeed, one might wonder whether the value of species can be separated from the value of ecosystems. These questions were discussed at length, but no consensus was reached."[79] However, since the term *biodiversity* focuses attention both on individual species and on the ecosystems they maintain (or that maintain them), it requires no consensus to be reached.

Yet if focus on species is politically expensive and practically infeasible, and if it is ecosystem or habitat that is more important, why not focus efforts on ecosystem or habitat conservation? Why has *biodiversity* become the key buzzword in conservation circles in the 1990s? Vickie Funk worries about "tying the fate of the environment to a passing sort of concept." She'd "very much like to see it less decentralized under the concept and more in touch with what's actually involved. Like habitat preservation."

But *habitat* and *ecosystem* lack precision as scientific terms. Endangered species, for all the shortcomings of the concept, are at least (usually) definable, concrete entities. One can identify grizzly bears, count them, and prepare some plan that purports to preserve them. Bureaucrats can grasp their ontological reality. As Falk observes, "At least we can be reassured that one organism from the beginning of its life to the end is going to remain a member of the same species." Ecosystems, on the other hand, "are abstract entities, with boundaries that mesh and blend with each other," Constance Hunt notes. Orians does not think "endangered ecosystems" are the answer either, inasmuch as "we have no accepted system whatsoever about ecosystem classification. . . . Any system that you attempt to set up is going to be highly political, for a whole set of reasons." Malcolm Hunter concurs that ecosystems "are not even very tangible to ecologists, who have no precise way to define them."[80]

The problems of focusing on ecosystems do not end with definition difficulties. Ecosystems, Hunter observes, "have no big, brown eyes to endear themselves to the public." Noss offers pithily: "You cannot hug a biogeochemical cycle." Woodruff believes that "you get more leverage from focusing on charismatic megavertebrates or on a few higher plants than you do by saying, 'The next forty-eight miles of forest, or the next thirty-eight miles of grassland'—that all look the same to the general public—'are very important in maintaining life support systems on the planet.' I think people went that way because that's the way they were trained, and because they discovered they had leverage, political leverage."

Biodiversity gets you this leverage, while maintaining the life-support systems as part of the package. Rolston gives a more complicated explanation of the shortcomings of an ecosystem focus:

> Unlike higher animals, ecosystems have no experiences; they do not and cannot care. Unlike plants, an ecosystem has no organized center, no genome. It does not defend itself against injury or death. Unlike a species, there is no ongoing telos, no biological identity reinstantiated over time. The organismic parts are more complex than the community whole. More troublesome still, an ecosystem can seem a jungle where the fittest survive, a place of contest and conflict, beside which the organism is a model of cooperation. In animals the heart, livers, muscles and brain are tightly integrated, as are the leaves, cambium, and roots in plants. But the so-called ecosystem community is pushing and shoving between rivals, each aggrandizing itself, or else seems to be all indifference and haphazard juxtaposition—nothing to call forth our admiration.[81]

So what is the biologist concerned about so much more than species to do? No conservation biologist exists in a vacuum; he requires societal support for his research and conservation goals, and people may be reluctant to embrace poorly defined entities with which they cannot identify emotionally. "All human institutions are transient expedients, and the conservation systems that are fashionable today will certainly undergo many changes in the next century. Opportunism and tolerance must be the watchwords of the science, the politics, and the art of nature protection," Soulé notes.[82] The biodiversity concept allows for this kind of opportunism. Those who want habitat and ecosystems and the interconnections between different levels of the biological hierarchy will find them under the term's aegis. But since it is difficult to identify emotionally with habitats or ecosystems, the concept of biodiversity includes pandas and the other charismatic species that are real, recognizable, embraceable entities for citizens and policy makers. If it is difficult to define precisely, that is biodiversity's best asset as an opportunistic conservation tool.

With biodiversity "we can distinguish very different components and select those that are suitable for different kinds of objectives," says Reed Noss. Dan Janzen spells this out clearly:

> "People get caught up in missions, and they look for vehicles to carry the mission forward. And species are a lot easier to hang things on than habitats are. And it's a lot easier to talk about extinction of species than extinction of habitat. You get into these huge terminological arguments about whether you can ever distinguish a habitat. How do you distinguish a marsh? Well, I don't know, but you can distinguish a whooping crane. And so the single-species focus, which is a disaster for conservation, keeps coming back and gets used as a device. Well, biodiversity is a list of species in some people's minds. Its very easy to talk about it and say, 'Well, what's biodiversity?' And you list the ten species. And you say, 'Why should you care about it?' Because you should care about these ten species. So it's pragmatically comfortable and easy. And what that translates to is reporters can pick it up easily. TV can pick it up easily. And they can—you can take photographs of objects in biodiversity—i.e., species or individuals—very easily. Put them on the television set, put them on the wall and all that kind of stuff. Habitat is very hard to photograph. To most people it looks like a green blob."

The neologism *biodiversity* crystallizes a new approach, a new perspective for conservation, one that requires biologists to appeal to the media in the first place. Don Falk talks of "the increasing sophistication of the conservation movement." He says that

> "any serious biologist has always viewed endangered species as just the tip of the iceberg. That's the place where a lot of impacts on land and ecological processes become visible. It gives you something to aim at, it gives you a target. But it's not necessarily the whole objective. And so in talking about biodiversity more broadly, that is, taking in ecosystem functions, community processes, and genetic diversity within species, and so on—taking in the whole picture, it begins to approach a more, I think, a richer and a more realistic view of what we're actually trying to protect. The danger with species conservation . . . is that it's at risk of becoming isolated from these other elements of diversity—that you can literally save a species without saving the ecosystem or community."

Use of the term *biodiversity* represents greater sophistication both in how we conceive of conservation and in how we promote broader conservation goals. In his foreword to *BioDiversity*, E. O. Wilson discusses the new word, "which aptly represents, as well as any term can, the vast array of topics and perspectives covered during the Washington Forum."[83] This broadened range of multidisciplinary concerns is embodied in this consciously created new term. It is the label for a new, synthetic discipline devoted to conservation. The word represents a new approach, but not necessarily a new entity: the terms *biological diversity*, *natural variety*, and *nature* have been around for quite a while. Under the rubric of *biodiversity*, these terms are repackaged to unite amorphous, diverse endeavors in a streamlined, do-or-die conservation effort with biologists at the helm.

NATURE AND BIODIVERSITY

"Biodiversity is no frill," Noss warns us, "it is life and all that sustains life." Aplet et al. provide an apt summary definition: " 'Maintenance of biological diversity' can be thought of as another way to say 'maintenance of *everything*.' "[84] When reading and listening to the definitions biologists apply to biodiversity, it is hard to imagine what in nature does not fall under the rubric of the term.

Appeals to nature are appeals to external reality. Not only scientists endeavor to use the lability of nature to justify their claims; not only scientists act as mouthpieces for nature. Nature validates the arguments we all make to others in a heterogeneous society. It is only "natural" for humans to engage in free trade, communism, democracy, cooperation, competition, and so on. Rather than being the transparent carrier of external reality, *nature* becomes a rhetorical vehicle that authorizes a speaker's purpose.

This is not a lapse into relativism, for external reality certainly informs our ideas about nature and portrayals of it. But *nature* is so polymorphous a term that what one attributes to it may say more about the speaker than it does about the natural world. Raymond Williams has noted that in the word *nature*, the "real complexity of natural processes has been rendered by a complexity within the single term." Nature gives us much room to maneuver: each of us can look to nature and find examples that will validate our point of view. Nature, so often defined as beyond human artifice, is shown to be quite human-made.[85]

Conservation biologists do not often go to bat for nature per se; they do not often describe nature in their writing. According to Neil Evernden, "The environmental advocate sits on the horns of a dilemma: the time honored technique of invoking the authority of nature has been essential to the presentation of a persuasive argument, and yet that technique is now vulnerable to charges of fraud."[86] The term *nature* not only carries a multiplicity of confusing, often self-serving, meanings; it also carries the taint of association with bleeding-heart liberal tree huggers. To be considered a "nature lover" is not a compliment in many quarters. So rather than running to nature, biologists flee from it. Instead, they describe and defend biodiversity. It maintains an aura of scientific respectability while still meaning so many different things to so many different people, without having yet acquired the notorious etymological reputation of the word *nature*.

BIODIVERSITY VERSUS NATURE

It is difficult to imagine what aspect of nature is excluded from the concept of biodiversity. In my interviews with conservation biologists, I asked specifically how the terms *biodiversity* and *nature* differed in meaning, and what in nature fell outside the interviewee's definition of *biodiversity*. Compare the responses below with the same respondents' definitions of *biodiversity* above.

PETER BRUSSARD: "I don't know. I'd have to think about that one. . . . I think nature in people's minds, once again, gets tied up with sort of wilderness, nuts-and-berries kinds of lifestyles and things. I think there's a negative connotation in there too. I've said a number of times that if we're going to make any headway in conserving biodiversity, first of all—I mean everybody knows if we all ate tofu and rode bicycles, the world would be a better place. Take that as a given. Go out and tell that to a bunch of middle-class Americans and they're going to go nuts. So I think you have to approach it differently. You have to figure out ways we can reduce our impact on the world. And make them aware of the value of these things in both the short and the long term. And the fact that certain lifestyles are more impacting on this than others. And try and get at it from that sort of approach, rather than wearing hair shirts."

PAUL EHRLICH (in response to what in nature is excluded from his definition): "Anything that's dead. [Laughs.] Basically."

THOMAS EISNER: "When you talk about nature, you're talking also about the feeling, the nonscientific feelings that you might have toward biodiversity. You might think of the spiritually uplifting impressions you can get from beauty. When you say *biodiversity*, you're already loading the term in the sense that you're about to talk about something that might be threatened, or something that we owe—we should be scientifically aware of. But I use them interchangeably."

TERRY ERWIN: Good question. I think probably it's just a practical—*biodiversity* is a buzzword. I'm not sure exactly when it started. Probably in the early 1980s . . . people started picking up on it, and it just became a buzzword for a lot of the stuff that we do. Of course, people have been working with biodiversity for generations, if not millennia. Nature . . . probably is all-encompassing, includes geology, universal factors, and so forth. Biodiversity is diversity of life."

JERRY FRANKLIN: "Well, biodiversity, even as broadly as I defined it, which includes genetic structure and process variability, is still narrower in my mind than nature. Maybe if I thought about it for a while, I wouldn't feel that way."

K. C. KIM: "Well, biodiversity . . . is simply components of nature. And does not necessarily indicate the synergism involved in species interactions. In other words, biodiversity in some sense is more of a concept of evolutionary—how should I say, diversity of living things. Nature, however, is more holistic, and more in a functional whole, which includes species interactions, as well as interactions in given environments."

THOMAS LOVEJOY: "I don't know. *Nature* is a very, very general term. And you know, I suppose you could look at all of nature as equating to all of biological diversity. But when you then try to get specific, . . . you can have biological diversity of some sort in situations that people wouldn't think of, conceptually, as nature. People don't necessarily think of an agricultural field as nature, yet there could well be biological diversity there—probably in the soils. Nobody's paying any attention to it. So, I mean, I think there's a good reason to talk about biological diversity aside from normal concerns about nature conservation. Nature conservation's got to include things like landscapes, those kinds of aesthetic values. It's got to include sort of—well, it just seems to me that it includes a bunch of other values."

JANE LUBCHENCO: "I think the difference is, in using the term *biodiversity*, one focuses on the organisms as well as the functions that they serve. As opposed to something that's much more nebulous, which is what nature is to me."

S. J. MCNAUGHTON: "Well, *nature* is some sort of encompassing term for what's present in the world. Biodiversity is a characteristic of that."

REED NOSS: "That's a good question. And I have thought about that some but probably not enough. To me, again, biodiversity and nature are probably equivalent. The difference is that with the term *biodiversity*, we can recognize a distinct spatial and temporal hierarchy. And we can start to pull out particular indicators, measurable attributes that we might want to pay attention to over others. . . . No one has attempted to do that with *nature*. It's a very, very general term, more general than *biodiversity*. . . . But you know, to me, in a philosophical sense, they're equivalent concepts. But in a practical sense, as a scientist, biodiversity is something that we can distinguish very different components [in] and select those that are suitable for different kinds of objectives."

GORDON ORIANS: "I have to confess I haven't thought about it in exactly those terms, so I'm having to think a bit here. I suppose in some ways, now that you ask me, it does sound somewhat like a jargony way of saying what we've always said. [Laughs.] But it does have somewhat more precision, I think. Nature is very, very ill-defined. And I think biological diversity, the notion, introduces some precision, although not an enormous amount, I must confess."

DAVID PIMENTEL: "Well, in a sense, *nature* is even a more vague term than *biological diversity*. Biological diversity I really think of as . . . species diversity. Nature I take it as being all of that and maybe some of the interactions. . . . I think *nature* is coming closer to *ecology*. It's a very vague term that people respect, just as they do the environment. . . . Many people don't know what they're talking about. But I'm glad they respect it."

G. CARLETON RAY: "That's interesting. I hadn't thought of it that way. Really not much. I think it's a little more concrete. . . . I would simply think it's more specific. Nature to me would include everything that one paints in a painting, a Van Gogh or something like this. I think *biodiversity* is basically a scientific term. *Nature* isn't. Nature is much broader than that, much more inclusive. So I would not put [it] in the same framework. I think Thoreau was certainly a natural historian. But he certainly was more on the nature end. . . . And *biodiversity*, after all: a word that ugly couldn't be anything *but* scientific."

MICHAEL SOULÉ: "None. *Nature* is a little more inclusive; it includes the physical and chemical aspects of biodiversity, such as waterfalls and the Grand Canyon as a geological phenomenon, strictly. But since the biosphere pretty much covers much of the surface, there's a lot of overlap between *nature* and *biodiversity*."

DAVID WOODRUFF: "I think they mean the same things. I think, however, the word *nature* was so disparaged by the scientific community in this country in the 1950s and 1960s, for any scientist to come forward and say, 'I'm a naturalist,' or 'I want to be the Baird Professor of Natural History,' rather than the 'Baird Professor of Science,' would be suicidal. Once science, once biological science became highly reductionist, then the word *nature* became anathema to the practitioners, to the people who controlled it. And so they steered away from it. And so instead of doing natural history, you did behavioral ecology. It's natural history. But in this country, people found that word was like stamp collecting in scientific circles. . . . And so we lost the utility of a word like *nature*. I'm very comfortable with it, but I have to be careful when I use it who's listening. Otherwise people very quickly write you off as being a flake, meaning a tree hugger, or being an 'ologist,' somebody only fixed on birds or cats. I think once you're identified that way, you lose a lot of flexibility and a lot of clout. So I've avoided using [terms] like *nature*, or *natural history*." "Ultimately it's *nature* or it's *biodiversity*. It's more than *biological* diversity,' because it involves the processes."

The term *biodiversity* incorporates the conservation goals toward which many biologists really aim—preservation of intact ecosystems and biotic processes—while still allowing the public to maintain an emotional grasp on charismatic icons. It puts a scientific spin on *nature*, while avoiding the word's negative connotations. See Brussard's answer to this question. He starts by concurring that perhaps there *is* no difference between *biodiversity* and *nature*. He describes the negative connotations the word *nature* holds. He suggests that people must be convinced of the social goals of biodiversity advocates. The word *biodiversity* is part of a *convincing* strategy—that is, it is designed to convince and has been quite convincing thus far. By scientizing the concept of nature, biologists aim to convince you both that the biodiversity crisis is grave and that they have special expertise in understanding and addressing it.

Eisner, while admitting in his response that he uses the terms interchangeably, adds that *nature* includes what scientists traditionally exclude from their discourse: emotions and beauty, for example. Lovejoy also says that the term *biodiversity* excludes values; we shall see later that these *are* included under *biodiversity*. Eisner also hints that *biodiversity* means action, is meant to prompt us to remember that nature is threatened. It is a biologist's way of eradicating emotion, while adding objectivity *and* conservation goals to the discussion of nature.

For Noss the ecophilosopher, *nature* and *biodiversity* are equivalent. But for Noss the conservation biologist, *biodiversity* gives latitude for manipulation depending on what one seeks to achieve. It is not that one could not do this with the term *nature*. It is just that biologists have found it more expedient to use the scientifically formulated and scientific-sounding term *biodiversity* for conservation purposes: this is what it was designed for.

Kim lists specific biological differences in the meanings of *biodiversity* and *nature*, and Pimentel's contention that *biodiversity* is narrower fits with his equation of it with species diversity. Franklin, Erwin, Lubchenco, McNaughton, Orians, and Ray assert that the meaning of *biodiversity* is narrower in some ways, although their responses lack specificity. Erwin says that *nature* includes "geology"— which is to say, that which is not living, although he has also said that "for biodiversity to form the basis of a pluralistic conservation strategy, 'bio' must be interpreted broadly to include both patterns and processes of nature as well as much of what is technically considered 'abiotic.'"[87] The term *biodiversity*, like *nature*, becomes so all-encompassing that it may include even what is not living.

Lubchenco limits *biodiversity* to "organisms and the functions they serve," although her definition above seems broader than that. Ray asserts that *biodiversity* is "basically a scientific term," and he's right—scientists created it and promoted it—although that does not necessarily make it easier to pin down than *nature*. Ehrlich, Eisner, Noss, Soulé, and Woodruff pretty much concede that *biodiversity* and *nature* are synonyms. Woodruff adds that *nature* and *natural history* have fallen victim to biological snobbism. *Biodiversity* more befits biologists' usage: as Ray puts it, a term that ugly *has* to be scientific.

Scientists have coined this new word, which scarcely differs from *nature*, so that we will add it to our working vocabulary alongside other neologisms we have come to hold dear. As Alfredo Ortega, a Mexican newspaper columnist, notes: "Biodiversity, or natural riches, if you prefer, is a new term that describes something very old. . . . We shall incorporate it into our dictionary along with democracy, ecology, free trade, and other complicated products of modern civilization."[88] Eisner comments on the success of *biodiversity*: "I've never run into a situation where we got hung up because we were using different meanings of the term. . . . I mean, people respond to the term now. . . . People have a feeling for it. . . . There's a collective noun that represents a lot of different things. . . . So *biodiversity*, for those of us who have become somewhat expedient in our urge to have an impact in conservation, is a term that is more useful in that context."

If *biodiversity* is blurry and all-encompassing that is, in part, why it has been so successful as a conservation buzzword. When Noss thinks "about biodiversity, the images I get are just *everything*, the full richness of life on this planet." *Biodiversity* has entered the dictionary, people respond to it, it works, because each of us can find in it what we cherish: the biologist can find her study subject or conservation goal, the policy maker can isolate the fragment of the biological world that demands legislative focus, and the rest of us can conjure up any icon from nature that we wish to see endure. What is it you most prize in the natural world? Yes, biodiversity is that, too. In biodiversity each of us finds a mirror for our most treasured natural images, our most fervent environmental concerns.

BACKLASH?

In a taxonomy of the ways in which we value nature, Michael Soulé defines *biodiversity* as "the living nature of the contemporary Western biologists. It is the most concrete of all these perceptions, which makes it the most vulnerable."[89] On the contrary, I argue, little is concrete about most definitions of *biodiversity*, and it is this imprecision that makes it so vulnerable. Indeed, by representing it as all things to all people, the seeds of a backlash against biodiversity may already have been planted. Eisner and other biologists may share a tacit understanding of what *biodiversity* means, but those who oppose conservation do not. Eventually, they may call the conservationists' bluff.

David Ehrenfeld calls *biodiversity* "a horrible word. . . that, because it's vaguely defined, has a broad appeal, like motherhood. But it sometimes can cause more trouble than it's worth." Recognizing the potential for backlash from this buzzword, some biologists do not use *biodiversity* but have reverted to the term *biological diversity*, whose unwieldiness and history may lend it more scientific credibility. But the same problems of nonspecificity remain. For example, the National Biological Diversity Conservation and Environmental Research Act, a bill Congress kicked around for a few years, defined *biological diversity* as "the full range of variety and variability within and among living organisms and the ecological complexes in which they occur, and encompasses ecosystem or community diversity, species diversity, and genetic diversity." Testifying against the bill, David Ford of the National Forest Products Association recognized the broad sweep of this definition: "Today there is no single definition of biological diversity that can be agreed upon by scientists. There are numerous definitions. Some are very general, and some are very, very specific. . . . an overly broad definition of biological diversity would create an opportunity for unintended court challenges."[90]

It is not just that such a law would open a Pandora's Box of litigation: how would we set concrete, enforceable conservation policies or goals based on this definition? As expressed by George M. Leonard, associate chief of the Forest Service, "a requirement that we simply describe . . . the impacts on biological diversity, absent some guidance as to the level which is expected, would simply open us up to the continued challenge that 'You haven't gone far enough.'"[91] A friend of mine who is a wildlife biologist for a large National Forest (and who prefers to remain anonymous) confirms that it is quite difficult to fulfill her mandate to manage for biodiversity: "Nobody wants to decide what it is or how we're going to measure it." She can do little more than stick to using charismatic species as indicators.

THE ARGUMENT FROM IGNORANCE

In his encyclopedic review of the U.S. environmental movement, Samuel Hays argues that in recent times, environmentalists have placed far greater demands on the scientific establishment than scientists have been able to meet. Hays claims that problems arose because the "overriding experience in environmental science was the degree to which much of what one wished to understand was not yet known. Each advance in knowledge seemed to expand what was not known more rapidly than it did what was known; society seemed to be faced with escalating ignorance."[92]

The term *biodiversity* symbolizes biologists' lack of knowledge about the natural world. Endangered species can be identified, counted, and planned for (even if a daunting task), but *biodiversity* stands for biological wealth and complexity whose depths biologists have scarcely begun to plumb. Such opacity is not necessarily problematic: when they employ the concept of biodiversity, biologists mean to turn the depth of their ignorance from a seeming weakness into a unique strength. They seek to use this ignorance as a lever, not only to promote their conservation goals, but to advance the privileged position from which they speak for those goals. Ignorance adds to the luster of the biologist's expertise; it becomes his domain, a source of his authority.

How many species are there on Earth? No one really knows, despite well-publicized guesses. Somewhere around 1.7 million have been described. British ecologist C. B. Williams estimated 3 million species in 1964; from there, citing new discoveries, biologists have hiked their estimates to 5 million, even to 10 million species, plus or minus a few million.[93]

Terry Erwin has dramatically raised the stakes as he has censused the scarcely known riches of the tropical rain-forest canopy. When poisoned by his chemical fogs, a staggering diversity of insects, especially beetles, drop into his waiting nets. He has found a high degree of host specificity—that is, he sees little overlap in arthropod species between the canopy of one tree and those found in other trees even short distances away. From his results, he extrapolated in 1982 that upwards of 30 million species of insects alone may lurk in Earth's hidden recesses. In 1988 he upped the ante to 50 million, and others have used similar startling gaps in knowledge to raise these estimates to as many as 100 million species of organisms.[94]

Why this numbers game? If you wish to save endangered species, then in so many specific cases, biologists' expertise is devalued in the face of so much that is so unknown. Biologists cannot formulate a plan to rescue a species if they know nothing about its ecology. They cannot tell you that something particular residing on this parcel needs rescue if no biologist even suspects its existence. But if you wish to save biodiversity, then you wish to save something that includes all that is known *and* unknown about life. Biologists can be your guide. The lead quotation in *Global Biodiversity Strategy* comes from Peter Raven: "We cannot even estimate the number of species of organisms on Earth to an order of magnitude, an appalling situation in terms of knowledge and our ability to affect the human prospect positively. There are clearly few areas of science about which so little is known, and none of such direct relevance to human beings."[95]

That is not a credibility sink for those who would speak for biodiversity; rather, this lack of knowledge becomes a source of power for entomologists and other biologists. Your society wants to conserve biodiversity? To know what you have, to know how biological systems work, biologists say, we must figure out how many species there are. Only biologists, particularly entomologists, can tell you what biodiversity is, how much of it there is: "Probably 30 million species of insects on one continent alone and less than 2 million animal species described in the entire world! What a challenge for taxonomists! What a problem for conservationists!" Marc Dourojeanni exclaims.[96]

Bruno Latour, who studies how scientists accrue power, portrays them as generals marshaling allies to do battle against competing knowledge claims. A scientist wishing to stake a claim will attempt to enroll as allies both all those scientists who have gone before and those who are at work now, their instruments, and their organisms: "Doubting a spokesperson's word requires a much more strenuous effort because it is now one person—the dissenter—against a crowd —the author."[97] When the author speaks for biodiversity (a.k.a. all of nature), that is quite a crowd to come up against. And if speaking for 5 million species of things is not enough, try speaking for 50 million, or 100 million!

If we don't know how many species we have, we don't know what or how many we're losing; the more we have, the more we have to lose. A few years ago, Wilson provided a "conservative" estimate of 17,500 species lost/year; more recently he upped this figure to 27,000, and then to 50,000.[98] Myers estimates that we are losing more than 100 species/day and stand soon to lose perhaps more than one-third of all extant species. As noted earlier, these estimates are contentious. But they also disquiet the listener. Erwin, who launched species numbers into orbit, points out that "such estimates may be so far off the mark that they are worthless, or they may be 'just so' stories, or they may be close to accurate and thus very scary!" To find out, he proposes "an agenda for massive, but achievable biotic inventories that will not only yield voucher and study specimens for museums and basic systematics, but also set up a system for long-term monitoring of species-collectives, that is, co-occurring species and their communities, in their naturally integrated habitats."[99] That is to say, fears of biodiversity loss should prompt society to support biologists' efforts to understand that loss.

For this to happen, society must want it to happen. Erwin calls his original 1982 paper "kind of just a punch in the awareness." In 1991, he wrote: "Now that the problem has been recognized and a bleak future forecast if policy changes are not made, we must all do our part in changing policies and stopping the carnage. In terms of potential extinctions, if the weight of higher species numbers is greater on the collective human conscience, then so be it. For this reason, I'd rather err on the high side."[100] Not knowing how many species inhabit Earth thus becomes a high-profile problem, put forward by biologists to raise public awareness, and only to be solved with biologists' expertise.

If readers find estimates of how many species we have already lost and are losing scary, they may not be pleased to hear more from Norman Myers. He wonders how we shall know when species loss reaches catastrophic levels: "We simply do not know the answers to these momentous questions. Indeed we are very far from knowing even how to get to know them. Rather than learning to supply the right answers, we have yet to learn to ask the correct questions. In light of this gross uncertainty, is it not the path of prudence to err on the cautious side by saving a maximum, rather than an optimum, number of species?"[101]

Who is in the best position to figure out what such catastrophic loss means to humanity? Iltis asks, "Who is to say what is useful and what is not, especially about species not yet discovered that, unknown and unstudied, fall prey to plow or cow? And who can predict the value of a monkey, a butterfly, or a flower? Or of intact ecosystems, to which we are inseparably linked, whether we acknowledge this or not?"[102]

Who is to say? This is not a rhetorical question. The answer is: biologists will say. Simultaneously they create our worries and pose themselves as palliatives to those worries. And their answers always require us to pay more attention to their own expertise and save more of what they want to save—that is, as much biodiversity as possible.

If biologists only dimly suspect how many species inhabit Earth or how fast they are disappearing, they are even more in the dark as to what functions organisms play in ecosystems. "And the unfortunate thing is," says Erwin, "we don't know yet how many players it takes to keep a system going and how many systems it takes to keep the biosphere going." To complicate matters, some biologists opine that the roles given species play in specific ecosystem functions may be overrated; species may have eco-equivalents. While biologists may agree that systems need some level of redundancy, no one knows how many species or what kind of species are needed to perform various tasks in different ecosystems.[103]

Why are so many species uncommon and so few species abundant? What is the ecological function of abundant species?[104] Who knows? Biologists urge us not to regard the answers to these questions as mere esoterica; they are offered as crucial knowledge if we are to know how ecosystems support humanity. So that we may come to know all that biologists don't know, but wish they did, and come to preserve all that we don't, but that biologists wish we would, they strive to shift the focus away from individual species conservation toward conservation of aggregates of species and the systems they maintain. That is to say, biologists move us toward conservation of biodiversity.

One of the three major research recommendations outlined in the Sustainable Biosphere Initiative (SBI) is to "address both the importance of biological diversity in controlling ecological processes and the role that ecological processes play in shaping patterns of diversity at different scales of time and space." Its distinguished authors from the Ecological Society of America, led by Jane Lubchenco, intend the SBI as a "call-to-arms" for ecological researchers to address pressing ecological problems. The SBI also advertises to policy makers and funders the crucial role ecological knowledge plays in human survival. In the real world, according to the SBI's framers, "ecologists are increasingly asked to justify the benefits of biological diversity compared to the human benefits that might be derived from economic development. Ecologists will be challenged over the coming decades to evaluate the functional significance of genetic diversity, species diversity, and ecosystems diversity."[105] Several things happen in the SBI. First, as with other definitions, biological diversity is broadly characterized to include multiple levels of biological hierarchy and pattern. Next, present ignorance about the ecosystem functions of biological diversity is acknowledged. Finally, that ignorance is converted into a strength, as what society wishes to know can only be provided by the biologists, who do not know it now, but might if society were willing to pay.

Ignorance about biodiversity has more tangible consequences. Biologists advertise the latent value of undiscovered foods, medicines, and industrial products that we might yet unearth in the rain forest and elsewhere. Only by allowing biologists to discover and document the full range of species—the full spectrum of biodiversity—can we harvest the full bounty that evolution has planted for us. In 1988, Lovejoy addressed a group of biologists on the importance of "serendipity" in the uses of biodiversity. From what was not known in the past and was then discovered—who would have guessed?—we may extrapolate a myriad of useful things we have yet to discover. Biologists should treat this, Lovejoy contended, "as a strong argument in favor of conserving biological diversity."[106]

In light of our ignorance about the roles of organisms and species in maintaining ecosystem function, and about the practical uses of organisms to humanity, biologists urge us to heed Aldo Leopold's classic advice that "to keep every cog and wheel is the first precaution of intelligent tinkering." Or to use the Ehrlichs' updated metaphor, we should be preserving all the rivets, because we do not know which of them hold the wing on. Raven elaborates:

> "We know so little about biodiversity, the interchangeability of biodiversity in communities and all the rest, that we don't know what the limits are. . . . When human beings invented agriculture in several scattered centers about 10 or 11,000 years ago, biodiversity was at a maximum, at least for the past 65 million years, maybe a maximum for always. Now we don't know that you need that level of diversity. . . . We don't even know what the word *need* means. . . . We presume we can get by with lower amounts of biodiversity, but we haven't even got a grasp of the ways in which that might be done. We're not getting a grasp of the ways. We're not investing in it. And meanwhile it's disappearing extremely rapidly. For plants, there are about 250,000 species. If my predictions and other people's predictions are roughly correct, there would be 50,000 possibly extinct in the next 30 years or so. . . . Since they do form the backbone of communities throughout the world, it seems to me incredibly shortsighted that people have not said, 'Let's keep them.' Go back to Aldo Leopold: why should we let 50,000 of them become extinct? Why shouldn't there be a worldwide effort to try to save them?"

"Regardless of how ill-informed we are—and you know we're incredibly ill-informed—we're not totally mindless," Raven went on to say.[107] No matter how little they know, biologists know more than anyone about the biodiversity upon which civilization is built. If the foundation of life is not to crumble, we all must listen to biologists, fund their work, heed their warnings, conserve what they want us to conserve.

Iltis, too, cites Leopold's cog-and-wheel prescription for intelligent tinkering, saying:

"It is just enormously stupid to throw away the parts. I mean, why destroy something that is irreplaceable and just throw it away when we don't even understand what it's all about? . . . When I talk to a beginning class—I give these guest lectures, which is a way for a professor to get respite from giving a lecture every week, he gets off for a period by asking Iltis to come in and give them an extravaganza and shock the students out of their *wits*—I hold up an oak leaf from my backyard, a dry one, and say, 'Nothing we have ever done and nothing we will ever do in the future is anywhere near as complicated as what goes on in a single cell of that oak leaf!' I mean, the enormous complexity of even the simplest algal cell compared to a big computer . . . which is nothing compared to the biochemistry of the single cell. It's just absolutely *foolish*. It's stupid to throw things away."

Iltis scares undergraduates out of their wits by displaying our ignorance of the natural world. He aims to convert budding biologists to the goal of conserving every cog and wheel.

Biodiversity's promoters commonly appear as experts on ignorance. They know best that what we don't know *can* hurt us. Lubchenco, too, urges that we be cautious about impairing ecosystems that we do not understand, underscoring their crucial, and largely unexplicated, significance for human well-being: "It's pretty stupid to be destroying things that we may be dependent upon and that we can't recover." Daily and Ehrlich call it a "gigantic gamble with the future of civilization" to think that human survival does not depend on diversity of populations of nonhuman species. Wilson says that "the question I am asked most frequently about the diversity of life [is,] if enough species are extinguished, will the ecosystem collapse, and will the extinction of most other species follow soon afterward? The only answer anyone can give is: possibly. By the time we find out, however, it might be too late. One planet, one experiment." Above all, Wilson urges "prudence. We should judge every scrap of biodiversity as priceless while we learn to use it and come to understand what it means to humanity."[108]

Lovejoy further illustrates the uses of ignorance:

To approach the global problem from a scientific perspective, we are immediately confronted by the second problem commanding attention, namely, the limitations deriving from our relatively shallow knowledge of flora and fauna. Recent discoveries of insect diversity in the canopy of South American tropical forests warn us that we do not even know the extent of biological diversity on our planet to the nearest order of magnitude. Given such a poor inventory of life on Earth, biologists can say relatively little about which species occur where, which are in danger of extinction, where protected areas should be established, and where heavy environmental modification for development is permissible.

What is desperately required is a revitalization of the science of biological systematics, with all the ancillary strength modern technology and molecular biology can provide, combined with a crash program of biological exploration. A decade or two of intensive biological mapping is needed while development is halted, or at least severely curtailed, in areas that are evidently the richest but least explored.[109]

Scientifically speaking, by Lovejoy's reckoning, not much is known: there's a lot more out there than we expected, and biologists are poorly equipped to inform conservation decisions. But this is not a weakness. For if we are interested in making informed decisions about biodiversity, only biologists are in a position to help us; so we must help them. While their profession is revitalized and they search for the answers we all want, we must stop development in areas of potential biodiversity richness.

Here, in a concluding passage that illustrates the pattern we've been observing, is Reed Noss:

Given the immensity of the biodiversity crisis and the uncertainty of cause-effect relationships underlying it, what practical steps can we take to lessen the impacts? One key principle is prudence. In other words, we should err on the side of preservation when science is unable to provide clear answers. Reversing human population growth and resource consumption are long-term necessities. Scientifically credible environmental education must become a prominent feature of school curricula. Because the threats to biodiversity are so immediate—the battle may be won or lost in the 1990's—our imperative is action. Conservation biology is a contentious science, true, but there is an emerging consensus on certain elements of an active macrostrategy.

Noss alarms us and prescribes caution. But his further policy prescriptions are more severe: curbing population and consumption. We must also educate our children in a "scientifically credible" way— that is, teach them about the environment from the perspective of Noss and his peers. And we must act immediately, for otherwise it will be too late. The "macrostrategy" is a multipoint plan that conservation biologists must enact now; it includes the injunction "Be Visionary." Noss continues, "Maintenance of biodiversity must become our primary mission as a society, the principle that guides all resource uses." He asks, "Will public awareness of the crisis catch up in time? We cannot afford to wait for an answer."[110] Note that "we" has changed from a concerned general audience to the elite of conservation biologists who understand the dimensions of the problem, and the depths of our ignorance; these biologists must lead society to preserve what they care so much about, and what society, whether or not it yet realizes it, cannot live without.

The argument from ignorance can backfire. "We're being asked to take the entire scenario on faith," Julian Simon declares, and dismisses it as "guesswork and hysteria." The forest ecologist Ariel Lugo says that "no credible effort" has been undertaken to make concrete the figures and predictions espoused by conservation biologists. Such predictions are ridiculed as "bio-dogma." Zoologist Michael Mares derides conservation prophets' attitude of "data as luxury" when making pronouncements. Otto Solbrig declares that many of his peers' pronouncements lack scientific proof, and he chides them for crying wolf.[111]

But these nay-saying voices are, for now, crying in the wilderness. Responding to a *Science* article quoting those who downplay the seriousness of biodiversity loss and question the extent to which biologists' data support their dire predictions, Ehrlich writes:

> Mostly they are equivalent to saying that people should not be overly concerned about the burning down of the world's only genetic library because the number of "books" in it is not known within an order of magnitude and fire modelers disagree on whether it will be half consumed in a couple of decades or whether that level of destruction might take 50 years. Apparently a few scientists would never call the fire department

unless they could inform it of the exact temperature of the flames at each point in a holocaust nor, similarly, would they recommend beach erosion control unless every grain of sand had been counted.[112]

In our interview, Ehrlich added that "the 95–99 percent confidence intervals we're used to working with when the question is, 'Are the butterflies on one end of Jasper Ridge significantly longer than [on] the other?' just don't apply to these things. You're doing better if you can increase the chances of doing something right by 5 or 10 percent—that's worthwhile doing."

The all-encompassing nature of biodiversity guarantees us a degree of ignorance—we don't know everything about everything in the natural world. But the *bio* puts this ignorance firmly in the court of biologists. They tell us how little we know about the vanishing natural world and warn us of how severely we might pay for how little we know. They urge us to proceed with caution, that we ignore their ignorance at our own risk: therefore we must preserve as much of the natural world as we can. Biologists are thus the masters of ignorance about biodiversity; they accrue authority from it.

HOLISTIC VISIONS

The argument from ignorance presents two possible futures for the role of biologists' expertise. In one, if given enough time and resources, biologists can fill in the crucial gaps in knowledge. In the other, such lacunae can never be filled, because some things about the world are unknowable by science; biologists wish to make themselves spokespersons for these ineffable values that fall between the slats of science as it is currently defined.

"Conservation biologists are holistic; they do not believe that reductionism can satisfactorily explain ecosystem processes, and they are multi-disciplinary in their approach to practical problems. They are less concerned with the short-term profit of some corporation or government agency and more concerned with long-range viability of whole systems and species," Bill Devall writes in a treatise on deep ecology.[113] Devall conflates different explanations of what it might mean to be holistic. First, holistic conservation biologists are said to avoid reductionism in their scientific work; they would not believe that patterns at one level can be adequately explained by analyzing the next level down in the biological hierarchy and building up. However, conservation biologists certainly do use reductionistic scientific methods. Next, they are said to reach across disciplines to solve real life problems. Whether within narrowly defined science, or while working on extrascientific environmental issues, some conservation biologists work holistically as Devall suggests; some do not. Finally, Devall attributes a kind of holistic political commitment to conservation biologists, which some may indeed have; for example, Ehrlich told me he "would shift the burden to any development to show that this shopping center is more important than that stretch of chaparral. . . . And the presumption will always be that the chaparral is always more important than the shopping center." Others may not share this kind of political attitude at all.

Furthermore, according to Devall—and to practitioners like Soulé—some conservation biologists have normative and metaphysical commitments that we might label *holistic*. But supposedly holistic worldviews may, in fact, be quite atomized: worldview and approach to science may not overlap. It is perfectly plausible to have a holistic metaphysical or political worldview and do reductionistic science. Or one can be reductionistic in one's metaphysics and do holistic science. The environmental historian Carolyn Merchant conflates these two as well: "Although ecology is a relatively new science, its philosophy of nature, holism, is not. Historically, holistic presuppositions about nature have been assumed by communities of people who have succeeded in living in equilibrium with their environments. . . . Ecology necessarily must consider the complexities and the totality. It cannot isolate the parts into simplified systems that can be studied in a laboratory, because such isolation distorts the whole."[114] But ecologists do isolate systems and parts of systems all the time, ignoring the interconnectedness that Merchant believes informs the holistic, ecological, scientific view.

Thomas Dunlap, another environmental historian, recognizes that "ecology, which is intellectually holistic, often arrives at these ideas by severely reductionistic means, ones that emphasize the ethos of detachment, objectivity, and manipulation of the 'subject' for scientific knowledge."[115] Dunlap partially corrects the fallacies of

Merchant and Devall, but speaks of reductionism in scientific ethos, not practice. One can do holistic kinds of science with detached objectivity, or be quite subjective while using reductionist methods.

The kind of holistic paradigm said to epitomize ecology is used as a resource by others as well. Betty Jean Craige, who works in cultural studies, discusses "ethical holism" in both biotic and cultural communities. Echoing Leopold's oft-quoted dictum, she suggests that "the goal of the cultural version of ethical holism is comparable to that of the environmentalist version: to establish and maintain the integrity, stability, and beauty of the human community." She goes on to say that the "well-being of the system requires the well-being of all its parts." Yet this kind of holistic worldview could still lead to a nonholistic approach to both science and conservation, the very approach that partisans of endangered species often implicitly accept: study and save the parts to know and save the whole. Jennifer Daryl Slack and Laurie Anne Whitt partially correct for this when they urge a holistic perspective, "an eco-culturalist perspective which acknowledges ecological interdependence and offers a basis for intervening to resist the instrumental reduction of the ecosystem, including both its human and non-human constituents."[116]

Having issued these caveats, I would suggest that for many of its proponents, the biodiversity concept may represent a more holistic approach to scientific practice, to conservation interventions, and even to metaphysical commitments. Hutto et al. warn us that in the ecological world, "the whole is greater than the sum of its parts." Merely saving endangered rivets does not necessarily preserve the wing. McNaughton suggests the alternate directions conservation policy must take if the world "is reductionist, with only collective properties of importance, or holistic, with emergent properties of importance."[117] In McNaughton's holistic, ecosystems view, properties emerge at the ecosystem level that could not have been predicted from an atomistic dissection of lower levels of the biological hierarchy. Such interactions that define ecosystems, that make them function, can neither be identified by studying ecosystem components nor preserved by preserving individual species alone, especially if removed (e.g., through captive breeding) from their functioning ecosystems. Focus on the totality of biodiversity more

accurately reflects McNaughton's holistic approach to science and
to conservation.

Erwin writes that biological diversity "is in fact *the product of
organic evolution*, that is, the diversity of life in all its manifestations. Biological diversity is thus holistic, and this is indicated in
the very nature of the word's root meanings; it is the sum of earth
species including all their interactions and variations within their
biotic and abiotic environment in both space and time."[118] Here,
Erwin uses *holistic* as a synonym for *all-encompassing*, for the
totality of nature's organisms and processes that biodiversity represents. It is also possible that the multitude of interactions subsumed
by the term suggest holistic, or emergent properties that one could
not predict from studying isolated particles of biological diversity.

Lubchenco suggested a different use of *holistic* to me: "There are
lots of very serious chronic problems. I think that what needs to be
done is to view them holistically and to emphasize that a focus on
just short-term kinds of things—for example, stimulating the economy—to the exclusion of looking at the longer-term consequences
of how that's done, the longer-term consequences to the environment, is inappropriate." So, for example, when we attempt to save
endangered species on a case-by-case basis, we avoid looking at, and
trying to avert, the more global determinants of species loss. Too
many people with too many wants and needs are wreaking havoc
on the environment. Focus on endangered species is focus on brush
fires, when what really needs to be extinguished are the flames of
unfettered human growth, wasteful styles of life, and values opposed
to conservation.[119] *Biodiversity* is a macroterm to focus attention
on a macroproblem. Biodiversity symbolizes a more inclusive, more
holistic approach to political problem solving; it makes us think
long-term, makes us consider how all our sociopoliticoeconomic actions will affect the future of the natural world.

Donald Worster identifies a persistent, if sometimes nondomi-
nant, strand in ecological science of an "ecological ethic of interdependence." Organicism stresses the unity and interconnectedness
of all of nature, and has "a way of gaining a foothold on even the
most unpromising surface."[120] In the ideas packaged in the biodiversity concept, holism again gains a foothold on the rocky shoals of

science and the slippery surface of conscience. Biodiversity presses us to consider nature as a whole and to evolve conservation policies that reflect this rather than disarticulating nature into endangered species and habitat parcels to be preserved in isolation.

Noss calls *biodiversity* "the most popular buzzword in conservation these days, and rightly so. Here at last is a chance to expand beyond a fragmentary species by species approach and address environmental problems holistically."[121] Rather than managing for single objectives, he urges "management for multiple values, multiple species and processes at the same time." Broad definitions foster broadmindedness. As opposed to *endangered species*, the term *biodiversity* embraces this multiplicity, and also embraces humans and our interrelatedness with the rest of the natural world.

Kim uses the concept of biodiversity as a conceptual vehicle to convey a holistic, ecosystems approach to conservation to landowners and policy makers: "I think this biodiversity issue brings them a more holistic perspective on their properties . . . so that they still can cut the tree but in the context of preserving habitats and what's in it, the plants and animals." Kim uses *biodiversity* to bring distant environmental problems home, literally and figuratively, to teach holistic messages about interrelatedness. While reflecting biological understanding of interconnections in nature, *biodiversity* also brings people in a democratic society together toward common corrective actions.

The environmental philosopher Warwick Fox promotes a way of thinking based on the "this-worldly realization of as expansive a sense of self as possible." In the absence of a transcendent God, the world around us is the sum total of reality. We can do nothing greater to expand our sense of self than to seek to understand Earth and all its creatures. The wider the circle of the diversity of life we come to know, the greater our chances for self-expansion, and the fuller our sense of self.[122] This metaphysical holism runs through definitions of biodiversity. Raven's studies and notions of biodiversity have, for example, led him ineluctably to an expanded sense of self. "Peace, social justice, human order, the protection of biodiversity, the production of or promotion of a stable biosphere are all inextricably interwoven," he notes. "And since I see that all as a unity, I can present that any way that I want for a particular audience." Raven focuses on biodiversity rather than endangered species because he supports "dealing with everything together, dealing with regions, dealing with whole systems." The all-encompassing view of biodiversity that stems from his science also informs his greater world picture, one in which things must be viewed as interconnected, a unity, a whole.

Woodruff's "holistic view of biodiversity" embraces all levels of biological hierarchy, as well as how these levels affect the life-support systems of the planet. He says his holistic hierarchical definition of biodiversity parallels a holistic worldview, "one that isn't nationalistic; it's international. One that isn't locked into any one small society or any one religious group; but it's one that recognizes interconnections. And interdependency." Ray told me he believes "there is something about diversity per se, rather than an individual species living in a vacuum, that is quite different emotionally, economically, mathematically, than the individuals. It's the sum-of-the-parts principle."

Michael Soulé draws the parallels between ecological and metaphysical holistic worldviews:

"Each individual, each element of the universe is an infinitely faceted mirror that reflects all other individuals and entities in the universe. We're all constantly reflecting everything else. So it's like a hologram. Each part contains the whole. That's very mystical, but—speaking as a scientist—our genetic material remembers the days in the organic soup when we were basically a few simple biochemical pathways—they're still there. They haven't changed hardly at all from the bacterial days. So in a sense, even a scientist has to admit that we still are a part and parcel of everything else, everything else organic on the planet. Regarding the question of 'What is intrinsic value?': I'm not a philosopher and I haven't figured it out. But intuitively, when I'm asked, you know, 'Should we save this species or that species, or this place or that place?' the answer is always 'Yes!' with an exclamation point. Because it's *obvious*. And if you ask me to justify it, then I switch into a more cognitive consciousness and can start giving you reasons, economic reasons, aesthetic reasons. They're all dualistic in a sense. But the feeling that underlies it is that 'Yes!' And that 'Yes!' comes out of the affirmation of being part of it all, being part of this whole evolutionary processes. And agreeing with Arne Naess that each species, each entity should be allowed to continue its evolution and to live out its destiny—It's not ordained or anything, but just do its thing, as we say. Why not? And the 'why not' is there's too many people."

The holism in Soulé's approach to the study of biodiversity parallels the holism in his approach to the conservation of biodiversity. The sense of the interrelatedness of all things inherent in his worldview and in his definition of biodiversity bespeaks the obviousness, the "Yes!" of the need to conserve as much of the natural world as possible. His science is imbued with this interpenetrating view—how can an ecologist not feel this way?—and it makes him an activist. He is not an isolated, "objective" laboratory scientist; rather, he is reflecting everything else. His view of interconnecting, interreflecting entities obliges him to reflect back what he sees and what he feels. This holism is part of his science both because he is an ecologist, and must by definition consider the interrelatedness of things, and because he is obliged to proselytize out of responsibility for those other entities that are not really "other," but are part of him emotionally and evolutionarily.

Granted, Soulé is an exceptional biologist. At one point, he took a break from academic biology to direct a Buddhist studies institute. Roderick Nash describes the role that Eastern religions have played in the American wilderness movement:

> The key element in these religions was the assumption that a web of kinship unites all things. Man's task is to discern these interrelationships and submerge his ego in the concept of universal community. Worship consists of feeling a oneness with the living as well as the nonliving components of nature. The approach is pantheistic. Even mountains and waterfalls and soil are sacred. This was a vision that transcended ecology, in its scientific aspects, to probe toward the mystical concept of oneness.[123]

Even for those biologists who disavow any connection with religion, Eastern or otherwise, the biodiversity concept may reach toward oneness, toward a unity of all living (and nonliving) things. Like Zen devotees, ecological biologists seek to understand how the world works while appreciating the interrelatedness of all living things, while cherishing both the parts that comprise the whole and the whole itself (which may be greater than the sum of its parts). Biodiversity is both the parts and the whole of the natural world that biologists study and fight for. They feel this "web of kinship" in the entities they study, and they feel their connection as part of that web. The idea of biodiversity scientizes this ecological and holistic unity, this evolutionary and metaphysical kinship.

CONCLUSION

The term *biodiversity* makes concrete—and promotes action on behalf of—a way of being, a way of thinking, a way of feeling, and way of perceiving the world. It encompasses the multiplicity of scientists' factual, political, and emotional arguments in defense of nature, while simultaneously appearing as a purely scientific, objective entity. In the term *biodiversity*, subjective preferences are packaged with hard facts; eco-feelings are joined to economic commodities; deep ecology is sold as dollars and sense to more pragmatic, or more myopic, policy makers and members of the public. Biodiversity shines with the gloss of scientific respectability, while underneath it is kaleidoscopic and all-encompassing: we can find in it what we want, and can justify many courses of action in its name. It reflects the interrelatedness of all living beings, of humans with the rest of the living world, of our ideas of nature with nature itself. By promoting and using the concept of biodiversity, biologists hope to preserve much of the biotic world, including the dynamic processes that shape that world, while simultaneously appropriating for themselves the authority to speak for it, to define and defend it. If biodiversity is a much more complex and dynamic focus for conservation efforts than endangered species, it likewise offers a much more complex and dynamic role for biologists in society at large.

EXAMINING

BIODIVERSITY:

SCIENCE STUDIES

MEETS

ENVIRONMENTAL

HISTORY

 4

During the North American summer of 1988, temperatures soared, Yellowstone burned, ozone thinned, waste fouled the shores, and public interest in environmental issues bloomed.[1] Delivering the keynote address to the American Institute of Biological Sciences (AIBS) that summer, Thomas Lovejoy proclaimed: "This is in many ways our moment in history as biologists."[2] Lovejoy is an assistant secretary of the Smithsonian Institution, a conservation biologist, and a tireless promoter of environmental causes. He and a cadre of like-minded biologists were armed and ready to stoke the flames of public concern over perceived threats to the integrity of the natural world. At the forefront of their arsenal was the recently coined term *biodiversity*.

MAKING HISTORY

"I think there are the seeds of something among some that could change the course of global history," says Smithsonian tropical biologist and biodiversity advocate Terry Erwin. "I've always kind of been ahead of the game or ahead of the thinking in my own little field of entomology. . . . I think there will be individuals who can influence the direction of change. And I'd like to be among them, for whatever little part I might be able to play." Biologists are "making history" (jargon for the novelty and significance of what they're attempting) by making history (changing the course of events so that human and natural history will continue to their liking). And they are making history by making nature: a socially constructed idea embedded in and indistinguishable from the term *biodiversity*. Furthermore, by making history, they are making nature in more concrete terms of biopolitical boundaries of parks carved from public opinion, species that are allowed to thrive or disappear, ecosystem functions that can be maintained in their integrity or compromised.

How can we come to understand the relationship between the riot of life that biodiversity represents and the sometimes extravagant claims made by biologists on behalf of biodiversity?

I shall attempt to answer this question by employing ideas from environmental history and science studies. My goals here are numerous. I wish to bring the tools of these two academic disciplines to bear on biologists' ongoing attempts to define, analyze, and solve the biodiversity crisis, which, through biologists' promotion, has emerged as one of the key elements in the discourse on global environmental change. In so doing, I hope to make everyone—especially biologists—more aware of the political grammar of environmental rhetoric, and therefore more effective in advocating accurate, reasonable, just conservation strategies. In the process, I hope to offer some examples or guidelines for analyses of other complex environmental dilemmas. At the same time, by putting environmental history and science studies in the service of understanding biodiversity, its social construction, and the claims made about it, perhaps I may help subtly transform these interpretive disciplines.

ENVIRONMENTAL HISTORY: AN INTRODUCTION

Environmental historians charge themselves with the task of discerning the roles the natural world has played and does play in human history, and the roles that humans have played and do play in natural history. As a result, environmental historians have provided cogent analyses of the dialectic between nature and culture. In *Changes in the Land*, which chronicles mutual transformations of the New England landscape and New England cultures of Indians and settlers, William Cronon observes:

> An ecological history begins by assuming a dynamic and changing relationship between environment and culture, one as apt to produce contradictions as continuities. Moreover, it assumes that the interactions of the two are dialectical. Environment may initially shape the range of choices available to a people at a given moment, but then culture reshapes environment in responding to these choices. The reshaped environment presents a new set of possibilities for cultural reproduction, thus setting up a new cycle of mutual determination.[3]

The dialectic between land and humans can transcend the physical. In her study of the battle to save the California redwoods, Susan Schrepfer says that "ideas do mobilize material forces; this is especially true of environmentalism."[4] Environmental historians can find fertile ground in examining ideas as agents of ecological change, spotlighting the dialectic between the Earth and our ideas of it. Ideas can cause ecological impact by reshaping how we view, and thus how we treat, nature. We must examine who creates and promotes ideas about the natural world. From where do these ideas arise—how much from nature and how much from culture? How do a society's ideas about nature shift, and how do our changed attitudes toward the Earth remake the Earth itself? And how does the physical manifestation of our ideas continue the cycle? In particular, I am interested in how scientists forge conceptual tools that produce environmental change, and how changed environments remake scientists' ideas of what they study.

"For the historian, the main object must be to discover how a whole culture, rather than exceptional individuals in it, perceived and valued nature," Donald Worster says.[5] On the contrary, I wish to show how a few "exceptional individuals" are trying to change how the whole culture values nature, while they simultaneously try to change the culture of science. Whether lecturing to ladies' garden clubs, testifying before Congress, or bringing Jane Fonda and

Ted Turner to sleep in hammocks in the heart of an Amazonian rain forest, biologists are attempting to transform how people in our society view the natural world. Realizing that ideas don't change— they are *changed*—biologists are proselytizing on behalf of nature via their vigorous promotion of the concept of biodiversity and its attendant values. While attempting to alter cartographic boundaries to preserve habitat crucial for the sustenance of biodiversity, biologists are attempting to redraw normative boundaries to include more space for biodiversity's importance. Simultaneously, they are redrawing the boundaries of what it means to be a biologist.

IDEAS OF NATURE

Virtually no one denies that nature exists out there, something Neil Evernden calls that "great amorphous mass of otherness that cloaks the planet." But that amorphousness makes nature so hard to contain, so hard to define, so hard to know. Trying to understand what, exactly, nature is all about is like trying to nail jello to a tree. Worster points out that humans constantly engage in conceptual reorganization of the world around them, and that a society's view of nature may say as much about the society as it does about the natural world. What we consider to be nature is a multiplicity of constantly shifting views: "Nature," says György Lukács, "is a social category." Like "art" or "morality," nature can be thought of as a complex web of ideas that expresses the views of a society.[6] And, of course, each individual has her own personalized view of nature as well.

Historians have provided fascinating histories of the idea of nature, or its close relative, wilderness. Clarence Glacken's *Traces on the Rhodian Shore* was a landmark and a benchmark, a monumental attempt to trace this dialectic between nature and our ideas of it from antiquity to the nineteenth century.[7] Glacken's intellectual descendants often pursue the agenda that if conceptions of nature can be manipulated, we should choose the ends of our manipulation carefully, or at least pay careful attention to where past manipulations have led us.

For example, Linda Graber illustrates what the cultural construct *wilderness* represents to those who invoke it in defense of nature, showing how it is imbued with meaning by its devotees. Roderick Nash, too, traces the shifting history of this term, revealing it to stand for "a resource . . . defined by human perception." Max Oelschlaeger updates Nash's history in search of a postmodern view of wilderness, or an "old-new way of being." While he would agree with Nash and Graber that the true meaning of *wilderness* is open to interpretation, he traces a strand of wilderness worship back to prehistoric times, depicting its devotees as sages, whom we must heed if we are to become reenchanted with hidden, sacred nature and thus survive as a species. In fact, Oelschlaeger sees the very process of searching to understand how we have constructed, do construct, and could construct our ideas of nature or wilderness as a way out of the conceptual prison that shackles us, as modernists, to value nature only in a utilitarian, and therefore destructive, way.[8]

In *The Social Creation of Nature*, Neil Evernden asks "an apparently simple question: what is this thing 'nature' that we hasten to defend?"[9] As portrayed by humans through the ages, "nature" is a morass of contradictions. At times we mimic and obey it; at other times we strive to control and transcend it. Nature is something to be worshiped as wholly apart from humans; at the same time, nature is a product of the human mind. Evernden decries all attempts to define nonhuman nature as acts of control. His project resembles Oelschlaeger's: through bona fide attempts to understand wildness, an aspect of nature not under human physical or conceptual control, we can begin to cross over to some nonhubristic, nondomineering relationship with nature.

Dubbing *nature* "perhaps the most complex word in the language," Raymond Williams provides several instantiated definitions of it and shows how it serves as a resource for its manipulators. We appeal to nature to eschew solipsism, to justify our actions and beliefs in terms of a realm beyond humanity. Langdon Winner concurs: "Anyone who examines the range of meanings that have been attributed to nature in Western history must be impressed by their number, diversity, and glaring contradictory implications." Like Williams, he concludes that since real-life nature is protean, nature can justify any action, any belief.[10]

Opinions about "what nature is or wants" may not say much

about either biology or the zeitgeist. Rather, they may serve the purposes of the definers. N. Katherine Hayles urges that social justice and epistemological accuracy would be better served if the powerful few privileged to define nature became aware of the cultural biases that shape their constructions. Elizabeth Bird declares nature a social construct, traditionally shaped by those in power; at best, we can hope for a more democratic process in defining nature, incorporating as many interests as possible. Thus action taken in the name of nature, or attempts to define and solve environmental problems, would more likely result in justice for a broader cross-section of society.[11] Even those sympathetic to this goal might counter that how we view nature is at least partially constrained by the natural world itself. Yet this compromise epistemological position still leaves nature malleable, able to be bent by those with the power to speak in its name.

IDEAS AS FORCES OF NATURE

So we reveal little new when we highlight nature's cultural constructedness. But it may be new that with biodiversity, biologists are deliberately acting to shape our social construction of nature. Those who would never question the concrete truthfulness of the representations of nature they produce in their labs seem aware that cultural conceptions of nature are shifting, subjective, and can be molded; and they act on their intuitions.

Ideas can act as forces of nature. They can have tangible ecological repercussions. They can reshape how we view, how we value, and thus how we treat nature. The battle to sway human value judgments, human constructs of nature, may be decisive in the war over biodiversity. I take an environmental historian's standpoint to tell the story of biodiversity biologists, who claim special insight into nature based on the copious time they have spent surrounded by it and on the singular way of understanding the natural world afforded them by their science. They portray the natural world currently revealed to them as damaged and in danger of collapse. In seeking to remake our perceptions to match theirs, conservation biologists hope we shall help them heal nature.

This book is about the dialectic between two natures: nature, the real world that surrounds us, and "nature," how we portray that world. Or, rather, it is about the dialectic between biodiversity, the notional totality of life on this planet, and *biodiversity*, the term biologists have concocted as an approximation for that totality: a scientized synonym for *nature*, imbued with the values biologists cherish. The term *biodiversity* is only a decade old, but it stems from *nature* and *wilderness*, notions whose roots burrow deep into human history. In a rare opportunity to watch the conscious creation and dissemination of a new paradigm of our conceptions of nature, we are able to examine how and why biologists have concocted and promoted the word.

Donald Falk, director of the Center for Plant Conservation, notes that "to go from seeing nature as something upon which we are utterly dependent, and which sustains us physically and also psychically and spiritually, to go from that to thinking that nature is a kind of weekend amenity that we can protect in little bits and pieces to the extent we feel like and to the extent that we care to— I mean, that is an enormous change in how we view nature." Biodiversity biologists realize that modern, industrialized people come to know nature as much through words and images as through direct experience. Our physical and conceptual distance from nature allows those who would interpret it for us more room to maneuver. Given this distance, Evernden asks, "To whom are we to entrust the engineering of the *concept* of nature?"[12] Of course, conservation biologists would have it be them.

If one "is able to control the communication over a phenomena, then one controls how it is to be understood," Gary Lease observes; "A contest over what is allowed to represent reality—and, that is what intelligible access is all about—is a struggle over reality itself."[13] Similarly, Warwick Fox notes that to control discourse is "a common tactic that is employed in ideological battles over significant concepts: winning the semantic battle puts you well on the way to winning your particular war—and not just at a theoretical level."[14] Biologists have been largely successful at controlling the biodiversity discourse because we have invited them to be. We grant them cognitive authority to speak for nature. According to Donald Worster, in our post-Darwinian, nihilistic world, members of society have turned to scientists—especially ecologists—for moral guidance. Thomas Dunlap concurs. Our culture worships at the temple of science: "We have made its practitioners our oracles."[15]

Dunlap suggests that when we attempt to make order of wilderness, we invent "myths"—stories that provide a road map, a set of tacit directions that guide our way in the world, a complex adaptive paradigm to which we adhere, often without realizing it. Science, particularly ecology (and now conservation biology), has provided one such set of myths. Of course, ecologists and conservation biologists draw their own images from both a rapidly changing natural world and a rapidly changing society: science is embedded in both. Catchphrases and metaphors such as "balance of nature" and "diversity yields stability" act as synecdoches for the abstruse or the barely understood, making concrete our conceptions of the natural world: this holds true for both ecologists and laypeople. Each culture, including the culture of ecologists, has its nature myths, and each individual interprets these myths in her own way, navigating the byway between humans and nature slightly differently. People take from the science of ecology—and from nature—what they need to make peace with an ever-changing, ever-industrializing world. Ecologists and conservation biologists take from culture and from nature what *they* need to make peace with that same world and what they require in their attempts to distinguish ecological signal from noise.

Environmental historians have paid scant attention to how scientists' ideas and actions have been shaped by the natural world, and how they have, in turn, reshaped the natural world. "Most historians, because of the educational system, regard science the way the stereotypical American tourist views foreign countries: exotic places, possibly dangerous, filled with natives in peculiar costumes, speaking new languages, and practicing strange rites," Dunlap observes.[16] Those environmental historians who do enter the strange world of science often paint a picture of ideas trickling down or diffusing out from science to provide common ecological myths.

But the story of biologists and biodiversity is not a story of diffusion, of scientific terms adapted by the perplexed layperson seeking a mythic road map. Rather, biologists are attempting active mythmaking. Carolyn Merchant suggests that one of the tasks of the environmental historian is to discern how a society comes to know the natural world. Noting that a "society's symbols and images of nature express its collective consciousness," she declares that "scientific, philosophical, and literary texts are sources of the ideas and images used by controlling elites" to shape our norms with respect to nature.[17] In the case of biodiversity, scientific elites formulate and promote their own mythic images. A nature perceived as depleted has reshaped the culture of biologists who attempt to remake broader culture so that more people will come to appreciate the Earth's biotic bounty.

Ironically, biologists use the power gained by the deepening wedge between humans and the natural world to urge us to return to nature—both return to valuing and respecting nature as we allegedly used to and return in the physical sense, so that nature can work its persuasive charms on our psyches. "Like men of every age, we see in Nature what we have been taught to look for, we feel what we have been prepared to feel," Marjorie Hope Nicholson says apropos of how our aesthetic perceptions of mountains have changed.[18] Biologists are coaching us for our return to nature, hoping that we shall see and feel what they have, wanting us to experience the very things they have experienced that make them value it so much. Although Evernden and Oelschlaeger warn against any attempt to control nature or our ideas of it, biologists are using their control to have us recapture wildness, some essence of nature that can transform our lives, our values, our actions. Furthermore, in promoting something as various as biodiversity, biologists harness the wisdom of those who tell us how vast nature is, how we can see anything we wish in it. To speak of biodiversity is to appeal to the multiplicity of nature; we can see in it what is important to us. The multiplicity of reasons touted to preserve biodiversity find parallels in the multiplicity of images the term evokes, in the range of meanings one can find in it.

So the question: can we ever know what nature really is or what it really wants? Worster tells us that "the foremost philosophical challenge of this age, in my view, is to escape the state of nihilism, relativism, and confusion that modernistic history and modernistic everything else has left us in. That requires an ability to step out-

side ourselves, our dreams, artifacts, and domineering drives, to discover and acknowledge another, objective reality that we have not
created nor ever fully controlled . . . acceptance of the unconscious,
unplanned, unsuperintended wisdom of evolution."[19]

This theme infuses this preeminent environmental historian's
work. But who should interpret this wisdom? Evolutionary biologists? Environmental historians? And does Worster fall prey to the
naturalistic fallacy—in other words, is appeal to the "wisdom of
evolution" just a modern-day way of confounding the "is" of nature
with the "ought" of humans?

For Daniel Botkin, "solving our environmental problems requires a new perspective that goes beyond science and has to do with
the way that everyone perceives the world," a perception that "depends on myths and deeply buried beliefs." Like many of the authors
cited above, Botkin feels we must break our conceptual shackles
and see nature in a different way. He calls for us to change our paradigm of nature from a static entity to one in which flux and change
prevail: "The way to achieve a harmony with nature is first to break
free of old metaphors and embrace new ones so that we can lift the
veils that prevent us from accepting what we observe, and then to
make use of technology to study life and life-supporting systems as
they are."[20]

Botkin believes our metaphors blind us to how nature really is:
by dusting away the cobwebs of old paradigms that obscure our view,
of theoretical presuppositions not grounded in hard "facts," we may
see the essential nature. Apparently, Botkin feels he has done this;
his computer models can provide us with an accurate, nondistort-
ing mirror for nature. This is dubious on several levels. First, how
can we trust Botkin to tread firmly where apparently everyone else
has found quicksand? Next, the technology that he proposes might
save us is itself a result of past constructions of nature; it is impreg-
nated with a history of socially influenced theory and metaphor-
laden observation.[21] Finally, as Peter Taylor points out, it pays to
nurture a healthy skepticism of those who claim to hold the key
to a value-neutral, universally beneficial social order, particularly a
social order where the claimant would be poised to exert control.[22]

Are the biologists promoting biodiversity doing what Botkin is trying to do? Are they attempting to use the authority granted them by society to speak for nature, to tell us what nature wants, and to control the process whereby we remake the Earth by remaking our ideas of it? According to Evernden:

> Ecology is today's official voice on natural matters, an institutional sha-man that can be induced to pronounce natural whatever we wish to espouse. Ecology is, in this sense, simply being used as a blunt instru-ment to help implement particular life-styles or social goals. . . . What matters is not what ecology is, but how it functions, how it is perceived and used—and perhaps *why* we seem compelled to assert such assump-tions of "naturalness" at all. What ecologists do or think they do is, in this context, irrelevant. What is interesting is that ecology, and science in general, should be called upon in defense of differing social ideals.[23]

Yes, this is interesting. But what ecologists do or think they do is not at all irrelevant, especially if we wish to understand why they are called upon as arbiters in societal matters. Scientists promote themselves as objective observers of the natural world; they can therefore claim to know nature and what it wants, can act as mere conduits for nature to reveal its facts, its importance. Biologists who have the preservation of biodiversity and an expanded role in society for themselves as their goals promote their own right to speak for nature. This should be important—and unfortunate—to Evernden, because by controlling the discourse on biodiversity, they perpetu-ate the relegation of nature to human control. And they also exert social control, since the policies to preserve biodiversity that they promulgate may also dictate where and how people may or may not live.

SCIENCE STUDIES: AN INTRODUCTION

Here is one of the many points of entry for a science studies perspective. This discipline descends both from the science, tech-nology, and society programs that sprang up in the 1960s to re-direct research and the products of science toward juster, more peaceful ends and from movements to study the history, philoso-phy, sociology, and politics of science. Today, scholars in science studies watch scientists construct scientific knowledge; they aim to explicate how scientists negotiate and defend their cognitive and social authority in society. Their laboratory animals are scientists in action, their resources the spoken and written traces left by scientists in and out of the lab. Their philosophical commitments are usually relativistic, their insights fascinating, their impact on science as practiced small.[24]

If environmental historians have been reluctant to pursue the elusive scientist, so, too, science studies scholars have generally avoided environmental issues. Scholars—including most environmental historians—who analyze environmentalism tend to embrace a set of normative commitments to specific environmental objectives. They seek to help effect change in the world that will further some of the varied goals that fall under the very broad umbrella of environmentalism. Donald Worster, one of the most respected environmental historians, asks: "How can we survive as a species without undermining or degrading the planet Earth and its fabric of life, the very means of our survival? This predicament gathers force year by year . . . historians are beginning to add their voices to its resolution."[25] Environmental historians wish to use history as a moral compass to guide us in our current thinking and, therefore, in our current actions.

Science studies' practitioners may lack such normative commitments to real-world change, in part because such engagement is difficult to defend in a relativist universe.[26] Where scientific truth and scientific institutions are whatever we all agree they are, biologists' or environmentalists' values may be little more than a resource for science studiers to deconstruct and explain away. Furthermore, science studiers are attempting to carve out a niche for themselves as a respected discipline; they are attempting to black box their own authority in the halls of academia. Activism or overt normative commitments might blur the boundaries that they are scrambling to erect between the academic and the nonacademic. The sociologist of science Steven Yearley tentatively suggests that environmentalism might nudge science studiers away from detachment and toward engagement with moral positions. He does not indicate, however, how this might or could happen.[27]

Yet science studies does offer a comprehensive methodology and a deep understanding of science that environmental history lacks. A science studies standpoint can also correct for environmental historians' nonreflexivity in treating scientific and nonscientific claims. Often lacking a perspective from which to view science as a particular form of social activity and subscribing to a positivist view of scientific knowledge and practice, environmental historians may treat the words and beliefs of scientists with unwarranted reverence.

Environmental historians may have sound political reasons to leave science in peace. Steven Yearley points out that green arguments often hinge on the preeminent authority granted science by broader society.[28] Science studies disallows this kind of knee-jerk allegiance; claims made by scientists seeking power in environmental disputes are assessed as claims made by anyone else would be. Science studies provides the mechanics for looking at the maneuvers of scientists as they seek to speak for nature. It allows us techniques to examine how knowledge claims about the environment are constructed, or what resources are marshaled to create durable webs of expertise and status.

But when analyzing environmentalism—in this case, biodiversity—deconstruction is a high-stakes contest. What if conservation biologists' warnings are not just rhetorical weapons to get us to save what they like? What if we really are standing by mute while a holocaust progresses? What if the awareness of imminent peril, combined with a deep-rooted love of the natural world, that prompts biologists to take to the hustings is not merely an idiosyncratic bias— and it is a bias no matter what they say—but an *adaptive* bias? If we shared this bias, might we enable our survival, or at least maintain a more interesting, beautiful planet? We must take deadly seriously the possibility that some of the claims made by biologists are correct: they may be passing on information that is vital to each of our lives, to the lives of billions of organisms, to the perpetuation of millions of species and the ecological processes that sustain them and us.

Brian Martin warns that science studies critiques may play into the hands of the opponents of environmentalism or other social goals.[29] Without certain normative commitments, when science studies meets environmentalism, the resulting analyses can easily become resources for those who oppose social or environmental justice. However, it is precisely the alleged gravity of the situation—a gravity conservation biologists take great pains to show, thus adding ponderous weight to their pronouncements—that should give us pause. For conservation biologists have, in a sense, created "the biodiversity crisis," and the cataclysmic urgency they attribute to our current predicament preys on the fears and interests of the audiences they are luring. If effected, some biologists' prescriptions would usher in sweeping, radical, and sometimes coercive and harsh changes in how people everywhere live their lives. And, like so many other claims made on behalf of the environment, the claims makers seat themselves in the center of a huge resource web, seeking to control information and power. Without a science studies perspective, those analyzing environmental affairs or making environmental policy may all too easily be captured in the resource net that biologists are casting. Obeisance may be as analytically and practically hazardous as deconstruction.

REDEFINING SCIENCE

Not all scholars working in science studies have avoided environmental topics. Exceptions include Sheila Jasanoff, who has examined how environmental regulatory agencies such as the EPA use science, and how science's authority shapes perceptions of environmental risk. Peter Taylor engages in social analyses of environmental change; he searches for the political and social commitments that underlie environmental knowledge and, especially, the policy pronouncements that scientists claim follow ineluctably from such knowledge.[30]

Steven Yearley, who also writes extensively on environmentalism, believes that the subject is "of special interest to science studies because of its claims to be a scientific social movement. In particular, negotiations over the kind of certainty which this scientific basis lends, and over the degree of confidence which science gives to practical decision-making, are of central importance. More attention needs also to be directed to the question of why 'ordinary' people support this social movement; it cannot just be—as some seem to think—because of the strong scientific arguments supporting it."

Or can it? It may be that these "scientific" arguments build on values scientists take from society and then sell back—consciously and unconsciously—to the public. Of course they would resonate with us! Furthermore, what is "scientific" is subject to negotiation. Elsewhere, Yearley notes that "many social movements in this century have been based on moral or religious claims. I shall argue that the green case is unusually dependent on scientific authority and that this fact lends the movement peculiar weaknesses as well as strengths."[31] Yearley demarcates what is and what is not science—in other words, science includes no moral or religious claims. Yet if scientists may be empowered to stake such claims as freely as they now stake factual claims about the real world, then the question of what constitutes a scientific argument takes on new meaning.

Conservation biologists are attempting to redefine what constitutes a scientific argument. For science and its workings are not ineluctable, ahistorical phenomena, handed down from the gods. What science is and what scientists may properly do are contested and shifting turf. Science studiers talk of "boundary work," a phrase coined by Thomas Gieryn to describe the processes scientists engage in as they continuously negotiate and redefine the borders of science according to context-specific needs.[32] Nonscientists may simultaneously struggle over the same patches of contested ground, attempting to circumscribe—or even expand—science in some way. I say "expand" because environmentalists try to pull scientists into debates, expand scientists' role, capitalize on science's cultural cachet. This may happen even when the scientist wishes (and I stress this is a choice each scientist makes) to remain "above the fray," out of the politics she prefers to deem "extrascientific." Thus what science "is" at a given time and place results from complex negotiations among scientists and those allies whose allegiances they would enroll, or who would enroll them, and those opponents whose authority they would discredit or who would discredit them.

Conservation biologists attempt to redraw the boundaries of science in several ways. They describe themselves as "mission-oriented"; their mission is to proclaim the imminent, cataclysmic loss of the Earth's myriad species and ecological processes, and to enable action to halt that trend. To fulfill their mission, they attempt to expand the space where scientists may reasonably speak and act, from the laboratory or field out to congressional hearings, foreign villages, television networks, best-seller lists, and church pulpits.

Biologists will also make policy recommendations and ethical prescriptions based, they say, on their "scientific" understanding of what biodiversity—and humans—need. Soulé calls conservation biology a "crisis discipline," in which "one must act before knowing all the facts," basing such action on "intuition, as well as information." Biologists' recommendations, then, will not spring forth indefeasibly from their data analyses, which, as we have seen, are incomplete and tendentious. Intuition may well reflect the same values that made biologists conceive of and promote the concept of biodiversity in the first place.

Biologists may recognize this. They attempt to speak for values that go far beyond what one might think of as falling within their realm of expertise. Note that conservation biologists take some of these values as part and parcel of the science itself. Built into conservation biology's foundations are normative principles, or, according to its founder Michael Soulé, "value statements that make up the basis of appropriate attitudes towards other forms of life— an ecosophy." This ecosophy affirms that biological diversity, ecological complexity, and evolution are "good," and that biological diversity has value exclusive of any potential use to humans. Many biologists I have interviewed subscribe to these normative precepts, including the notion of biodiversity's intrinsic value.[33] These norms may then shape all aspects of their science—including observation and data collection—while simultaneously dictating that an inextricable part of being a conservation biologist is acting out those norms to the fullest—that is to say, acting in society to see that those norms are recognized and heeded. Internalizing these norms forces biologists out into the world.

Given the ecological destruction they perceive, biologists choose to risk the public's image of the scientist as value-neutral expert— an image scientists themselves have long burnished for public viewing—by the very act of public advocacy, by speaking about a host of heretofore extrascientific values, by promoting a conservation mind-set they consider an ethical imperative. Should they succeed in their advocacy, they will heighten public consciousness about the values of biodiversity, so that people's images of nature come to match their own. Simultaneously, biologists empower themselves to be mediators of human values. Furthermore, considerable personal and professional profit and power in setting public priorities comes to those to whom we grant authority to speak for nature. But the carefully tended reputation of scientists as objective experts is on the line, as is the effectiveness of the biodiversity conservation movement.

BOUNDARY WORK

In his explication of boundary work, Gieryn demarcates "essentialists" from "constructivists." Essentialists find "unique, necessary, and invariant qualities that set science apart from other cultural practices and products, and that explain its singular achievements (valid and reliable claims about the external world)." Constructivists—a label that defines many, if not nearly all, of those in science studies—see that "no demarcation principles work universally, and that the separation of science from other knowledge-producing activities is instead a contextually contingent and interests-driven pragmatic accomplishment drawing selectively on inconsistent and ambiguous attributes." That is to say, either you do or do not believe science is somehow distinct and special. Either you believe that scientists deserve some of the cognitive authority they have been granted by dint of their objectively derived discoveries of verifiable facts about nature, or you believe that scientists have arrogated this authority to themselves by deft marketing and skillful negotiation.

"Essentialists *do* boundary-work; constructivists *watch* it get done by people in society," Gieryn continues.[34] The dichotomy strikes me as problematic in several ways. First, it suggests that constructivists (science studiers) are somehow the neutral observers, with no political or ideological program of their own. Yet they, too, wage fierce battles to draw boundaries as they try to redefine science, just as they say scientists have always done. If science is contested space, science studiers are certainly among the groups who try to demarcate where its boundaries do, will, or should lie.

Next, Gieryn's dichotomy is false. One can certainly believe in the notion of boundary work, can describe how it works, and yet have essentialist tendencies. The world is very seldom black or white, as protagonists claim. It is usually some rich shade of gray.

Scientists, and those who adhere to a positivistic, realist's worldview, portray the process of science as one of discovery of facts about nature. Scientists are neutral transmitters of nature's wisdom, engaged in the process of building an ever-expanding corpus of facts, an ever more complicated picture of nature's processes, coming ever closer to core truths about the natural world.

Nonsense, say relativists or constructivists, who rob nature of agency in the portraits scientists paint of it. Facts are social constructions, built on socially derived metaphors, theory-laden observations. Truth may be said to exist, but truth is social convention; truth is whatever we all agree on, or whatever becomes too difficult or too expensive to contravene.

Martin Rudwick uses debates in nineteenth-century geology to depict science as neither about the discovery nor about the construction of truths. I follow Rudwick in believing that scientific knowledge is *shaped*. Constructivist sociologists of science have convincingly shown that theory shapes even apparently neutral observation, that culture constrains framing of questions, appropriate attitudes, likelihood of accepting or rejecting facts, what counts as reasonable evidence. Yvonne Lincoln and Egon Guba show that values from our personal backgrounds and disciplinary paradigms inescapably affect our methodology, even how we perceive the world.[35] That much-vaunted ideal of "objectivity" cannot be met.

Yet at the same time, nature intransigently insists on challenging our portraits of it. To say that truth is purely a social construction is to slip down the well-greased slope to cultural or personal solipsism, to deny that we can ever really know anything about the real world. Rather, I believe, like Rudwick, that science is about the shaping of knowledge. Using a core of natural reality, scientists mold verifiable knowledge. Of course, the palate of grays with which the world can be painted is rather polychromatic. Any given knowledge claim may have virtually nothing to do with the real world it pretends to portray, or may represent some profound truth about the real world. Scientists have much room to negotiate, and we have much room in which to interpret their negotiations.

The same may be said of values. Many conservation biologists not only present themselves as neutral transmitters of biodiversity facts but claim to be neutral transmitters of biodiversity's values too. They even claim that biodiversity has intrinsic value that can act on us; such a locus of value would inhere in biodiversity and not in the value system of the human observer. Of course, the constructivist will argue that value is in the mind of the beholder: we see and value what influential members of society prepare us to value, or what it is in our best interest to value. Here, too, I hold a compromise position, discussed more fully below.

Based on their self-representations as neutral proponents (itself an oxymoron) of biodiversity's facts and values, biodiversity biologists make policy prescriptions that say, in effect: "Listen, this is what nature wants; this is what biodiversity really needs. And this is what humans must have, too." Constructivist observers, on the other hand, will say that biologists impose their values on nature and seek policies to preserve what they find most valuable—that their policies amount to defenses of a value-ridden, socially constructed idea of nature.

The middle ground—terrain I stake out here—says that if biologists have some epistemological handle on the truth, then they may well be in a reasonable position to advise on policy decisions and shape our ideas of the natural world. And if the time they have spent studying biodiversity plausibly makes them experts on its values, that gives further credibility to their claim to sit at the policy table; it gives us further reason to take their ethical prescriptions seriously. But we must also heed the constructivists' warnings and remain skeptical about accepting biologists' voices as the only credible ones to speak for nature.

The middle ground requires us to search for the values underlying policy recommendations, the ideological commitments behind ethical prescriptions. By admitting the possibility that biologists may speak, not only for nature's facts, but for nature's values, it expands their role in policy making; but it limits their role to the extent that we make space for others who may have equally plausible access to facts about the natural world and its values. In the intermediate terrain between realist and relativist, we thus simultaneously make biologists' voices more powerful by allowing values into the scope of biologists' expertise and less powerful by downplaying the paramount role of scientific facts in conservation and policy prescriptions.

What of how we view "science in action"? This phrase comes from Bruno Latour, in whose writings scientists muster Machiavellian moxie to widen their influence by whatever means necessary. They enroll allies—other scientists, equipment, machines, government officials, biological organisms themselves—to create networks that allow them to gain resources, to expand their boundaries. For Latour, science is nothing special. There is no "science" and no "society." Rather, all society is shifting alliances, power winners and losers, changing nodes in resource networks.[36]

To apply a Latourian worldview, conservation biologists are tying the traditional epistemological and cultural cachet of science—a cachet itself established by the fierce and long-running accumulation and defense of a diverse, but now firmly united, coalition of allies—to a new set of actors whose interests they seek to capture. They are speaking not only for the 10 (or 30, or 50, or 100) million species of organisms out there. They are also convincing all of us, the empowered and the barely powered, of the importance of each of these organisms for each of our lives. Biologists would have us grant them authority to speak for all of nature's facts and values; they would have us listen to the many prescriptions that maintain their supremacy at the center of a complex and increasingly powerful resource network.

Biologists will not recognize this portrait of themselves as generals marshaling forces to grab power. They claim to speak for nature, and for all of us, in order to save the things that give meaning to their lives and that each of us requires to exist. And so they may. I do not doubt biologists' sincere belief in many of the values they champion for biodiversity. And I have no doubt that their lives— as well as mine and yours—will be diminished or even endangered with the diminution of wild places and wild species. But I also do not doubt the essential wisdom of Latour's model of science and society. Power and authority come to those who tie your interests to theirs, who make themselves and their resources an obligatory point of passage en route to your goals and interests. But what Latour cannot account for is the possibility that some of the arguments are stronger because they are valid; that we should sometimes heed the partisans of biodiversity because they may be right; that sometimes scientist-generals win battles, or at least wage wars, because they have the force of truth behind them.

What of "science"? I am not a strict constructivist, as Gieryn would have me be, but I concur with his notion of boundary work. Science is not predestined; it is what we all agree it should be. And a group of biologists at the cutting edge are trying to expand the boundaries of biology to respond to what they see as the exigencies of conservation. Yet . . . I concur that some external justification exists for science's current negotiated position in society— namely, that science has produced demonstrable facts about the natural world, and that its far-flung boundaries stem in part from this. Believing that they are to some extent justified and necessary, I tentatively applaud the attempts of conservation biologists to expand those boundaries, to change the definition of what it means to be a scientist.

Which brings us to biodiversity: is it an objective and clear representation of natural reality, or a social construction designed to manipulate and accrue? A little of both, naturally. The concept of biodiversity shoulders the burden of all that biologists find intriguing and wonderful in the natural world. It is a creation of a specific time and place and creator. Yet it is informed by the richness of the natural world. It is the multitude of real-world organisms, species, and processes commingled with biologists' factual, emotional, political, aesthetic, spiritual, and ethical values of the natural world, all combined to shape public perceptions, actions, and feelings.

This compromise view of facts, values, science, and nature may leave me little constituency. Scientists may see me as a traitor who casts doubt on the objectivity and value-neutrality of the scientific enterprise, the very positions (as we shall see below) to which scientists retreat, because their power has traditionally sprung from them. The science studies scholar may, conversely, see me as having been captured by the constituency I am portraying—lured by the siren call of the scientist, ensnared in his resource web.

I engage in boundary work of my own as I study—and participate in—the struggle to define what science is and where it may go, what it may do. Harry Collins metaphorically relates scientific facts to ships in bottles: they appear to belong there organically, but by careful scrutiny one can see the labor involved to make so much effort seem effortless.[37] I do something similar for the idea of biodiversity. I am not concerned so much about how operational facts about the natural world are made to seem like irrevocable, permanent, seamless knowledge claims. Rather, I am trying to show how biologists attempt to have a new paradigm of nature, with all the values that inform it, appear like a ship in a bottle—so natural it seems as though it has always been there, could never be extricated or denied—that there could never have been such strenuous effort to put it there in the first place. Simultaneously, they try to have their roles as experts on these diverse values accepted as the biologist's unquestioned privilege.

MAKING CONSTRUCTIVISM CONSTRUCTIVE

Biologists are beginning to take notice of academic debates over the social construction of nature. Michael Soulé has organized a conference and co-edited a book that both attempt to counter "the social siege of nature." He asserts that when scholars with various political agendas wage attacks on objectivism and scientific knowledge, they provide covert, ideological cover for more overt attacks on the natural world. Soulé does not see these debates as esoteric and cloistered; rather, he believes those most likely to be influenced by arguments about the social construction of nature are those most likely to be making policy decisions on the future of biodiversity, "sometimes to the detriment of living nature."[38]

Indeed, by showing the labor involved in constructing the biodiversity concept and in redrawing maps of authority, I may be damaging the reputation of conservation biologists by removing the guise of inevitability from the various layers of their enterprise. Soulé fears the undermining of objectivism and therefore of scientific expertise, particularly in matters of conservation. But by analyzing how and why conservation biologists do what they do, I may in fact strengthen their hand. The feminist scholar Sandra Harding advocates "strong objectivity": by recognizing the societal and disciplinary cultures in which each of us is positioned, and that therefore cannot help but mold our scholarship, we can take steps to becoming more objective. In other words, by highlighting the "positionality" of conservation biologists, I may simultaneously help them become more self-aware and lend credence to many of their expanding boundary claims.[39]

Other scholars attempt to transcend the realist/essentialist–relativist/constructivist debate, to move beyond polarized arguments over whether science is all about fact or merely fiction. Sergio Sismondo parses the metaphor of "social construction" in part to show that many scholarly enterprises bearing the "social construction" label are quite compatible with a realist's view of scientific activity: "We have to come up with a theory of science that doesn't make successes (and failures) into miracles." N. Katherine Hayles writes of "constrained constructivism"; like Harding, she notes that we are all situated in social positions whose allegiances, assumptions, and metaphors invariably shape the knowledge we produce. However, this does not mean that the natural world does not constrain our portraits of it. The trick, according to Hayles, is to allow a large chorus of differently situated voices to produce knowledge; by so doing, we shall come closer to producing views of the natural world that more closely resemble that real world. In this ecological epistemology, Hayles stresses the interrelatedness between the viewer and the viewed, and between the viewer and all other viewers. And it admits and asserts the value of all other forms of life, for their points of view must also be respected and appreciated.[40]

Peter Taylor's research program, which he calls "heterogeneous constructivism," dovetails nicely with Hayles's ideas. Partly inspired by Latour, partly by Gieryn, but mostly by a heterogeneous band of antecedents and contemporaries, Taylor takes examples of environmental knowledge and uncovers the diverse natural, social, and ideological strands that went into making that knowledge. After demonstrating how certain choices were made and others ignored, he shows how the resultant knowledge facilitates or constrains other choices, other decisions, other knowledge claims.[41] By adhering to this way of thinking, one can abandon the positivist position without necessarily slipping down the slope to absolute relativism. For Rudwick, Harding, Sismondo, Hayles, Taylor, and myself, nature may inform knowledge products; nature may make some of the heterogeneous elements that contribute to and stem from knowledge more plausible, more like truth.

My work is itself an example of heterogeneous constructivism. As do the biologists I portray, I pull together a multiplicity of diverse elements to tell a story, to construct a piece of knowledge that I hope will exact compliance from you, one that may facilitate certain future ways of thinking and disable or debilitate others. It is up to you to analyze whether I have chosen pertinent strands, to find ideological holes or commitments, to enable or disable the portrait of science in action I serve up to you.

THE PROSELYTIZERS

Biodiversity did not attain buzzword status without vigorous promotion. I hope to show that this promotion has been intentional, and that biodiversity proponents feel that they and their fellow biologists bear a *responsibility*—indeed, that it is part of what it means to be a biologist—to proselytize on behalf of biodiversity. Paul Ehrlich speaks of the "National Rifle Association Principle"—you don't have to persuade 100 percent of the people; if you get 2–3 percent riled up, they'll do your work for you: "They're willing to do something about it, propagandize others, get out there." Ehrlich and his colleagues use biodiversity and the values it subsumes to entice people, scare them, and *convert* them into missionaries who will fight to save the things biologists love.

As we have seen, biodiversity biologists are not the first to promote ecological ideals to society at large. In 1970, Leo Marx suggested that as environmental fervor grew in the late 1960s, most scientists were content to sit on the sidelines, believing that a handful of their peers were screaming loudly enough to convey the conservation message effectively.[42] After Marx wrote, biologists watched the public's interest in environmentalism wax but then wane. Biologists have lined up behind biodiversity to reverse this trend. I believe the movement behind biodiversity is unprecedented in how widespread it is among biologists, in how many of the top environmental biologists participate, in how many different kinds of venues they choose for their message, in how strongly they urge fellow biologists to join them, and in how many different kinds of values they presume to speak for.

Thomas Dunlap chronicles the transformation of wildlife conservation from an emotional crusade by private citizens into a bureaucratic, bloodless march where "academically qualified scientists have displaced naturalists and nature-lovers." He warns that "wildlife protection is, more than most of its defenders like to think, a matter of organization and management, and bureaucratic plans are as vital as nature essays." Biodiversity biologists bridge this gap. They have infiltrated the bureaucracy with their data, their plans, and their neologism. But they simultaneously promote love of nature; they write the nature essays, launch the emotional crusade. Dunlap writes that "science is a body of knowledge and a source of authority to which people appealed, a myth, and a social activity. While it is supposed to guide wildlife policy, that activity unfortunately involves choices and values more than decisions of fact."[43] And biodiversity biologists attempt to influence those choices by having us adopt their values. The biodiversity movement is unique in how many different hats prominent biologists attempt to fit simultaneously on single heads, how many different boundaries they try to expand and defend, and how far into uncharted territory so many of them seem willing to sally.

Thomas Lovejoy, who accurately claims to be "probably *the* organismal biologist closest to government," finds it "very hard to understand why every single biologist of that sort isn't up in arms." Biologists do not just work on behalf of biodiversity; they urge all their colleagues to join them in their self-appointed mission.

In a 1992 *Conservation Biology* editorial, Kamaljit Bawa and Garrison Wilkes asked, "Who shall speak for biodiversity?" Their answer: the Society for Conservation Biology, at least in government and policy forums, for "in order to speak directly to the merits of biodiversity conservation and seek a consensus, we need to enhance our capabilities to inform the public at large and policy makers in particular, and to communicate among ourselves more effectively. This will require new mechanisms and an expanded role for the Society." A few years earlier, in a more contentious *Conservation Biology* editorial, Noss had asked, "Should the conservation biologist try to be a scientist and activist at the same time? This is tricky, but I suggest that the answer is yes." This editorial is a call to action: "We need to stop arguing over esoteric details, stop declining to comment when we do not have all the data, and pull together to offer strong guidance on how to save the Earth." This is imperative, according to Noss, if conservation biologists are to control the discourse: "If conservation biologists fail to respond, there will be plenty of economists, developers, industrialists, timber executives, livestock barons, and others jostling to offer their advice."[44] The Society for Conservation Biology now heeds such warnings and has begun to issue specific policy prescriptions.

Some biologists had previously tried to urge their peers to combine science with activism. In 1981, Eisner and a group of colleagues (including Wilson, Raven, and the Ehrlichs) outlined the current crisis of species extinction and urged scientists, and especially biologists, to take more responsibility for tropical conservation. Iltis has been cited for his "indefatigable pursuit of converting the inconvertible." Among those difficult to convert have been his fellow biologists. He warns biology teachers who neglect to inform their students of impending eco-doom: "May all the fates help us and you! For then you will be guilty—guilty of being shepherds leading innocent sheep to slaughter."[45] As early as 1970, in a call to action to his fellow scientists that was rejected by several major journals, Iltis proclaimed:

> For the preservation of this diversity we must work and fight. *For this we must become politically involved*, for all effective conservation is ultimately decided by political judgment! . . . The trouble is that too many of you, who are perfectly able to get involved, are unwilling to fight. Many of you are apologists, and worse, *many of you refuse even to mourn*. You are embarrassed to be sad, much less to become angry and act. And your refusal to mourn, to get angry or involved, will cost you your self

respect and the power that you now have by virtue of your knowledge and profession. You will lose, then, the *right* to criticize others, to have any voice, political or otherwise, in what is being planned.[46]

Iltis believed the knowledge base and institutional clout of scientists *required* them to venture forth into society in the name of conservation.

Why didn't they listen then, and why do so few of them listen now? Arne Naess provides a perceptive list of reasons (see below) why scientists have been reluctant to go public and support the conservation movement; but "then came conservation biology!" Naess's fellow deep ecologist Bill Devall praises conservation biology, where "some scientists are speaking up, *witnessing* as it were."[47]

Peter Raven is a self-proclaimed proselytizer for biodiversity. When he addressed the American Institute for Biological Sciences, he urged its members to join him, to unite in the name of biodiversity: "At any rate, fellow biologists, let us resolve to put aside our differences, and some of our narrow views, and to use and to improve this organization and make it a constructive force for accomplishing what we want, and what our nation and the world require of us." Raven notes that political leaders were "hungry for authentic expressions of opinion from informed people, and we need to take part in the political process at all levels to make that process work." He concluded:

> We must begin to learn to give credit to our colleagues who do speak out, often making severe professional sacrifices in the course of doing so. We need to approach the media, the politicians, one another—anyone who will listen—and try to improve the sustainability of the world. Ecological problems generally, with an emphasis on the preservation of biodiversity, will dominate our agendas during the next century, but the steps that we take now will be of critical importance in establishing the contours of the world of the future.

Raven urged his colleagues to preach to anyone who would listen, and at very least to stand aside and facilitate attempts to speak for biodiversity. His message was echoed to me by Paul Ehrlich, who wishes the 70,000 members of the American Institute of Biological

Sciences "could get themselves organized—and the AIBS has failed totally to do so—if they get themselves organized, they'd have an *enormous* impact on people's attitudes towards biodiversity."[48]

This proselytization runs all in the family. Many of the biologists I interviewed have been inspired by their fellow biologists' conservation activities. Or they prod one another on in the conservation game. K. C. Kim talks about a colleague who along with "Peter Raven, Tom Lovejoy, [and] a number of other guys—has been giving workshops at the National Science Board, basically brainwashing board members on biodiversity issues." Why had Daniel Janzen attended the National Forum on BioDiversity? "[E. O.] Wilson asked me." Raven cites Ehrlich as formative in his interest in real-life conservation issues. Ehrlich himself seems to be a force of nature. Soulé says of his own compulsion to fight for conservation: "I was Paul Ehrlich's graduate student; I come by it naturally." Wilson credits Raven with getting him into the conservation business:

"But it was Paul Ehrlich who was making noises all along. And yet Paul was sort of a professional agitator, in addition to being a very fine scientist. . . . But then Peter Raven, who is a plant systematist, and head of the Missouri Botanical Garden, *really* began to yell about it. This was the late 1970s. And making inflammatory speeches on the loss of biodiversity. And I said to myself—and I've told Peter this—I said, well if someone like Peter Raven feels that it's necessary to get up on the barricades, maybe there's more of a need for someone like me to be involved as well. Maybe these conservation organizations aren't doing the job. Raven seems to be the guy who's doing all the talking in public, you know, the strong statements about biodiversity. I had made that statement in *Harvard Magazine*, which got a lot of attention, too, in 1979 on the greatest folly that will be least forgiven us by our descendants. And I got so much attention that I said, that surely indicates that leadership from scientists in this area is desperately needed. So I took the plunge in 1979-80."[49]

Raven corroborated this story: "One of the greatest compliments I ever got is that E. O. Wilson told me that the reason he got back into this preservation thing or into it heavily in 1980 or 1981 was because he heard me give a speech. So you see we all reinforce one another." They do indeed. They form a support network, urging one another on to breech science's value-neutral walls, to stand on the front lines for biodiversity.

G. Carleton Ray wants "the universities to get off their high horse a little bit. . . . Most of us in ecology are out there testifying or one thing or another. I wouldn't say most—a lot. A lot are. Not enough." Peter Brussard, who has been president of the Society for Conservation Biology, feels that "biologists have done a terrible job in informing the lay public about biological resources, and the fact that all our renewable resources are biological." It's clear to Brussard that the public doesn't know the basics, and "I think we can just take the blame for that right square on ourselves." Inculcation of appreciation for biodiversity is the biologist's job; so, too, is more explicitly political work. Brussard believes it is critical that conservation biologists educate environmentalists and economists "about ecological realities." And in the inaugural issue of the Ecological Society of America's *Ecological Applications*—a spin-off of *Ecology* designed for practical research (much of it difficult to distinguish from conservation biology), Brussard advises, "to be blunt, preserving biodiversity is clearly in every ecologist's own self-interest, and he or she should pledge to spend a measurable portion of time working to accomplish this goal. The first step is a commitment to the political action necessary to enact an appropriate national policy on biodiversity. . . . Every ecologist should lobby for [the National Biological Diversity Conservation Act's] passage."[50]

In the pages of *BioDiversity*, the collection of essays on the subject edited by E. O. Wilson in 1988, several of the contributors consciously sought to launch a movement. Ehrlich urges, "We must begin this formidable effort by increasing public awareness of the urgent need for action." The scientific community should "aggressively participate in writing and executing the contract" between managers of wild lands and society, Daniel Janzen declares. "Conservationists spend entirely too much time talking to each other," Tom Cade says. "We do not need to convince each other of the importance of what we are trying to do. We need to convince the vast majority of other folks!"

Worldwatch President Lester Brown had taken his crusade against soil erosion on the *Today Show*. According to him, to win the battle to save the Earth, "scientists are going to have to become activists." He suggests that the Club of Earth (formed at and for the National Forum), "the purpose of which is to bring scientific attention more quickly to important but neglected environmental problems, should write the White House and explain why it is important to restore full support for the UN Fund for Population Activities in order to protect biodiversity." Brown continues: "We've got to move the issue from the scientific journals into the magazines and the popular press, so that maybe someday Jane Pauley will say, 'And today we have a scientist who's going to discuss biodiversity. THAT'S RIGHT, biodiversity.'"[51]

Soulé told the same audience that for conservation biologists to be effective in their mission, they had to embrace tactics few would recognize as scientific:

> Biologists wish to convince others of the importance of protecting biodiversity, including ecological and evolutionary processes. The problem is that very little thought and research has gone into the best ways to accomplish this vital goal. . . . Though it may sound heretical, our primary objective as conservationists (not as educators) should be to motivate children and citizens, not necessarily to inform them. Research may show that the two objectives are incompatible . . . the new motivators for nature might take a page from the advertiser's book. . . . We must learn from the experts—politicians and advertising consultants who have mastered the art of motivation. They will tell us that facts are often irrelevant. Statistics about extinction rates compute, but they don't convert.[52]

To Soulé, these are dire times: biologists must *act*. For them to engage their listeners' hearts before their minds is far from inadmissible; it is necessary if biologists want to promote biodiversity effectively.

In his textbook *Conservation Biology*, Soulé notes that "not many scientists, during the few decades of their careers, are able to commit their minds and labors to an epochal task like saving the planet." He warns of "The Nero Dilemma," and urges his fellow biologists to dedicate themselves to research that will help inform conservation decisions, to get this information out to relevant decision makers, to proclaim one's love of nature far and near, and, as noted above, to attempt to convert people using any means necessary: "It is simply that it is approaching 'high noon' and we should use every (ethical) tool at our disposal to minimize the damage to this planet."[53]

Others join Soulé to advocate that biologists use any method that works. Alvaro Ugalde urges use of advanced communications techniques to get the message across. "I always had very good relations with reporters," Eisner told me. In fact, he and the science writer Natalie Angier published a piece in *The Scientist* (a weekly newspaper for scientists) entitled "Use the Media for Your Message," in which they explicated the nuts and bolts of communicating to the press. Hal Salwasser urges that to sustain biodiversity, conservation managers must implement "innovative marketing to build awareness, create a more informed constituency for endangered species and biodiversity goals, and secure the resources needed to meet those goals."[54]

Which arguments should these crusading biologists use? Many of those I interviewed discriminate among which values of biodiversity they voice to whom. Many say the correct message is the effective one. "I always play my audience," Erwin says. Brussard agrees that "obviously you have to play the audience a little bit. . . . I'm all in favor of using whatever arguments work, to tell you the truth." "Whatever works to sell it," urges Pimentel. "You use every argument you can," advises Woodruff. Ray replied, "When you're asked to talk to the public, you're trying to use any intent to convince them that something is important." Wilson finds three major compelling reasons for preserving biodiversity, "and I tend to emphasize the one or two or combination of all three, according to the audience." Falk talks about the need "to win over as many people as we need to by using every possible argument we have." He feels that no one argument should be primary, "Because you're really talking ultimately about values, what people's values are, and what you need is an argument pro or con at each level of values. And then you've got the kind of diversity in your own argumentation that enables you to persuade people."

So biologists urge polemical diversity to reflect the natural diversity for which they're arguing, and to stimulate the human diversity in motivation and interest. Like Falk and Wilson, Raven finds a multiplicity of valid reasons for preserving biodiversity, so he simply tries to "modify my own ways of presenting and thinking about them depending on the audience that I'm dealing with." Iltis says, "The point is: we are part of nature. We are totally dependent on it. It's just foolishness to do what we are doing. I don't think it's wrong to use every conceivable argument to argue for the preservation. I mean, I've used everything I could think of." The common thread for these biologists is to *use anything that works* to convince people to adopt their values.

The responses cited above vary in opportunism. Most speakers seem to believe in the veracity of all the arguments, although they maintain the pragmatic view that one needs to suit the message to whoever is listening. The ends may always justify the means, even in those cases when the biologist may not necessarily agree with the means. For example, when asked about deep ecology, Eisner replied: "I'm not familiar with that. Let me tell you where the opportunist in me would come in. I mean, my feeling is so strong about the need to preserve biodiversity that if I had a feeling that that, as a political course, became constructive, I could easily see myself backing that. Because I find most of my political moves are partial reconciliations of my principles with reality."

Biologists realize the term *biodiversity* packs clout: it aids them in their various arguments. Eisner told me "the term has proven its usefulness because of its gutsy meaning. I mean, people respond to the term now." Ehrenfeld sees it as part of "television mentality. Gee whiz! mentality. And so it's caught on very well. It's very easily conveyed to reporters. A lot of our top conservationists, including some who are really biopoliticians, but top conservationists, and including some who are very good, have found it very convenient to use it. They've latched onto it." So the term itself is part of a multiplicitous, opportunistic conservation strategy where biologists use any means at their disposal—including using a shibboleth expressly created for promotional terms—to arouse public interest in nature and its values, or in nature and biologists' values.

Peter Brussard, Dennis Murphy, and Reed Noss explicitly recognize the role that language, specifically buzzwords, plays in shaping society's view of nature. But they want to change the operative word from *biodiversity* to *wildlife* when pitching to the general public:

> Just what is it that conservation biologists want to do about the biodiversity crisis in the United States? We suggest that a morally and biologically defensible goal would be "no net loss of native wildlife in the U.S., particularly on our public lands." Why do we specify "wildlife" and not "biodiversity"? It is simply because most people have no idea of what biodiversity means and are unlikely to learn in the near future. "Wildlife," in contrast, is meaningful to almost everyone. The best strategy for preserving biodiversity is to convince the general public that it is a good thing to do, and making biodiversity and wildlife synonymous in the public mind may be the most expedient way to do this.

They go on to discuss the "tactical challenge" conservation biologists face if they wish to convert public land policy to something more benign for biodiversity. They still use the term *biodiversity*, but only for the right audience: "So rather than confusing the issue, let's use 'biodiversity' among biologists and educated laypersons and use the strategically powerful term 'wildlife' (emphasizing its broadest meaning) when trying to communicate our concerns about the loss of biodiversity to the general public."[55]

Of course, the term *wildlife* is "strategically powerful" for many of the same reasons *biodiversity* is: it has come to mean virtually all of life and prompts listeners to conjure up their most cherished natural images, particularly those of charismatic megavertebrates. According to Brussard, Nevadans see *biodiversity* as a newfangled environmental ploy; the term *wildlife*, on the other hand, has a tradition behind it. While the two terms may be fungible for him, using *wildlife* is a purely pragmatic strategy to win over public opinion: "I think I would find it a lot easier to get up and explain to a bunch of realtors why it's important to preserve wildlife than why it's important to preserve biodiversity."

Along these lines, Noss sees a danger that biodiversity will be perceived as "somehow a smokescreen that environmentalists are trying to use to lock up land as wilderness." For biologists to proselytize effectively, they must use the right word for the right occasion, and *wildlife* appeals to more people: "It's simply a question of directing your—using appropriate terminology for the audience."

Soulé is aware of this, too. He takes "an intuitive approach" to biodiversity when speaking to public audiences, because more "scientific" terminology would bore and confuse his listeners. Biologists recognize that to gain support for their conservation goals and values, people's concepts of nature must be changed. Accordingly, biologists fine-tune their conservation strategy, choosing terminology carefully to make sure the words they use convey the most propitious messages (i.e., the ones that best represent biologists' values and goals) to the most people.

CONSERVATION BIOLOGY AND RESPONSIBILITY

Reed Noss, who now edits *Conservation Biology*, stands at the cutting edge of conservation biology—perhaps not in his research, but in his efforts to interest people, to arouse us, to change the way we think about biodiversity, to get fellow biologists to fight for it. Noss bases his essay "Biologists, Biophiles, and Warriors" on his conviction that "the Earth is going to hell." He argues passionately that it is unthinkable for biologists to step aside and refuse to battle for biodiversity: "Environmental policy is too important to be left to the policy-makers, most of whom know little and care little about all that ecologists do and love. Those individuals who know and care about the biota have a moral obligation to act in its behalf." Fighting for the Earth is not merely desirable; it's mandatory, a "moral duty" for the biologist.

"Defending wild Nature and biological diversity, in my view, is the highest calling for biologists," Noss says. "Most biologists do not think of themselves as soldiers. But war has been declared against wild nature, and we best acquainted with that marvelous web of life have no moral choice but to defend our nonhuman friends and relatives, the innocent victims of human greed, ignorance, and arrogance. Our defense is not contingent on probabilities of winning or losing; it is an absolute obligation." Noss's prose brims with defense metaphors. For these biologists, a war rages. On the front lines of biodiversity's defense, biologists must not cower from battle; they must not shirk the responsibility of fighting for biodiversity.[56]

Dan Janzen certainly stands among those who feel this way and who feel no compunction about stating their views resolutely. The gravity of the impending diminution of Earth's genetic heritage is clear: "If nuclear winter threatens all of us and those things we work to save, then the nucleotide summer is surely on the other side of the coin."[57] The future of tropical ecology is placed squarely in the hands of biologists: "It is this generation of ecologists who will determine whether the tropical agroscape is to be populated only by humans and their mutualists, commensals, and parasites, or whether it will also contain some islands of the greater nature—the nature that spawned humans yet has been vanquished by them." Janzen continues:

> Engineers build bridges, writers weave words, and biologists are the representatives of the natural world. If biologists want a tropics in which to biologize, they are going to have to buy it with care, energy, effort, strategy, tactics, time, and cash. And I cannot overemphasize the urgency as well as the responsibility. Within the next 10–30 years (depending on where you are), whatever tropical nature has not become embedded in the cultural consciousness of local and distant societies will be obliterated to make way for biological machines that produce physical goods for direct human consumption. In short, biologists are in charge of the future of tropical ecology. If the tropics of the world go under, biologists of the world will have no one but themselves to blame. We can see very clearly what is happening, what will be the irreversible consequences for biology and humanity, and how the solutions must be constructed. An active as well as a passive audience is there. It is up to us to make the world conscious of its interactions with the tropical living world. If we cannot set aside our personal interests, research, and development, and put our entire effort to affixing permanently some of tropical nature, then we have sold the tropics' long-term fitness for a handful of instant gratification. We are the generation for whom the only message for a tropical biologist is: *Set aside your random research and devote your life to activities that will bring the world to understand that tropical nature is an integral part of human life.* If our generation does not do it, it won't be there for the next. Feel uneasy? You had better. There are no bad guys in the next village. They is us.[58]

Janzen pulls out all the emotional stops here—the situation could not be more urgent, the self-centered biologist could not be more culpable—and all the literary ones, too. Compare this to any of his (or anyone else's) more conventionally scientific papers. Alliteration ("writers weave words"), neologisms ("biologize"), potent metaphors (agricultural workers as "biological machines"), virtually no passive constructions, use of the personal "I" as subject, cute grammar ("They is us"), use of the imperative, varying sentence lengths, italicized points driven home. Even the subject evolves to envelop his audience: first "biologists," distant and elsewhere, then the inclusive "we," and finally "you": if you are a biologist, the responsibility rests on *your* shoulders. The centerpiece of biology as we've known it is marginalized here, even sneered at, as "random research."

For Janzen, Noss, and many biologists working for biodiversity, the status quo of science must change. The boundary lines of the profession must be redrawn, expanded farther out into "the real world." To do so is not a matter of choice; it is an obligation. During interviews, I asked what each person felt was the responsibility of biologists to work for biodiversity conservation to be. Here are some of their responses:

PETER BRUSSARD: "First of all, our raw materials are disappearing right, left, and center. Or are about to." [DT: "By raw materials, do you mean . . . ?"] "The stuff we study! The stuff we study. We might like to think of an analogy. But suppose you were into collecting antique automobiles. You wouldn't sort of stand by passively while these all went off to the car crusher. And I think biologists are doing exactly that."

DAVID EHRENFELD [in response to another question; I didn't ask him about responsibility directly]: "I think people in the academic world, with a few exceptions, are not doing very much for conservation. It's part of the general collapse and moral collapse of the university." In a letter to *Science*: "Many specialists in a host of fields find it difficult, even hypocritical, to continue business as usual, blinders firmly in place, in a world that is falling apart. They publish in *Conservation Biology*."[59]

PAUL EHRLICH: "Oh, I think it's absolutely enormous. . . . biologists are totally dependent on biodiversity for their own discipline. So if they want to have something to study, and this includes, for instance, when you think of all the gene splicers that we're training; nobody that I know in molecular biology thinks they're going to be manufacturing organisms from scratch. . . . I think it's critical for the world as a whole, and we're the group of professionals most knowledgeable about it."

THOMAS EISNER: "Yes. Absolutely. Who else is going to do it?"

TERRY ERWIN: "I think—this is a really difficult one because, why do I feel responsible to vote in the Democratic primary? Why do I feel responsible to mow my lawn so my neighborhood looks good? It's part of the human family activity. . . . It's on every person's conscience how much they give to it, how much they're going to get involved. And so basically there is no real responsibility in that sort of traditional sense of how we take our societal responsibilities. However, if we want to talk about it in really abstract terms, I think the human species has to take a huge responsibility on what we're going to do about this planet, and living space on this planet. And there it may become more of an obligation rather than something that you just do as you feel. And I think probably by the turn of the century that's going to be pretty obvious to everybody." [Does he promote this among his colleagues?] "Yeah, I try to get everybody to think along these lines."

DONALD FALK: "I believe that science, like art, should be able to progress on its own terms . . . the notion of responsibility should not be attached a priori to either art or science. . . . in other words, it should be the prerogative of any scientist or any artist to spend their time doing something completely useless if they do so choose. Because art and science, as embodiments, as expressions of the human spirit, have intrinsic value. They require in my view no other justification. . . . That's my basic principle. Although I'm not a practicing Quaker, when I was growing up I spent a lot of time in Friends meeting, and that's been one of the big influences in my thinking and in my moral sensibilities, if you will. And there is, in Quaker practice, the very important principle of witness . . . which is that you don't decide for other people what they're going to do about a situation. You only make sure that they are conscious of it, that they are presented with it, that they see what the situation is, whatever it is. . . . I would go on to say that the elements of society that make decisions about allocation of resources are making a pathetic and potentially tragic mistake by underinvesting in research into biodiversity because of its survival value. So I would not go to any plant systematist and say, 'You know you really ought to work on groups of plants that are threatened with extinction.' But if anyone comes to that decision on their own, I think society ought to be prepared to encourage and support that work because it has societal relevance."

JERRY FRANKLIN: "I think the responsibility is to educate as many people as possible with as many sound facts as possible. And I've always taken the position that it was not my position to tell people."

VICKIE FUNK: "I would say that every biologist has the responsibility to think about how they can take what they're already doing and make it— you know, look at ways that it can be applicable to other things. And then take some initiative to make what they already know how to do more applicable to the general public. And I'm not asking biologists to change what they basically are interested in doing."

K. C. KIM: "[The] entire human species has a responsibility for biodiversity conservation. And every biologist has responsibility if they want to have relevance. Reason is, ultimate argument is this: when human species becomes extinct, nothing really matters. . . . even taxonomists, they have to get busy, to produce the better data, learn more about the species, and that data be available to the conservation people, and used to develop strategies for conservation of particular species, or particular communities, particular systems. So I am not saying that systematists should not be doing what they've been doing. Likewise ecologists now should put their energies, not just to learning principles or theories, but trying to apply those things into how are we going to sustain the system."

THOMAS LOVEJOY: "I, in fact, find it very hard to understand why every single [organismal] biologist . . . isn't up in arms. It's no different than when Florence was flooded in the late 1960s and all the art lovers in the world got all upset. It's no different from that at all! It's, you know, it's no different than being concerned about all the books printed on acid paper, you know, slowly crumbling away. And yet, there are large numbers of them who don't even think it's important to care about. I mean, it is so bizarre. The very *stuff* on which their science is being built is vanishing while they play their little games."

JANE LUBCHENCO: "I think that's a very individual decision. And that each individual ecologist has to make that decision for himself or herself, depending on what they feel comfortable doing and what their own motivation is. I think that what the SBI [Sustainable Biosphere Initiative] does is provide justification for people, encouragement for individuals to devote more of their energies to conservation kinds of activities. But I think it's inappropriate to tell people they have to do that. Many, many people have emphasized to us, and written to us, and said they were delighted to see the SBI because they have felt personally that the kinds of things that they were doing were not recognized as valid by the rest of the ecological community, and this gives them—it validates what they felt needed to be done anyway."

S. J. MCNAUGHTON: "I'd never make that sort of value judgment. I don't know that biologists have a responsibility to do anything except find out the truth about whatever problem they work on. I wouldn't say they have any responsibility beyond that. They should be kind to other human beings; they should do their work well."

GORDON ORIANS: "Well, I don't know that I want to lay a moral trip on a profession. In the sense that I don't want any biologist who doesn't want to do that to feel guilty. Sometimes the way I frame it is: would the world be better if Mozart had been a political activist? There are talents and there are talents. And there is a real role, an importance, for understanding basically how living organisms work, how ecological systems work, just the basic science. And good basic science is and will continue to be tremendously important. And I have no negative feelings about anybody that chooses just to do that. And, in fact, there's lots of pressures on people to do that. And with young people, I mostly say, 'Lay low until you get tenure.' Because a lot of this, there's no real reason for you to sacrifice your careers to this."

DAVID PIMENTEL: "Well, I think, of course, they do have a responsibility to do this. . . . Trying to change the system is what it's all about."

PETER RAVEN: "The point is: everybody has to be responsible. Scientists don't have a unique need to be responsible. . . . You have a responsibility, you should speak out, you have to speak out responsibly. You have to give the reasons for what you do. But just because you're talking about things that people don't want to hear doesn't mean that you're wrong. The degree to which people might be penalized, so to speak, for speaking out, has gone way, way, way down over the last thirty years."

G. CARLETON RAY: "We're people in society. It's as simple as that. And you can't forget that. . . . I've spent a lot of time in Australia in the past couple of years. They have nothing equivalent to the National Science Foundation, and no private foundations, virtually none. So they've got strong scientific programs, but we in this country are so lucky to have the public that supports Congress to give money to the NSF and also to all these foundations. That's *amazing*. That people would assume that they can just do whatever they want and money is just going to be handed out by some prestigious organization like the NSF without any effort to give any of it back. And you give it back by writing articles and talking in front of ladies' clubs. And it's a very small thing to do—or teaching, obviously. I think that the only answer to your question is something smacking, not so sappily put, as social responsibility. That's all it really amounts to. And I feel it fairly strongly."

MICHAEL SOULÉ: "Oh, I do. It's always amazing to me that all ecologists aren't also conservation biologists. But it's only a minority of ecologists that would consider themselves to be that. . . . Freud said something about work and love. And the relationship between work and love. . . . if we don't love the objects of our work—that is, if we're a mason and we don't love working with rock, or a carpenter and don't love wood—there's something wrong. We probably shouldn't be doing that; we probably should be doing something else. And so I think it's true, most biologists *love* plants or animals—they love different ones. Some like lizards, some like grasses. But there's a certain affinity we have and even identification we have with the objects of our study. So it's hard for me to imagine why a person would not want to protect the diversity of those entities in the group he or she is interested in. So maybe the answer to your question is: they all do. All ecologists and biologists are conservationists, but they don't know it yet. Or they feel constrained by social pressures not to become applied, for example, because anything applied in the academy is beneath one's dignity. So maybe it's that they're all closet conservation biologists, and it's just a matter of coming out of the closet. . . . If you could go back in time and ask the nuclear physicists who developed the atomic bomb the same question, they would say, 'Well, you have to consider the context that our civilization, our values, everything we stand for is in jeopardy if we *don't* do this. Because what is the world going to be like if it's run by Hitler?' And I think today, ecologists and conservation biologists and geneticists can argue the same way. We are living in extraordinarily difficult times for life on this planet. And we're obligated as a citizen of the planet, and as the only species capable of undoing the damage that our species is doing to do that, to prevent a massive holocaust. But whereas 100 years ago, I wouldn't have felt any obligation to devote much of my professional time to this field, because there wasn't any jeopardy at that time—relatively speaking."

E. O. WILSON: [The question was approached obliquely.] "What I'm saying is . . . that in addition to doing their great stuff on the classification of spiders, or phylogeny of the birds or whatever, that they, this kind of biologist [systematists] especially, should take a larger role in presenting their group and what they know about their group as a result of scientific research, to the remainder of the public. Certainly at least the scientific public, and perhaps the public at large. . . . And one thing I was trying to do is put some spine in the systematists who are gathered like forlorn Bedouins around a dry water hole. [Laughs.] You know, they have a very important role and they have a destiny. And they should take that more seriously."

DAVID WOODRUFF [although he feels great personal responsibility]: "I have a mixed answer to that. I feel some of them shouldn't. I feel some of them are just dangerous when they do step out. And I've had some negative experiences where I've taken scholars of world renown out into the public, sat them down in an arena, and they've made absolute—they really set us back because they didn't have the skills to deal in that other arena." [Later:] "I don't want to leave you with the impression that I have any misgivings that one person can make a huge difference in this game. There are people, you may have met some of them, who really do become personal crusaders. I think you have to be a team player if you're interested in biodiversity conservation. And you play your strengths."

While all of these biologists feel a sense of personal responsibility to work for biodiversity conservation, a few hesitate to preach to their colleagues to join them. For example, while Orians feels this responsibility, he notes the peril posed to advocates' careers by those committed to defending the boundaries of science as traditionally—but not inevitably—practiced. Falk won't preach either, but seems to share Orians' concern, and urges that society should reward those who expand the role of science. Lubchenco is evenhanded in her treatment of her colleagues, yet her leadership in the Sustainable Biosphere Initiative reveals that she, too, feels compelled to urge her colleagues to act, and works to support and reward such expansion of biologists' role: the SBI provides "justification" for those who would do research that brings the biologist into conservation decisions.

Brussard, Ehrlich, Lovejoy, and Soulé note that biologists' self-interest dictates action on behalf of biodiversity: biologists' responsibility is, in part, a responsibility to themselves, for if they do not work to preserve biodiversity, they will have little left to study. Soulé notes that applied work has been considered beneath one's dignity—an attitude that has perhaps been one of the weapons scientists have used to defend the boundaries of science as usual, to rein in unruly colleagues who would breach the walls that have successfully defended scientific privilege. Soulé speaks of "coming out of the closet" about conservation issues—about loving what one studies and seeking to defend it. This love is portrayed as something deep-seated or innate, like sexual preference. Having to hide it suggests that it is somehow seen as shameful—"objectivity," like "heterosexuality," has been constructed as a powerful norm, which is defended in the interest of the majority's goals. By revealing your sexual preference, or your affinity for living creatures, you liberate yourself, free yourself to work to change society's values so that they look more like yours, work to make others accept and even acclaim that which you hold most dear. It is a potent metaphor.

Many biologists link the gravity of the biodiversity crisis to a moral obligation to work to mitigate it. For example, Funk sees it as the most serious problem we face, after global warming. Wilson puts it second to the threat of nuclear war. Lovejoy, Raven, and Orians pronounce it our most serious long-term problem. Brussard says that "ultimately the loss of biodiversity is probably going to do us all in," and Erwin declares, "It's the key factor in whether we're going to survive as a species." The Ehrlichs announce: "A substantial portion of the life that shares Earth with us is now doomed to go extinct. Partly as a result, a billion or more people could starve in the first few decades of the next century, hundreds of millions of environmental refugees could be created, the health and happiness of virtually every human being could be compromised, and social breakdown and conflict could destroy civilization as we know it."[60]

For Ehrenfeld, it is part of the "moral collapse" of the academy that in the face of such grave peril, so many biologists decline to act. Ray derives this responsibility from the money society invests in the scientific enterprise. Janzen was clear on this:

"By being involved in society, in fact you're paying—you're giving a return on the bill society pays for your existence in the first place. Scientists don't eat air; they get salaries. Somebody's tax money paid that salary. And that tax money was not given because they just like this scientist and they would like to give him a nice friendly free ride. . . . Now a faculty member, a typical scientist says, 'I do my work independently of that contract. Therefore I don't have a conflict of interest. Society says "Find truth" so I find truth.' The problem is if all you do is find truth, society comes to realize—and sometimes it takes awhile—that it's not getting much for its money. . . . So I would argue that my 'advocacy,' if you like, adds up on the positive side of the ledger there."

In Janzen's view, science's boundaries have to be continuously negotiated with the forces of broader society. Biologists are obliged to repay a social debt and can concomitantly use the services they provide society—which for Janzen include advocacy—to negotiate for a stronger position, to accrue further support.

Biologists need reinforcements in the negotiating process, and they need backup troops to carry the banner of biodiversity into the public fray. Many of the biologists who sing the praises of biodiversity in public, who call for spiritual revolutions and policy upheavals, are older and well-established. Their scientific credentials (as we traditionally think of them) are beyond reproach. Having accumulated enough capital, they can afford to risk some of it on a gambit that could simultaneously help preserve biodiversity and help expand their own purview in society. But what about the future? Eisner finds it "scary" that "all these active people are getting older fast and I'm dying to find out whether the twenty-year-olds will be able to pick up the ball." Recognizing that "you can no longer be in the vague area of biology that is called organismal—you know, ecological, behavioral, evolutionary—and disassociate yourself from the body politic," Eisner has taken a small step at Cornell to translate student interest into action on behalf of biodiversity. At CIRCE, the Cornell Institute for Research in Chemical Ecology, students get involved in research, education, and conservation: "The emphasis is chemicals, but biodiversity is the entire driving force. And [to] the chemists that I work with, you know, the term *biodiversity* was [an] 'Oh yeah, third page in the *New York Times*' type of term. I mean, they've really bought it hook, line, and sinker." Capitalizing on the "conviction in young people that this is an area where one can do something," Eisner is training students to be renaissance conservationist scientists.

Soulé maintains an upbeat attitude on the future: "Those of us who are concerned are going to go ahead and do what we think needs to be done the best way we can—pass on the torch to the next generation. They will do the same. And there'll always be enough people around, sort of a fifth column, that will be protecting life for this difficult passage." An NSF official explains the decision finally to fund work in conservation biology: "'We're trying to stimulate the fearless biologist—one who is solidly rooted in ecology or systematics, for example, but who has no compunction whatsoever about running off to find a computer scientist or a molecular biologist to learn a new technique.'"[61] Or, he might have added, a policy maker, or an ethicist, or a fellow biologist who will teach him a new technique in how to effectively reach a general audience.

Also seeking "the fearless biologist," but not quite so confident of finding her, Wilson, Ehrlich, and others are taking proactive measures. Upon receiving the Crafoord prize, a distinguished environmental biology award, Wilson "suggested to Paul that we publish a lead article in *Science* called 'Biodiversity Studies' to . . . anoint a new discipline, which I saw would inevitably rise to prominence in universities." This new field would go "all the way across from biology to economics, government, ethical philosophy." Wilson pictures "biodiversity studies as something that is both of and apart from other disciplines in biology and social sciences, and which will be rigorously defined as an academic field, but at the same time become a catchword to the public at large, just like ecology has. Ecology is a pretty well-defined and mostly now rigorous discipline within academia. But it also has a broader meaning in the public eye and can be used to instantly communicate to almost everybody. And I hope biodiversity has the same effect, and biodiversity studies." Setting up such programs has been discussed at Harvard, Duke, and other universities.[62]

But from informal conversations with professional ecologists, I sense discomfort with the vulgarization of ecology. They feel it somehow blunts the rigor or smirches the purity of their science. Conservation biologists, on the other hand, vigorously promote the popularization of *biodiversity*, which represents their complex view of a chaotic, interconnected, and aesthetically and even spiritually infused natural world. Silverback biologists would design biodiversity studies programs to train the young polymaths who, they hope, will carry on the torch of mandatory scientific activism on behalf of nature.

BIOPOLITICIANS FOR BIODIVERSITY

I have briefly noted some of the ways in which biologists are moving biodiversity into society. They do so not only by making esoteric research more applied. They also disseminate this research to citizens and policy makers here and abroad. They lobby and testify before government bodies. They write and speak to the public. And they try to shape how we experience nature, which they believe will determine our environmental ethic: how we feel about biodiversity dictates how we treat biodiversity.

Recall that Ehrenfeld referred to some of his colleagues as "biopoliticians." Biodiversity biologists' activities are growing increasingly political in two ways: (a) in the sense of being involved in the mainstream political process, and (b) in the Latourian sense: "If by politics you mean to be the spokesman of the forces you mould society with and of which you are the only credible and legitimate authority."[63]

Wilson, Lovejoy, Raven, Eisner, Stephen Jay Gould, and others have testified before Congress on biodiversity's values in hearings on the Endangered Species Act and the proposed National Biological Diversity Conservation Act. Brussard urges ecologists to lobby Congress for passage of the latter. Orians notes "the despicable display of the U.S. government in Rio" at the UNCED Earth Summit. Soulé predicts, "It won't be long before many conservation biologists are spending more time at community meetings than in the field or laboratory."[64]

Jerry Franklin, who serves as an advisor to congressional committees on old-growth-forest issues, does so because

"I want all the information out there. And if only half the story is being presented, I'm going to get up there and make sure the other half of the story is presented. Like with old-growth forest, if society chooses not to save very many, that's society's call. But *by damn* I'm going to make sure that everybody understands there are values there, that those forests *do* things. And you know, we have a history, particularly with the resource management agencies, including the Park Service incidentally, of only presenting those facts which serve the decisions that they wish to make."

Franklin feels his facts are *facts*, untainted by policy goals. Lubchenco urges biologists to educate people about biodiversity, ecosystem services, and how science works, to "respond with information that is relevant to policy, that is relevant to the kinds of decisions that need to be made." These principles also underlie the Sustainable Biosphere Initiative.

Orians also testifies in court on controversial management issues: "I fundamentally just talk science as a witness. There's just a law, and the law says the Forest Service shall manage to preserve populations of the vertebrates. And therefore it's under a mandate, a legal mandate, to maintain viable populations of the owl, for example. And all I have talked about is: will Forest Service policies do that?" He notes that "some of the most excruciatingly difficult hours of my life are in court being grilled by hostile attorneys."

Janzen has been working to help remake Costa Rica as a haven for biodiversity, and for the humans who depend on it, by restoring large areas of tropical dry forest there (I discuss this further in Chapter 6). Erwin and Raven had the ear of the former Mexican president Carlos Salinas de Gortari, whom Erwin helped establish a national biodiversity commission. According to Erwin, "We're starting to see some kind of movement and interface between us, the scientific community, and the political leaders that are savvy. So, something is happening that's new." It's new that biodiversity is a hot political issue in whose name new alliances are forged, on whose behalf heads of state turn to biologists as political advisors.

Vickie Funk has spearheaded a program at Guyana's national university that will combine education, conservation, and science. Although she is the driving force, "I keep trying to get the Guyanese out in front. You know, push the people at the university to make statements and prepare them for it, you know, give them the information they need, but let the statements and everything come from the university as a very supporting thing." David Woodruff adopts a similar strategy in Thailand: "They resent, even in poor countries, rich Americans coming in and telling them how to do things. So my strategy has been sneaky. You go there, you make friends with influential academics. You help them do what they'd like to do in the way of training people, in the way of writing reports, in the way of broadcasting their findings. And you build up the infrastructure locally." When Woodruff was dragged into a conflict with proponents of a potentially destructive dam in Thailand,

"we put them all on notice that a few snail watchers, ill prepared, never having done this before, hung over from the night before, can really screw things up but properly. And if I had the opportunity of doing that again, I could cream those guys. Those documents are so easy to take apart. But that was my first entrée into affecting policy on that level. And I discovered that with doing a little bit of homework, it's trivial. It's easy to play the game at that level. And you do have to own a suit. Now as a result of that experience on my last trip to Thailand, I commissioned the construction of a dark blue suit. Why? Because everyone in the room wore a dark blue suit. Now the problem this morning [at a meeting about a $1 million grant] . . . is that I don't own black shoes to go with it. So one day I'll get some black shoes and I'll be equipped to play the game that Lovejoy and Wilson and Eisner and Ehrlich can play."

The rules for this biopolitician game are still being defined. "We live here only because we keep people elsewhere in the state of poverty, in the state of malnourishment, in the state of ill health," Woodruff notes. "And that's a direct consequence of policy actions here. If you take a global view, then I think you have the responsibility to point these things out to people who haven't heard that. What people do with that information—I'm not a politician, so I don't try to bring about change. I try to provide people with information and then step back." But, of course, he is a politician! He wields knowledge and the power that comes with it to mold society so that the world will be more to his—and biodiversity's—liking.

Rocky Gutiérrez, one of the biologists most deeply involved in the spotted owl imbroglio, has lobbied and preached to many audiences. His talks are filled with political barbs and jabs, but he concluded one that I heard by saying: "We as scientists depend upon your support—because I don't consider myself an advocate. I'm a scientist."[65] Like Woodruff, he tries to distinguish between what he does and what people labeled *advocate* or *politician* do, while acting in ways that are political at multiple levels.

Eisner has been involved in political deals throughout his career. He is currently spearheading the Endangered Species Coalition, which is attempting to persuade Congress to retain the Endangered Species Act in full force. Eisner's efforts to broker a deal between Costa Rica's National Institute of Biodiversity and the pharmaceutical multinational Merck & Co. (discussed in Chapter 6) brought together the National Academy of Sciences, the National Cancer Institute, megacorporations, Third World countries, the John D. and Catherine T. MacArthur Foundation, heads of state, and Costa Rican *campesinos*. Eisner labored to forge connections that would change the foci of power. When he presented a talk on the Merck deal at a conference, an audience member asked if he was concerned about the political implications of, for example, taking land from people to make preserves. Eisner responded: "I'm not a politician; I'm a biologist."[66] When I asked him about this, he replied:

"Yeah, I remember that answer. And I regretted it the moment I said it almost. . . . what I clearly should have said is, yes, indeed, you know, conservation doesn't make any sense unless you couple with that a form of generating economic livelihood for those who are disenfranchised. But the solutions that I'm bringing in are, in that sense, long-range. The problem is what do you do in the immediate sense if you suddenly put a fence around something and say, you can't gold mine there anymore—you know, as the Brazilians should do but are unable to do. . . . So, in fact, it was disingenuous to the extent that I said I'm not a politician; I'm not an actual politician but I'm very conscious of politics."

Eisner started working in the public domain "later than someone like Ed Wilson because I didn't sense my political power until later." Whether they admit it or not, Eisner and his peers are becoming consummate politicians. Again, they are political not necessarily for personal gain—although fame and fortune may come—but for biodiversity's gain. They use their authority to create a demand for biodiversity preservation, to mold the forces of society here and abroad to work with them to change policy and power structures.

AIMING TO TRANSFORM

"Solving our environmental problems requires a new perspective that goes beyond science and has to do with the way that everyone perceives the world," Daniel Botkin argues. Our perspective on nature, he says, "depends on myths and deeply buried beliefs."[67] Biodiversity biologists wish to help create this new perspective, to unearth old beliefs, to develop new myths.

To do so, biologists proselytize for biodiversity and its values to whatever audiences will listen. Erwin urges a comprehensive biosurvey "simply to raise the awareness of what we might lose. And it's a ploy, a ploy to reach the human conscience." He uses his experiences as a field biologist to talk to conservationists, to "charge them up with stuff that *they* don't usually get first hand . . . and you know, especially at cocktail parties and stuff like that, you get people really excited. . . . So give them that and then they start thinking about animals and plants and continents and whatever."

He is certainly not alone in attempts to raise awareness, to get people thinking and feeling, to get them out in the natural world. Although he believes he wouldn't be effective on the *Tonight* show, McNaughton does give public lectures—for example, to the National Science Teachers Association, "And by conveying this information to teachers, who then translate it back into the classroom, I can do much more than I can do in other ways, I think." Brussard believes his most important contribution to biodiversity conservation has been in educating government workers in the Park Service, BLM, and Fish and Wildlife Service. "I think all of these agencies are now talking biodiversity," he says. "I think I probably did play some reasonably significant role in that. So that's probably my major contribution." Franklin has no question that he's been influential in educating congresspeople about biodiversity issues. Biologists are talking wherever and to whomever to convert others to the biodiversity mind-set.

The prelude to a *Nature Conservancy* interview with E. O. Wilson claims that "he has done more than any single individual to make the term biodiversity part of our language—and part of the national agenda. As a result, Wilson has become known as 'the father of biodiversity.'" How does Wilson deal with the nonbelievers? "Educate, educate, educate [laughter]. Keep talking, keep pointing out, keep explaining, keep demonstrating—show the wonderful beauty and promise of the natural world."[68]

Peter Raven is "optimistic because I believe so profoundly that individuals and individual actions make such a huge difference." He judges his greatest contribution to be "getting other people interested. I think any contribution that any of us make is in dealing with other people. That's all there is." Raven sees biodiversity as the key to future options for a sustainable human society, and he is "so convinced that individual action can affect that, that I want to do what I can to generate the kind of individual action that will do it." Paul Ehrlich has "talked to thousands of audiences about this sort of thing, both radio and TV and personal lectures and so on and so forth." "I'm a pretty good propagandist," declares Hugh Iltis. "I spent the better part of ten years going from one end of the country to the other preaching the fact that we have to . . . respect our genetic needs." These biologists have been on the front lines talking to people, getting them interested, excited, and, especially, back to nature via a conversion to biodiversity.

K. C. Kim directs Penn State's Biodiversity Center, which the university's vice president for research originally supported because "nobody else has such a thing. . . . [and so we'll] be well prepared by the time there are national institutes or center[s] for biodiversity . . . to get some funding." Kim hopes to "energize the intellectual community" and "sensitize the campus" through the Penn State Biodiversity Center and through a conference he organized to talk about biodiversity. "I have been a longtime missionary-oriented individual," Kim says. "I seem to be talking every day about the issues with somebody." The destruction of biodiversity, he believes, is an "ultimate symptom" that calls "the fundamental ways of Western culture" into question: "We need a new ideology, nothing less."[69]

Also to shape views of nature, biologists talk about the issues to students; and they include the feelings that make them want to talk about the issues. Woodruff advises that universities should be "training biologists differently . . . inculcating [in] them as undergraduates . . . the value of . . . the positive aspects of social responsibility" so that they too will work for biodiversity and related causes. Furthermore, Woodruff has "always taken the view that I would rather train one future lawyer or politician than ten biologists. . . . I mean, I've placed the biologists I've trained. But there aren't enough career opportunities out there for them in academia. So I take the view that I should be using my energies to train people for other professions. So I have always taught—and it was Ed Wilson that demonstrated to me the value of this—I've always taught the largest possible holistic natural history courses a university will let me teach." He has also fought a long (ultimately successful) battle at UCSD to institute an environmental sciences minor. Woodruff tries to win as many undergraduate hearts and minds as he can to convert them to share his values and to fight for them, too.

Peter Brussard has a similar view:

> "I know for one thing that biologists do an absolutely awful job of general education courses, getting people trained in biology: somebody who wants to major in criminal justice but discovers they've got to take a biology course. So what do they do? They get in this and they're taught the difference between mitosis and meiosis. I mean, what goddamned possible difference could this be to anybody? So this is where I would try to get some sort of appreciation of biodiversity, and get it inculcated into people and try to give them a glimpse of what the natural world is about. That's certainly a good place, and that's someplace where we've done a *miserable* job. Absolutely miserable."

Both Woodruff and Brussard add that biology professors should instill into students the value of biodiversity and the importance of caring about it as part of one's career and life.

In their book *Noah's Choice*, Charles Mann and Mark Plummer assert that in making tough choices about biodiversity, "the distinction between 'is' and 'ought' is vital. Science is a useful tool for answering questions about facts, but it cannot tell us what we should do about those facts."[70] Mann and Plummer do not see that conservation biologists, the scientists most intimately involved in making these decisions, attempt to bridge the is/ought dichotomy. Conservation biology not only supplies facts; it also offers a set of moral prescriptions that tell us what to do with those facts.

"An immediate—as opposed to a geological—solution to the problem of maintaining global biodiversity seems to depend on the collective behaviors and perceptions of people toward their habitat," David Challinor writes in the concluding essay in *BioDiversity*. Recall that Aldo Leopold, the spiritual and strategic grandfather to today's biodiversity advocates, observed: "No important change in ethics was ever accomplished without an internal change in our intellectual emphasis, loyalties, affections, and convictions."[71] Biodiversity biologists know this. Noss cites Leopold as his forerunner in ethical advocacy. Biologists, he says, "would naturally play a large role in articulating such an ethic and actually developing such an ethic, or refining such an ethic. . . . it comes from what we talked about earlier, the predisposition, biophilia or whatever that we have that also brought us into the training, brought us to seek that kind of training."

Warwick Fox proposes a "transpersonal ecology" in which if we truly, deeply understand the way things are, "then one *will* (as opposed to should) naturally be inclined to care for the unfolding of the world in all its aspects." A lasting and deep conservation ethic that makes it unthinkable to knowingly inhibit creatures from living out their lives, or to prevent the evolutionary process from unfurling unhindered, arises only when one knows and loves the natural world. Those who feel this already, according to Fox, must "respectfully but resolutely attempt to alter the views and behavior of those who persist in the delusion that self-realization lies in the direction of dominating the earth and the myriad entities with which we coexist."

Biologists are attempting to overcome such delusions and create a "steadfast friendliness" on the part of humans toward biodiversity.[72] Ehrlich believes that "we must, if we are going to solve this problem in any permanent way, create a feeling for other organisms that goes beyond what they might or might not do for *Homo sapiens*." The task of biologists, according to him, is in part to play a role in creating this feeling, to be out there "talking about it. Telling people how good it feels to get out in nature, how marvelous—you know, one of the things religion does for a lot of people apparently is [it] gives them a sense of wonder." Biologists attempt to create this sense of wonder to instigate what Ehrlich calls a "quasi-religious transformation" of feelings toward the value and wonder of the natural world.[73]

Wilson offers a similar opinion: "I believe that the more that you understand organisms, each species in turn, its natural history, its evolutionary history, behavior, the more involved people become with them. . . . Familiarity produces, if not love, at least affection. And affection produces a desire to salvage, to hold on, not to be reckless." As we shall see in Chapter 7, Wilson's heterogeneous activities on behalf of biodiversity are aimed not at producing this affection—for he believes it is already there—but at rekindling it. He attempts to arouse biophilia via a "style of writing [that] is a kind of poetry and science. And it is deliberately so."

Thomas Dunlap alludes to nature films of the 1950s that reflected new scientific ideas and "prepared Americans to 'appreciate' nature in new ways." Perhaps some of these efforts even instilled their deeply held views of nature into today's biodiversity advocates, who are in turn now preparing us to see nature as they do. Well before the current biodiversity movement started, Ehrenfeld wrote a textbook not only to educate readers but to change their ideas about nature and to get them out into nature guided with a sophisticated understanding.[74] Lovejoy founded PBS's *Nature* television series, because "there are just going to be millions and millions and millions of people each year reached by television who may not ever go out into nature."

Biologists' ultimate heuristic, though, is to have biodiversity itself effect people's conversion. Lovejoy has served on presidential science boards, written pieces for *Time*, and invented debt-for-nature swaps. Yet when queried as to the most important thing he's done on behalf of biodiversity, he cites taking people to the Amazon to spend a night in the rain forest. He has taken 10 percent of the U.S. Senate, various members of the House of Representatives, movie stars, and activists there in an attempt to get them to *feel* biodiversity, to have biodiversity work its persuasive powers "in a way words can't touch. . . . And it has never failed to be a truly touching experience for them. . . . And, you know, the Tim Wirths and the Al Gores and the John Heinzes, it was life-changing for them. I mean, John actually wrote in a copy of a book he sent me afterwards: 'You have changed my life.' And he really meant that." I asked Lovejoy how he prepared his guests for their experience. His response is worth quoting at length:

"Well, I do a soft sell. I mean, I give them stuff in advance to read. And we talk about it. But I let them experience it. And it was Bruce Babbitt who told me what it was really like. I really hadn't envisioned in my mind what it's like for these people. He said, 'You know, you've been going on this bumpy road in this truck for a couple of hours, toward Venezuela, and then you turn off on this side road for an hour or two. And you don't see another vehicle the whole time. You never quite know whether you're going to make the next hill. And it's getting dark. And then all of a sudden, in the middle of nowhere as far as you can see, the truck stops. And then there's this little path going into the forest. You don't know how long that path is, what it's going to be like in there. Or what's going to be there when you get to the end of it.' And so there's this—without my realizing it, there's this incredible amount of anticipation that's going on with these people. And then, there you are in a clearing in the forest. And you're living not very differently than Indians live in the forest. Some of the materials are different. But you're sleeping in hammocks, and if you want to wash up, you go down to the stream and wash up. And it's all tropical. And what I usually do is give them a spiel after dinner, not very long. Just sketch it in. How far are the Andes . . . how far to Venezuela; it's all unbroken forest. And how fast the forest is going. And then I have all these student types around who know about the frogs or the birds or the plants or whatever. So you have a very rich experience and many different voices talking about it. And we may do a little walk with headlamps out in the forest. And then the next morning, up at 5:30, out to see the birds in the bird nets. When you go into the forest, it's not what you think it is. It's not a bunch of animals just leaping around that you can't possibly miss. It's all this subtle stuff. Telling one tree from another is not a simple thing to do. It's the insects and the ants and the termites and the butterflies. And they will have spent the whole night listening to all these voices. That becomes biological diversity in their brains. If they don't spend the night, they don't get the experience."

Lovejoy primes the pump and then lets biodiversity speak and therefore convert these important people to a way of thinking, feeling, and acting that mirrors his own—he was also converted when biodiversity spoke to him.[75]

Orians concurs. People must be transformed: "This sense of value comes with contact and experience. Which is why Tom Lovejoy takes the senators down and shows them the Brazilian Amazon. Which is why the Organization for Tropical Studies [Orians has served as its president] runs courses for staffers, congressional staffers from Washington, D.C., to come down for an intensive week down there [Costa Rica], and they see the tropics, and they see where things go wrong. And they internalize it and they come back motivated and concerned about it." First, biodiversity transforms biologists. Then biologists bring important people to biodiversity so that they will be transformed and will work to see that it not be transformed.

Funk also brings congressional staffers and others to the Smithsonian's collections to show them what the museum does, to highlight the importance of its biodiversity work:

> "I take people through the National Worm Collection here; that's one of my very favorite things to do. Most people have no appreciation for worms. And they don't understand . . . the variety and the impact on our everyday life that worms have, the fact that they're on every part of our body. . . . And all these marvelous things that worms do, people have no clue. So I think taking people through a collection or maybe even showing them videos of worms would enhance their appreciation of different kinds of worms. But it takes somebody who knows about it and loves it. That can convey to that person their energy and enthusiasm. . . . You know, and you really feel like you've made a convert, no matter what they do for a living."

She believes you can take most anybody to see the Smithsonian collections or into the tropics "and explain things to them, and they'd be interested. But I think it takes somebody—in the beginning, it takes somebody to tell you or to explain to you why it's interesting. And that's why the education part is so important."

Biodiversity may convert, but for Lovejoy, Orians, and Funk, the conversion needs a guide to enhance observation and feeling, to tell you not only what you're seeing but why it's important. Orians maintains that "aesthetic values can be and are enhanced by education."[76] Franklin plans to write more popular articles to get people educated to enrich their aesthetic appreciation of biodiversity. Janzen told me that "for most people, you have to start them out somewhere else, and then prime the pump. Get them started on something and then they use the library. . . . It's like standing up and trying to describe a symphony. There are a couple of people in the audience that I can describe the symphony to and they'll get all into it. But the vast majority of people will not; they'll have to hear it. There has to be contact. . . . It's sort of like hearing good music on the radio in your car. Well, that's a start." When Janzen and the others lecture on biodiversity, they guide you to appreciate it. They may simultaneously tell you what to feel, which is what they feel. Biologists help biodiversity convey its values, injecting their own values in the process.

According to Worster, our enthusiasm for ecological science stems from our search for a fruitful marriage between ecology and ethics; we want ecologists to provide moral enlightenment.[77] This may or may not be the case. But we *have* granted scientists cognitive authority to speak for nature. We turn to them for facts about the natural world, for reliable knowledge that we can use to understand, appreciate, and even control that world. The biologists I portray here use their cognitive authority to promote more than just facts. They promote a conservation mind-set that carries with it, not only policy prescriptions, but moral and ideological prescriptions as well. They want us to reconceive our ideas of nature by discovering the concept of biodiversity, to return to nature so that it may work its persuasive, transformative wiles on us, as it has on them, and in so doing, change our feelings, morals, and values regarding the natural world. They whisper (or shout) *biodiversity* at us, and we listen in part because we see them as objective purveyors of truth; in part because they reflect back our own values—values that have been inculcated into them, as biologists, too; in part because their marketing strategies and techniques have been so adroit; and in part because what they say strikes a chord deep inside us, resonating as truth.

BOUNDARY WORK REVISITED: REDRAWING MULTIPLE BORDERS

I recently received a glossy brochure from the International Council for Bird Preservation advertising its book *Putting Biodiversity on the Map: Priority Areas for Global Conservation.*[78] Biologists are redrawing multiple borders to put biodiversity on the global political map—literally making more space for it—by putting it on the normative cultural map—making more space for it amid the values and norms that compete for our cognitive and ideological attention. They wish to make biodiversity a priority area for our affection and allegiance so that we shall agree to set aside more priority areas for its conservation and protection. In the name of biodiversity, they are expanding into the cultural space of policy, ethics, esthetics, values, ideology, and religion. As their territory becomes broader, they gain power to agitate for remade geopolitical and geo-cultural space for biodiversity the idea and biodiversity the natural object. The latter compels them to fight battles on many fronts, including a defense of science's traditional territory: the hard-won resources of objectivity and value-neutrality that have enabled biodiversity proponents to venture forth so far thus far must be guarded against capture by opponents, lest the conservation biologists be left vulnerable, isolated, with only one another as allies.

Some biologists show explicit awareness of their own need to do boundary work. Listen to Dan Janzen:

"I think what the scientist has to do is establish a negotiated position in society like every other subculture does. . . . I think we have to work out a negotiated position where we identify what it is, rather than sailing off into the blind unknown without the foggiest idea what is going to give any return to anybody, including me as the researcher. . . . So that my argument is that by being an advocate, science is learning how to come to the negotiating position. . . . The irony, of course, is that deans of research, directors of NSF, directors of NIH, certain congressmen . . . are constantly negotiating on behalf of science."

Note what happens here. Janzen demonstrates an awareness of boundary work in science while doing a very neat bit of boundary work himself. We can see from his earlier comments that Janzen clearly feels that it is the scientist's responsibility—it constitutes part of the scientist's job—to advocate on biodiversity's behalf. Janzen would expand the definition of what it means to be a scientist. At the same time, he argues that advocacy helps defend the scientist's traditional status, helps fund the basic research and the well-stocked labs. He is mindful that the scientists' position in society is not now, nor has it ever been, set; its bounds must be negotiated. It must be negotiated to include advocacy; advocacy is simultaneously one of the tools in the negotiation.

Arne Naess urges biologists to speak up for nature: "When biologists refrain from using the rich and flavorful language of their own spontaneous experience of all life forms—not only of the spectacularly beautiful but of the mundane and bizarre as well—they support the value nihilism which is implicit in outrageous environmental policies." Yet this avoidance of rich and flavorful language marks science; biologists *choose* this language to reinforce a perception of value-neutrality. Colorless suggests valueless, as Naess seems to recognize. He lists possible reasons why biologists do not publicly express strong views on conservation:

1. Time taken away from professional work.
2. Consequent adverse effects of this on promotion and status.
3. Feeling of insufficient competence outside their "expertise."
4. Lack of training in the use of mass media and in facing nonacademic audiences.
5. Negative attitude toward expressing "subjective" opinions and valuations, or violating norms of "objectivity"; reluctance to enter controversial issues.
6. Fear that colleagues or bosses think that they dabble in irrelevant, controversial fields, and that their going public is due to vainglory and publicity seeking.
7. Fear of fellow researchers, institution personnel or administrations; fear of the stigma "unscientific."[79]

Note that Naess himself engages in boundary work here: he wants to enroll the cultural authority of science as a resource to promote his deep ecology program. He entreats biologists to transcend their traditional boundaries to espouse deep ecology views. And the seven reasons he lists to explain why they do not normally do so all arise from previous and present negotiations about what science is or should be. The penalties that have customarily resulted for doing what Naess wants biologists to do are imposed by those who would maintain the constructed sanctity of science, who would defend what has been elaborately built up and maintained, and what has proven quite resilient in securing scientists' power, authority, and resources. That is to say, what constitutes "professional work" is what those in the profession decide it should be. "Promotion and status" are afforded by one's peers on a constantly shifting basis of what should count. "Expertise" is defined by the experts. The trainers choose not to train scientists to face nonacademic audiences, to cleave the sanctity of scientific walls.

What ranks as "subjective," "objective," "irrelevant," "controversial," or "unscientific" is not immutable; tacit or explicit recognition of its very mutability demands that harsh strictures and punitive measures be levied or threatened against those who would breach what has proven so successful, where what *successful* means is also a shifting resource. That is to say, the institution of science has proven "successful" in gaining large sums of money to continue its enterprise, in finding comfortable positions for its practitioners, in keeping the outside world from overseeing its day-to-day activities, in having its word accepted as a dominant source of authority in society.

Conservation biologists and their allies seek to expand what *successful* means in the context of biological science. The "successful" biologist would be she who takes this crusade furthest, who is most effective in changing the most minds, lobbying for the most laws on biodiversity's behalf, protecting the most habitat. What counts as "professional work" expands. "Promotion and status" would be more broadly bestowed. "Expertise" would cover, not just facts about nature, but its values as well. What is currently considered "subjective" would be brought under the control of the biologist and made "objective." Much of what is now viewed as "irrelevant" to biology would become very relevant, and what is "controversial" would become commonplace. Nonetheless, these biologists also wish to maintain their current perks: hence the delicacy of the negotiations in which they are engaged.

SOULÉ AND EMOTION

Michael Soulé is one of those engaged in these delicate negotiations on the present and future course of biology. In his contribution to the *BioDiversity* volume, Soulé warns that "statistics about extinction rates compute, but they don't convert." Biologists must, in addition, have "the courage to let ourselves describe our private, emotional experience of nature to our father-in-law" and to the public in general, he suggests.[80] I asked my interview subjects what they thought of Soulé's contention. (I did not put the question to Hugh Iltis and Reed Noss, because it seemed obvious from their writings that they would agree with Soulé.) Here are some of their responses:

PETER BRUSSARD: "I think that's fine. And if we're going to educate people on the values of biodiversity, I think we all have to indulge in that. I think if you read between the lines there, I think Mike really means that we need to convert to some sort of a religious experience having to do with biodiversity. And I just don't have it. But I certainly am happy to explain to people what I consider to be the very practical values of biodiversity. And if pushed, I will explain what my own personal values of biodiversity are. But I think they're rather personal things. . . .

[Asked whether he felt Soulé's talk of spirituality and emotional arguments reflected badly on other conservation biologists, and on the profession as such, Brussard said:] "Well, no. I really don't. Because I think that the message is important enough that there's plenty of room for a variety of approaches to it. Soulé doesn't bother me. Mike's a good friend of mine. And Soulé, he's delved into a number of different religions. He tried Zen for awhile. He was Jewish by birth. So he's tried a bunch of things. And it's obviously something he needs to make his daily life good. And sometimes—We were at a meeting in Wyoming having to do with desert tortoises in the mid eighties. And so here was a whole bunch of Wyoming Game and Fish types, all in cowboy boots with their big belts and whatnot, drunk as skunks. And Soulé gets up and gives this talk on deep ecology. And I said, 'Shit, we're all dead.' They *loved* it! They were just drunk enough to kind of eat it up: 'Hey, that makes a lot of sense.' So you never know what's going to work."

DAVID EHRENFELD: "I agree with that completely."

PAUL EHRLICH: "I think it's right. In a sense that's what I was doing when I—Actually, I've taken flak for that quasi-religious statement [see above] from some group of atheists. [Laughs.] I'm so used to getting flak from the religious right in one form or other, it was sort of amusing to get flak from atheists for mentioning the word *religious* in this connection."

THOMAS EISNER: "Sure. I'll go further. I think what Wilson has been saying is that the urgency is such that to say that we can't make decisions until all the facts are in is insane."

TERRY ERWIN: "If I understand it, yeah. . . . I don't like dry scientific literature. I think it sucks. I just get so bored reading some of this stuff; it's so hard. But if there was a little bit of humanism in there, I'd be much more intrigued. Obviously, where the hypotheses are laid out, they have to be very rigorously stated and tested and so forth. But for the introductory material, the concluding material, even descriptions—my professor up in Canada, George Ball, he always begins a taxonomic paper with some relevance on why he's even doing this study. And usually he works on carabid beetles in Mexico. He usually tells a great story about a trip to Mexico and the first occasion of seeing one of these beetles that's in this paper. And some thought process, some philosophy behind it. And based on all this kind of stuff, here comes the work. And that's how he got interested in doing this work. Immediately you're relating to this guy, and you want to read more. I'm tired of stuff like in *Science*. It's just *robotic*, absolutely robotic."

DONALD FALK: "Oh yeah! I think the reason people listen to scientists is because they know a lot. I mean, let's get right down to it. It's not that— It's because scientists are trained in a certain form of reasoning that's extremely powerful and helps us to understand and relate to the world as we experience it. So you can go to someone because of a wealth of knowledge and/or because of their reasoning ability—that is, their ability to respond to questions and say, if this, then what might happen? But the people who really communicate are the people whose reasoning ability is backed by passion. I mean, if Jacques Cousteau were a less passionate person, then tens of millions of people would be less interested about what goes on under the surface of the water, rather than whether a shark is going to come out and devour them. I agree, if I understand Mike's comment, and if I understand it correctly, I've read him say this: there's nothing antithetical about profound knowledge, highly developed reasoning ability, and a desire to view a situation as accurately as possible, *and* the sort of passion that drives all of us to do what we're doing. It's just part of—That's no more and no less than saying that scientists are human just like everybody else. Well, most of them [laughter]."

JERRY FRANKLIN: "Well, you know, I do that. Do I think we *have* to do that? No. Yet I think we erode our ability to influence outcomes by appealing to the emotions. I guess I am a believer that we have to primarily appeal to the rational in us if we're to be very effective in our political system. I'm not sure I'm right, but that's the path I've chosen. I'm going to be rational about this. I'm going to deal primarily with facts. I'll admit to you how I feel about it. I'm going to deal with that and that's going to be the way I'm going to influence outcomes the most. There are other people that are going to appeal to the emotions, but there's no end of emotional appeals that can be made for and against. And so I don't advocate that as a primary element in the strategy."

VICKIE FUNK: "Yeah! I think that gets a little bit about what I was talking about earlier, which is this gut feeling that most scientists have that what drives them is this desire to find out what's going on and to figure things out, and the wonder of the whole thing and how it all fits together. . . . and instead we make up all these other excuses, which are very real, about food and medicine and stuff. And we avoid this idea that what we're really trying to do, we're doing because we love it, we're doing it because we're interested in it and we're driven to figure out what's going on."

DANIEL JANZEN: "Well, what he keeps forgetting is that a lot of biologists don't have it. It's like—there's a lot of biologists who do, but a lot of biologists don't; they just do it as a job. That's the first thing. So they don't have anything in particular to transmit because they themselves are not really very involved with it. It's a nine-to-five kind of thing. Now for the ones who aren't, who are in the other category—the people who do have some connections to it, who can hear and see what's out there, I think it's basically a reward situation. Society isn't by and large rewarding the biologist for spending his time communicating those kinds of things. So they don't do it. . . . To me, blaming a biologist for not doing that, I think you should look a little more closely, not at the biologist but, at the reward structure. Why should he spend his energy doing that?"

K. C. KIM: "Well, I think that statement is basically what I'm doing. That's what his statement says. . . . If you're committed, your emotion shows. That does not necessarily mean you become emotional. Because I don't think you should become emotional, if you're talking about [it] in the context emotional vs. rational. You should not lose rationality. But explaining a rational argument of biodiversity conservation, you could become emotionally expressive. I read it in that sense. . . . Otherwise people, your in-laws or whatever, will not be converted."

THOMAS LOVEJOY: "Well, you know, I say each to their own. I mean, there are various ways to go at this. I mean, I have a pretty strong emotional attachment now to my research areas. It's sort of like going home when I go there. And in my advanced years, I now find that occasionally I need what I call a 'nature fix.' Which is what I did last weekend in Brazil. I mean, I don't think there's anything wrong with scientists seeming to have emotions."

JANE LUBCHENCO: "I agree with that. I have always been a strong proponent of talking about something that I think parallels this. And that is the *fun* of doing science. That when I talk to undergraduates or graduates, or give seminars, part of what I think is important to convey is just the personal satisfaction of doing science and of the intellectual challenges, etc. That's kind of the same thing that I think Soulé is talking about. . . . that we need to come to grips with the fact that there are factors other than just facts that motivate people. And part of that associated with doing science is the fun of it. Part of it that's associated, if you're trying to convert people to biodiversity, is to talk about personal satisfaction of being in habitats that are diverse. I think those are valid considerations."

S. J. MCNAUGHTON: "Oh, I think it's wonderful. I mean, I've been doing it for years. . . . I'm more likely to do it in a public lecture to a lay audience where they ask me a question that lends itself to it, that calls for some kind of a statement of my personal value system and how it relates to this. And I would speak to them just as I speak to you . . . I mean, should I be embarrassed because I have feelings?"

GORDON ORIANS: "I think Michael does make the separation. I've talked to him about it, too. And he's saying very much I think what I said in the beginning: there are big moral issues here—intergenerational equity, respect for the tremendous processes that have produced this world that we enjoy so much and feel a part of. There's no reason that we as scientists should pretend we don't have those feelings. And as I said, I do agree that unless there is also, in conjunction with the scientific data about extinctions, some of this ethical, emotional component, it ain't gonna sell."

DAVID PIMENTEL: "I don't, myself, agree with that philosophy. I think we ought to speak vigorously and support our positions vigorously, but I do not feel that we should get emotional about it. I feel we ought to stay with the facts; the facts, I think, are shocking."

PETER RAVEN: "I think there are a number of different comments that could be made on that. Obviously, I know Mike really well. I was on his committee . . . I've known him for thirty years, and I admire him very much because he is a deeply spiritual and very thoughtful and very intelligent person. Having said that, I think the first thing I would say is it falls into what I would call 'the Ehrenfeld trap,' this sort of magic-bullet trap: there is only one correct argument about species preservation, and it is your emotions. That's not the only correct argument. . . . You need all of those things. You need to be able to talk about extinction rates when you're talking to the United Nations or the UNCED meeting. You need to be saying, the quality, the character of life on earth is changing. On the other hand, I've already said that I firmly believe that people's attitudes, souls, have to change. But I think it's also wrong to say that by implication . . . talking about extinction rates or materialism or anything else isn't needed. It's *all* needed. It's all part of the same package. . . . It's human beings dealing with the Earth. These are just ways of describing it. We don't need to deny any of those ways to do what Mike is talking about. That's my first point. My second point would be: Amen! I mean I speak very passionately about these things. I talk to anybody and any group and any way that I need talk to them. I firmly believe that nothing will happen unless Americans and all people change their attitudes very profoundly. I'm going to talk to anybody that I can, whether it's my father-in-law, Mike Soulé, you, or anybody else in the way that I think is best for expressing my rage and my interest and my feeling about it."

G. CARLETON RAY: "I know Mike very well. . . . A statement like that, I take it in the context of the person who said it. . . . A lot of times it *does* convert. . . . If you say, species are going extinct forty a second vs. forty a year, for some people with that kind of mind, it does. So I don't think Soulé's quite right. . . . I can think of a lot of cases where emotionalism has been very definitely a distracting element. . . . It depends on who's doing it. If Meryl Streep is up there doing it, I believe her. But you know, if I'm up there, nobody's going to believe me."

MICHAEL SOULÉ: [I asked him "about the reaction you get to these kinds of comments from your fellow scientists and from the public. . . .

What happens to the public perception of science when you have emotional scientists, which a lot of people consider to be something of an oxymoron?"] "The only people who talk to me about those comments along those lines are those people who like them. Usually people who disagree don't say anything. So I don't even run into them. But those sorts of things appeal strongly to students, I've discovered, after the fact. Students are still at a stage where they haven't yet learned that they're not supposed to have emotions. And it seems obvious to them that my statement has some validity. It just seems to work. People don't make decisions on the basis of facts. Many of us who have been educating students for a long time find that they know a lot of the facts, but it doesn't change their behavior. That is, knowing about ozone, knowing about global warming, knowing about extinction, doesn't necessarily make people recycle, or become vegetarians, or whatever social personal lifestyle changes they are considering. It has to be something beyond knowing. It has to be feeling. And I think most experienced educators know that. And that's what motivation, that's what advertising, that's what that chapter's all about. It's about advertising, how you *really* convince people to do something, to go out and buy something. It's not with facts; it's by appealing to something emotional."

E. O. WILSON: "As you just described it, I agree with it completely. That's part of the expanding *role* of the scientist. Not the expansion of science, [but] expansion of the role of science. If artists were to be faced with a compelling need . . . to save art treasures housed in museums from being discarded, burned up, and so on—if they were faced with the compelling need . . . to explain more fully why they were artists, and why they feel about art the way they do, instead of just doing art, then I think exactly the same argument would hold."

DAVID WOODRUFF: "Soulé has always, as far as I know, had this ability to plumb the depths of his consciousness and see things in a more humanistic picture than the average scientist does. . . . I think his impact on biodiversity conservation has been extraordinary. And it's a tribute to his ability to engage people, and to write, that has brought him to this point of influence. . . . He appeals to people across a spectrum. He appeals to scientists across the spectrum of their experience. I think that's his personal strategy. . . . [At a meeting] the director of USAID thumped the table . . . and said . . . 'I don't want to hear one more statistic about the rate of species extinction.

It doesn't mean a thing to me. And it's not helping you. And I'm just getting turned off by it.' And it was a mind-set problem. They had just failed to
reach this policy political appointee. So I think Soulé's right . . . you have
to change attitudes of older people toward reality. I'm a pessimist about our
ability to do that. I think it probably can be done. Whether our society will
tolerate such retraining—I mean, how did the Chinese do it? Easy. Sit down
and reeducate people. Are we prepared to do that? I don't think so."

This is a mixed bag of responses. Pimentel strongly opposes using
emotional arguments, and Franklin thinks it is not necessarily wise,
although he seems to contradict himself: "Well, you know, I do
that." Brussard notes that being emotional is fine for some scientists, even though he doesn't do it, and Ray agrees that it's fine for
those who can pull it off; he does not count himself among those.
Ehrenfeld, Ehrlich, Eisner, Erwin, Falk, McNaughton, Wilson, and
Woodruff would go along with Soulé that "emotional scientist" need
no longer be considered an oxymoron. Raven, I believe, misinterpreted the quotation, thinking that Soulé believes *only* emotional
appeals make a difference; in fact, his response is very much in line
with what Soulé advocates and does. Lovejoy agrees, although a bit
more tepidly. Funk and Lubchenco both agree, although they turn
the question into talking about the fun of doing science, the excitement of finding things out about the natural world. Kim specifies
that biologists should be talking about rational arguments emotionally, and, like Raven, Orians thinks emotional arguments are fine as
long as they are expressed along with rational ones.

Janzen's response puzzles. Even if society doesn't reward the
emotional biologist, his reward should come in feeling that he is
helping to preserve biodiversity. Be that as it may, the reward structure of science—an elaborate construction designed to reward those
who play by the rules that reinforce the social authority of science
in society—does not currently reward for advocacy. Note Soulé's
remark that "students are still at a stage where they haven't yet
learned that they're not supposed to have emotions." In the same
vein, he said that biologists have to be "trained that feelings obscure
reality—that's what graduate school is all about [laughs]." Soulé

seems to recognize that training is about boundary work, including the necessity of teaching nascent biologists to repress their emotions, to conform to some idealized standard of detachment from the study subject and from the real world.

VALUE-NEUTRALITY

This is a major tension for the biologists I profile. The work of scientists traditionally depends—or has long been advertised as depending—on objectivity and value-neutrality. The scientist is not supposed to invest himself in his study organism, his methodology, the implications of his work. As the scientist detachedly discovers facts about the natural world, he gains cognitive authority, ideological clout. Or, as others have sarcastically put it, "When 'science' speaks—or rather when its spokesmen (and they generally are men) speak in the name of science—let no dog bark."[81] If scientists freely promote their values to society, or if values informing their work or advocacy are revealed, then some may question scientists' status as objective arbiters in societal or environmental debates. And so some biologists have urged their peers away from advocacy and politics.[82] Biodiversity biologists may be simultaneously drawing from and damaging the authority our society affords scientists.

Others maintain that this value-free self-portrait drawn by scientists is not only false; it may be counterproductive as well. Steven Yearley examines the role of scientific expertise in environmental debates, particularly when the self-promoted "inflated" scientific standards of objectivity clash publicly with the values that creep into environmental expertise. For example, environmental scientists called into Irish courts to provide "expert" testimony on standards for bog preservation had their credibility besmirched by lawyers who revealed the nonobjective values that informed their testimony. Thus scientists who wanted their word to count for things they cared about fell victim to the very standards of scientific objectivity they promoted to gain their positions of authority in such matters in the first place. In this case, it is not, as many scientists fear, that scientists' authority is discredited if they profess values; rather, it is because they pose as value*less* that their authority is discredited.[83]

Some observers agree with Soulé that expressions of emotion

may help the biologists' cause, while also recognizing the pitfalls attached to such expressions. Phyllis Windle notes the suppression of public displays of emotion by ecologists, but argues that in this "science of relationships," it is only natural that one grieve for lost species and places; for "people emerge from grief with new insights about their relationship to the deceased and renewed energy for loving again." David Orr acknowledges the explicit boundary work performed to suppress such expressions: "Excessive emotion about the object of one's study is in some institutions a sufficient reason to banish miscreants to the black hole of committee duty, or worse, on the grounds that good science and emotion of any sort are incompatible, a kind of presbyterian view of science." To Orr, "emotional bonding" and "love" are fundamental to the fight for biodiversity; "we have emotions for the same reason we have arms and legs: they have proven to be useful over evolutionary time. The point in either case is not to cut off various appendages and qualities, but rather to learn to coordinate and discipline them to good use."[84]

Biologists promoting biodiversity understand these tensions between objectivity and attachment, value-neutrality and value-ladenness, distance and love. They acknowledge it explicitly; or you can hear it in different responses given to different questions from a given biologist who will simultaneously attempt to justify expansion of science out into society and defend the very qualities that are said to demarcate science from society.

As noted earlier, many who sing biodiversity's praises in public are elder statesmen whose advocacy finds roots in a fertile, well-entrenched career. So it is not surprising that several (Kim, Ehrenfeld, Falk, Raven, Wilson, Ehrlich, Soulé) report no backlash to their careers from academic tensions raised by their "extracurricular" activities. Brussard notes that his career has actually been enhanced, although he feels he would have suffered backlash had he stayed in the Ecology and Systematics Department at Cornell, where his colleagues frowned upon "applied" research. But many warn that young biologists still need wait before they engage in conservation activities as part of their professional lives.

Ray cautions that the kinds of activities that Raven and Janzen undertake "can wreck your scientific career among your colleagues and can reduce your productivity." Young people face the greatest risk. "Lay low until you get tenure," Orians warns them, a strategy he says he followed himself, "so in a sense I couldn't be touched. The system couldn't hurt me all that much." He also continued his "basic research" to cover both terrains, the patches demarcated as properly belonging inside and outside of science. This productivity factor—the time consumed by conservation work—was cited as a potential or actual career impediment by Pimentel, Lubchenco, Iltis, and Funk. Rosen, Ray, Iltis, and Noss also revealed that they believed their conservation activities had hindered their career progress at one time or another.

Rosen organized the National Forum on BioDiversity at the citadel of the defense of objectivity, the National Academy of Science. He used the National Academy's imprimatur to hold a conference that expanded the boundaries of biology to include advocacy and launched the term *biodiversity*. He says, "had I been a little more conventional, a little more obedient, a little more of a company man, I might have been valued more highly there. . . . I have a feeling that I probably made people nervous about myself, and therefore you could say, in a sense, impeded my career." "We have to walk a very fine line," Wilson warns, and he is not the only biologist to use the metaphor of drawing lines; Lubchenco, Noss, and Orians do as well. The line demarcates science from nonscience. It is not only fine; it is movable. Wilson continues:

"And that's part of the reason I waited till around 1980 to plunge in. At that time, I saw the desperate need for scientists to get involved in a more activist role. One of the reasons for not getting involved in an activist role was a feeling that the role of scientist was the one that I properly filled. But it was just needed, that's all, in the 1980s. This was an area where scientists really needed to become activists. In 1980, I was fifty-one years old. I held the National Medal of Science. I was a member of all the leading scientific organizations. I had already accumulated a great many—Pulitzer Prize and all that stuff. I could—I had enough capital to risk. If a scientist goes in and becomes an activist at thirty or thirty-five, before they've built a reputation as a scientist, then they lose credibility rapidly. If you build a reputation as a scientist on into middle age, then your credibility is much less at risk."

Wilson had "credibility chips" (Jerry Franklin's term) to spare. Although he might disagree with my interpretation (when I asked, he replied "indoctrination is not my game"), his proposed biodiversity studies program would train scientist-activists so that they need not wait until they were "world class" to engage in a variety of boundary crossings on behalf of biodiversity.

Funk notes the need for biologists to work in society, adding, "It would be nice if they got a lot of support, peer support from the community to go out and do these other things, because I think in the end, science would be healthier." Like Wilson and Funk, Janzen would change the reward structure in science—in other words, change the definition of what it means to do good science. He is interested in helping incoming biologists do the things he does, and "the only resistance you get from young people is the perception, which is sometimes correct, that these old graybeards, who are people who are going to be deciding tenure, are not going to appreciate it, because they didn't—that wasn't the ritual they went through as graduate students. . . . if you want the young guy to feel free about going in that direction, what you have to do, people in my set, have to work on the upper level. They have to work on this reward structure up here so that it then rewards this young guy for going into it." Later, he talked about what happened to his curriculum vitae once he got involved in conservation activities:

"And from a pure, hard-science standpoint, it's not an impressive CV at all, especially not compared to the years before. So if somebody says, 'Oh, look, his career's gone all to hell,' well, there's two things there: one is, that depends on how you define your career. The other is that much of the stuff you do in this kind of thing doesn't express itself in publications that you list in a CV. If you bust your ass and you set up—or you help set up—an endowment fund for a national park: in work energy, that's worth 30–50 scientific papers. It takes, say, 4–5 years. The concrete outcome is that the park is still there—hopefully—is still there 20 or 100 years from now. But that doesn't—unless you reconstruct your

curriculum vitae, you invent a new kind of CV—it doesn't appear anywhere. . . . So the temptation people have is to not put it in there, so it's
just not there. . . . I don't think that conservation is intrinsically more
fragile, or perishable, let's say, than is science proper, but the identification of a finished product is a different process."

Janzen would institutionalize the practice of rewarding time and
energy invested in biodiversity conservation. He would have biologists' careers redefined so that conservation work shows up on the
CV, counts as part of the legitimate responsibilities or duties of a
biologist. Woodruff notes the same conundrum. His conservation
activities were numerous and diverse. "But your CV doesn't say anything about these community services," the chair of his department
said. "And I said, 'Well, there's not a space for them.' Well, now there
is a space for them. And I make a big point of listing all the things
I do on local, national, international committees. But that was held
against me."

By clearing space on his CV for conservation, Woodruff is clearing space in science and society for biologists' conservation activities. For Woodruff and his conservation activities, "coming out was
a long process that in part hinged on my career security." He would
hope to spare junior biologists the kinds of career backlash he endured by helping to redefine how the career of a biologist should
appear.

BALANCING ACTS

Others, too, recognize the threat of backlash to the hard-won
sanctity of scientific reputations when biologists venture out into
the deep waters of policy, values, and public representations of
nature. "It's a very, very dangerous game," acknowledges Lovejoy.
"I mean, it's sort of like tiptoeing around a volcano." Erwin notes
the tension between rational arguments scientists can make and the
temptation "to go circus with it"; but by doing the latter, "you lose
your credibility."

In a 1990 *Conservation Biology* editorial, Dennis Murphy attempts to draw some lines: "Conservation biology is not environ-

mental activism. While almost all conservation biologists are concerned with and involved in the resolution of environmental issues, the practice of conservation biology ends where science ends and where advocacy begins." One cannot just make a statement like this; the whole problem is that no one quite knows where those boundaries lie, and many biologists are involved in stretching them anyway. Murphy suggests that "we can eliminate these and other concerns and doubts about the nature and course of conservation biology if we step back and remind ourselves what scientists (all scientists, including conservation biologists) actually do, and how they do it." He then presents a positivist picture of scientific method, of generating and testing hypotheses.

By telling us what he wishes scientists did, or what he would like us to think scientists do, Murphy does not change the fact that "what scientists actually do" is . . . what scientists actually do. Conservation biologists do not merely generate and test hypotheses; they also write for the popular press, lobby legislators, and so on. Murphy suggests that reserve design and reserve management be treated as scientific hypotheses, experiments whose objectively studied results will guide future actions. But this is and always will be an explicitly political process. Try as he might, Murphy cannot depoliticize either conservation or biologists' role in it.[85]

A biologist must recognize the boundaries of his discipline, and he should cross them with caution in attempting to stretch them. Noss has experienced some career backlash because he has been perceived as too much of an advocate. "What most troubles many scientists is that advocacy jeopardizes our cherished credibility as objective observers, the quality that supposedly sets us apart from nonscientists," he writes. Were he to be "a flaming radical," Noss told me, "I'd lose my credibility very quickly, and I'd do a disservice to my profession." He continues: "So I think you have to be very careful. You have to be thoughtful. You have to know where to draw the line in your personal behavior, in your statements. Because after all, what is most important is getting the job done, not having some purist conception of what you should do as a person." That metaphor "draw the line" suggests Noss's awareness of the boundaries that exist, boundaries that he would change. Yet he dares not transgress too far for fear of not "getting the job done"—not being effective in saving biodiversity. Furthermore,

> "I've tended to hold back more often than I've made statements. But on the other hand, if biologists fail to make it known what their motivation is and what their personal as well as professional opinion is on some of these controversial issues, then we're not fulfilling our correct role as players in this policy debate. I mean, scientists are players in this. If we always say, 'Oh, we'll just analyze the situation. You make the decisions. We don't have any opinion or any recommendations on this'— if we always tell society that or policy makers that, then those who oppose biodiversity, and are making their opinions very broadly known, will win out, I think."

The "correct" role for biologists lies in the political process, where they should share, not only their factual material on biodiversity, but their emotional attachment to it and the values they adhere to: this is a cornerstone of Noss's view, the foundation that makes it so hard to build a structure that will not topple. He acknowledges that "unless we at least try to be objective when we wear our scientific hats, we might not be taken seriously by the public or by decision makers."[86]

What is a scientific hat? One way out of the "objectivity" conundrum is to counterpose the scientist as scientist against the scientist as human being, to declare that they function as separate entities. How can Noss be objective when he wears his scientific hat when he is so passionate about the cause of biodiversity conservation? He cannot; but he can do boundary work to try to convince you (or himself) that he can. If he wishes to avoid being misconstrued

> "in ways that would be harmful to our ultimate purpose or objective of protecting biodiversity or saving a particular roadless area, whatever it happens to be, then we have to be careful about the statements we make. Again, I think there's this balance between being philosophically pure and being pragmatic. And I try to achieve a balance between those myself. I think most people do. But some would just follow their gut instincts always. Others would sacrifice values perhaps to always be

pragmatic. I'm trying to find that middle ground; it's not an easy place. I don't know if I've found it, but I think it's a duty for us all to try to find it."

In this geographical metaphor, the "middle ground" Noss tries to find is that patch of terrain where he can have one foot in the camp of objective science and still venture out to move science into society in the name of biodiversity. He worries about stretching these boundaries to breaking point. Not just the institution of science risks being stretched out of shape; so, too, does the person representing that institution: "You have to be a whole person, and you can't be one kind of person one minute and another kind of person the next. . . . If you come into a courtroom as an expert witness, you're perceived as a scientist. . . . And in the society at large, how you're perceived depends on how the press quotes you, and displays you. And so a lot of it is beyond our control, what role we're put in. It's due to expectations on the part of others."

Noss may be disingenuous here. These quandaries do not merely arise from other people's mistaken expectations, to the roles that members of society desire scientists to play—although certainly if people did not want science as a source of authority, science would not have the clout it does today. Laypeople draw boundaries; but so do the scientists who create expectations by promoting the objective, value-neutral portrait of themselves that resonates in society. And no one forces them to appear in courtrooms and schoolrooms; they place themselves in roles other than those they have traditionally played.

Soulé, Noss's partner in advocating that biologists go public not only with their facts but with their feelings about biodiversity, says he also thinks that biologists can be objective in the process:

"The closer you get to humans and to life, the more people worry that emotions can interfere with objectivity, so-called. And they *can*; they certainly can. I'm the first to say that one has to know when one is attempting to be objective. And one has to know how and when, well when one decides to switch and go to another cognitive, or another conceptual, or another mental level and express things and feel things differently, more synthetically, more limbically or whatever. . . . I've had some training in doing that, in my Zen training. Where you're trained actually to go back and forth between a cognitive level and a more intuitive way of being. And you know, most people do that a lot anyway. It's just that they're not aware of the distinction. And it's a very cloudy distinction anyway, because there's always emotional interdigitation in what we think of as a purely rational, linear, cognitive process, and vice versa."

As we saw earlier, Soulé believes that biologists are taught that feelings obscure reality. The above passage suggests that biologists can "wear a scientific hat" and discuss things with some degree of objectivity, even though the boundaries are always blurred between rational and emotional ways of thinking and being. Soulé and Noss want it many ways simultaneously. They want biologists to be able to venture into public fora to discuss not only "factual" but also aesthetic and spiritual bases for biodiversity preservation. They want to present their emotions as part of the discourse, but believe they can, at least in part, switch their emotions off when called upon to speak "as scientists." They want to influence hearts as well as minds, but they have a stake in retaining science's traditional claim to objectivity.

Orians, too, expressed concern that by advocacy and by talking about values in public, biologists might weaken their platform: "It is something to worry about a great deal. And I do worry about it, and it's a very fine line. Because I very much feel that science is a very precious thing. And you don't want to undermine it and its credibility in the public mind." Yet Orians both cares deeply about the Earth and recognizes that "science is a particular form of human behavior, and it's tied up with everything else. And one tries to establish some standards in steering through this." Orians is not willing to say, "'I'm a scientist and therefore I surrender my right to be a citizen.' It just doesn't seem to me that that's going to make a better world than us taking the risks." Orians feels he can minimize the risks; he can steer around the boundaries. In speaking for biodiversity, he sometimes uses facts and sometimes feelings, and when he uses feelings, "I'm not going to try to tell you that the laws of ecology say this. . . . It's other feelings. So I try to be careful about this, and I think it is important . . . there's no reason that because I'm a scientist I've forfeited my right to feelings, or to advocate. It's just the base with which I claim to do that."

How does Orians separate the laws of ecology from his feelings? Science, even as traditionally practiced, never eliminates emotional input; the feelings a scientist draws from the broader social world help constitute how he views the world and therefore his science. When the scientist ventures forth into society, he may, like Orians, try to convince an audience—or himself—that he can separate these two forces, like a split personality. He can tell you he is putting aside his profession as scientist as the source of clout that justifies his words: "I do, in talking to groups, etc., . . . talk about values. But I just have to say what you're saying now: that I am now stepping outside what Gordon Orians ecologist-scientist can tell you as a scientist, and I'm now Gordon Orians as a person, who has feelings like the rest of you. And I know my feelings are influenced by the fact that I do these things as a scientist."

Like Noss, Orians tries to wear different hats, even when appearing on one stage before one audience. He claims to be able to enforce an unenforceable boundary. But audiences will listen to his feelings about biodiversity and give them extra weight precisely because he is a scientist. When he steps onto the public stage, he carries with him his lab equipment, his voluminous papers, his National Academy appointment, and the entire institution of science. When he speaks about the natural world, these invisible appurtenances add heft to his words that a nonscientist could not hope to wield. Orians extricates science from the boundaries of the lab, while nonetheless carrying those traditional boundaries with him. While attempting to reemphasize those boundaries—between Gordon Orians, rational scientist, and Gordon Orians, feeling citizen—he simultaneously expands them, becoming Gordon Orians, rational scientist/feeling citizen, for whom speaking out as a feeling, uniquely informed citizen is inextricable from the other roles of the scientist.

S. J. McNaughton roams these boundaries, too. He will discuss his feelings about the natural world "in a public lecture to a lay audience where they ask me a question that lends itself to it, that calls for some kind of statement of my personal value system and how it relates to this. . . . I mean, should I be embarrassed because I have feelings?" Of course not. However, the main reason people listen to him is because he's a scientist, and for the public, that means someone who is objective. By discussing values or emotions, he jeopardizes his role as a detached, value-neutral expert. So there's a trade-off. To this contention, McNaughton responds, "I don't believe that for a millisecond. As a scientist, I am not objective. I'm a human being; human beings are inherently subjective. However, I execute a process that is objective. That makes, if I do it well, that makes what I found out true—for the moment, until that truth is perfected by me or by somebody else, people who agree or disagree with what I find. The *process* is objective. As a person I'm not: I'm subjective." This is active boundary work, and it may genuinely be how McNaughton views science. Yet it is difficult to imagine how subjective people—who care passionately about both the natural world and about the processes and products of science—could ever transcend human fallibility to carry out a purely objective process.

E. O. Wilson engages in similar boundary work. I asked him if his proposed biodiversity studies program would be expanding the frontiers of what it means to be a scientist. His response: "Interesting question. No. I think that science itself will not change. What's changing is the role of science. You know what I'm saying. In other words, I don't expect the science that I do to be any more or any less rigorously conceived or any less rigorously judged than it has been all my career. Nor do I expect to see biodiversity studies as science treated any differently. But the role of science, scientists, in finding —how should we say—in seeking literary expression, in being engaged in public affairs, is expanding, at least in this end of biology."

I asked Wilson if his call for systematists to become conservation spokespersons for their particular groups did not expand what it meant to be a scientist, and he replied, "Again, it's expanding the role of the scientist. That doesn't expand the nature of the science." When I asked him about Soulé's views on biologists' emotions and his own assertion that "we need to encourage scientists to reveal ourselves," he responded: "That's part of the expanding *role* of the scientist. Not the expansion of science, [but] expansion of the role of science."[87]

By "science itself," Wilson means the hypothetico-deductive experimental science that has both found truths about the natural world and been used repeatedly by scientists to proclaim their cognitive authority. When he refers to the "nature of the science," he means nature as the essential element, the fundamental core. Scientists may do all manner of unusual things; but science doesn't budge. Contra Wilson, I'd assert that these actions—what he would have scientists do outside their labs—are just not separable from what scientists do in labs and how they're judged by their peers. It is all science. The authority to move out of the lab comes from what scientists do in the lab; and what scientists do in the lab lies at science's core. But scientists have placed this at science's core, have defended this center from which to expand as the context demands. A continuum of activities fall under the aegis of "science," but it is in the scientist's best interests to demarcate part of that continuum as "science" and part as something else (in Wilson's case, as the *role* of science, as opposed to some essence of science itself), something extra, something that draws on the considerable cachet of science's reputation to further scientists' goals, without jeopardizing that cachet.

Addressing this conundrum, Walter Rosen told me:

"Well, I think there's a paradox there. I think that the [National] Academy does, indeed, owe a lot of its credibility to its so-called objectivity, detachment, fairness, balance, all of those things. On the other hand, I think those same—when scientists use those same criteria to excuse themselves from getting involved in the public arena, and in policy issues—on the one hand, they're being true to their science. And on the other hand, they're copping out as citizens and as individuals. And I don't know how you reconcile that. I don't know how scientists can make value judgments without either consciously or unintentionally invoking their science as a source of authority. So the potential for abuse is built in. Science is supposed to be objective, yet I, who am a scientist, nevertheless feel and believe very strongly in this and that value. If I'm going to be listened to, it's probably because I'm a scientist, even though I'm making a nonscientific assertion. So I don't know what you do about that. Maybe if everybody was willing to do it, if all scientists or many scientists felt free to speak their *values*, the public wouldn't get confused about that. Maybe the public could be educated to differentiate between scientists speaking as scientists and scientists speaking as private citizens, informed by his or her science, but nevertheless private citizens. How's that? It's a speculation. I'd like to see it tried."

Rosen puts his finger on the problem: when scientists make value judgments, they draw their authority from their professional identity. Speaking off the cuff, Rosen proposes the kind of boundary work other biologists are doing—he tries to demarcate scientists as scientists from scientists as citizens, despite having just stated that such a separation is impossible. But if it is impossible in fact and deed, that does not mean that scientists could not do boundary work to try to make it seem as though it were a possibility.

Like Rosen, some other biologists have little problem admitting that science is not value-free. But they accompany such pronouncements with boundary work to maintain the institution's sanctity despite this seeming weakness. Eisner declares: "Science has never been value-neutral. That's a lot of horseshit. Science is scientific. I mean, if we're *supported* by the body politic, I mean, in a way, they pass judgment on what kind of money we deserve having—I'm sure we're not supposed to be silent in order to deserve that money. We're supposed to be scientifically good to the extent we can be. In fact, it's part of my job to say that we *cannot* make a scientific judgment on something—that data's not in—and to talk to the consequences of the data not being in."

Here Eisner limits the incursions of science into society to talking about values that are data-supported. Although he has moved into the political fray to support Zero Population Growth, INBio, the Endangered Species Act, and the preservation of several parks, to be consistent he would have to declare (and he might be right) that these actions are supported by the data that he as a scientist produces and best understands. Ehrlich affirms:

"In my view, it's preposterous for people who have spent their entire life immersed in a problem to present only a value-neutral thing. And politicians don't want you, ordinarily—they want not only to know what you think the situation is, they want at least suggestions on what society ought to do about it. And my view is, and I think people like John [Holdren] and Peter Raven and David [Pimentel] and so on, would all take the position that if you're standing in a building that's burning down, you don't just stand up and give measurements of the temperature and so on. You say, 'Let's get the fuck out of here' in addition."

So science is not value-free, and in Ehrlich's view, the public mandates that it not be so. Furthermore, the building—Earth's biotic treasure-house—is indeed burning down, and scientists cannot be value-free if they hope to douse the conflagration. Lubchenco offers a careful, measured interpretation of this dilemma. She, too, perceives the need to understand boundaries; simultaneously, she makes some of those boundaries herself:

"I think everybody has values. Everybody's activities and actions are colored by those values, whether they admit them or not. So I think that's the first point. It's probably impossible to have science be value-free. On the other hand, I think if one is being an active advocate of a particular position, one has to be particularly careful in drawing the line between what's fact and what's opinion. And that's a very gray line. And that's a very difficult decision right now, when in fact someone who is knowledgeable about a particular area feels very strongly that something is likely to be the case, but you can't prove it. So from a scientific standpoint, it's not really defensible in the traditional way that science operates. On the other hand, it's a fairly educated opinion, educated guess, an expert opinion, if you will, and I think that scientists right now are struggling with the way to act on and communicate that gray area because we don't have complete information, we can't always totally evaluate something, and yet based on our experiences, can make pretty good predictions."

Lubchenco feels some "people do cross the line. But I think it's something that we're all struggling with, for the reasons I just said." A continuum yawns between pure fact, which Lubchenco would agree can't exist, and pure opinion, which she would banish from the scientist's realm. It's that gray area between, the contested boundaries, that need definition if scientific expertise is to remain just that. As one solution, she proposes expert panels—for example, those under the auspices of the National Academy. Panel members would try to claim a part of the gray area between fact and value, isolation and incursion, and would be able to keep one another in line as they explored the gray area.

In the early 1940s, the pioneer sociologist of science Robert Merton highlighted peer accountability as one of science's fundamental norms. He called it "organized skepticism" and presented it as one of the ways in which scientists tacitly regulate themselves.[88] The current-day sociologist of science views a norm like "organized skepticism" as itself a tool scientists use to do boundary work. It not only serves to police the sanctified bounds of science from within; it is used as a rhetorical measure to convince the public that by policing themselves, scientists avoid subjectivity and falsehood. Soulé uses "organized skepticism" to tackle the objectivity problem:

> "What is objectivity? What is subjectivity? What does *value-neutral* mean? If you try to define *value-neutral*, you quickly find you can't. It's a meaningless term, because we all have values. . . . Because science has its own checks and balances built into it, unlike many other disciplines and endeavors, in that you present your data and then you sit back and wait for the tomatoes. You present your interpretation, too. But even data aren't collected neutrally. It's always been amusing to me and my colleagues that you can predict how a particular laboratory is going to react to a phenomenon on the basis of what they've already published. We can't avoid our biases; all we can do is be honest about our biases."

Soulé assumes that one's biases will be caught in the peer-review system. But what if all one's peers share one's subjectivity? What if you all have the same values? Scientists are on the whole interested in defending the institution's sanctity. The biologists I portray here all believe that biodiversity is important and that biologists should be empowered to speak about its value to society. So they are unlikely in the peer-review process to do things to disempower others who share their goals and values. They will make sauce with their tomatoes.[89]

We are not talking here only about factual claims about biodiversity, about the proper clade for this species of orchid or about the minimum viable population size for that population of tigers. These are peers reviewing one another's claims to speak about biodiversity's values, as well as reviewing the actual value claims themselves. I asked Paul Ehrlich about scientists being partly responsible for bringing about "quasi-religious transformations," for talking about ethics and aesthetics and beauty. He responded:

> "But I don't present them as scientific arguments. If you ask me, I'll tell you that it's not a scientific argument. One of the silly things is the idea that science is somehow separate from society. There is no value-neutral science. . . . The only reason, in my view, that science has any more objectivity than religion, or anything else, is because it's basically an adversary game. That is, if somebody can show, if I can show—much as I'm an evolutionist, if I can show that Charles Darwin was dead wrong, I'd be publishing it tomorrow, and so would any other ambitious scientist. And there are damned few unambitious scientists."

Ehrlich adds that peer review likewise applies to policy recommendations. But he does not say that so many conservation biologists share policy goals, share the values that lead them to make policy recommendations in the first place, share the methods that produced the science that makes the policy possible, share the clout that comes from enactment of policy that gives more power to the biologist, more credibility to his science, more cash for his laboratory. Few conservation biologists are going to throw tomatoes at Paul Ehrlich when he talks about values, ethics, and so forth, because they do not want to engage in public squabbles over science, thus weakening the institution. And they do not want to hurt the chances of successful forays by biology into society, because that might detract from conserving biodiversity and impede increased power for Ehrlich and all his peers. Ehrlich is not paving the way for Ehrlich alone; he is paving the way for all those others who would share the power to define public priorities that he seeks, whatever their motives for doing so.

TRANSCENDING SUBJECTIVITY

One can also skirt the value/objectivity issue by claiming that one's values are not values but facts, ineluctable consequences of what the natural world has revealed. Dexterity in demarcating the boundaries separating "value" from "fact" enables scientists to maneuver more freely in society on biodiversity's behalf.

Some claims seem relatively noncontroversial. Lovejoy told me that "if there is an inherent value judgment in what I'm doing, it's that biological diversity is important. But it's a value judgment based on a series of pretty solid convictions—provable convictions." It's no stretch for a biologist to make such a claim; as we shall see in Chapter 5, a plethora of evidence can be mustered to substantiate this mild assertion, and some of this evidence does fall into the traditional jurisdiction of the scientist.

Hugh Iltis takes strong stands on compulsory birth control, abortion, sustainable development, and other issues that he ties directly to biodiversity conservation. He derides his opponents as feminists or other special-interest groups, many of whom choose environmental goals that are "colored by the need to help people." For Iltis, "if you put people first, you're lost right from the start. It's a slippery slope from which you can't get up once you start sliding." Rather, biodiversity must come first, always. By his reckoning, this is not his opinion; it is fact. He has taken flak for this, for "if you take a very unpopular truth, you really get socked." He agrees ("Absolutely. Absolutely.") with my assessment that he believes that "when he expresses an opinion, it's an opinion that's very much based on scientific facts" and therefore would not harm the credibility of science in any way. Iltis claims his opinions are truths, unsullied; the opinions of his adversaries are tainted by subjective values, as when Stephen Jay Gould "for political reasons doesn't like to hear genetics introduced into our likes and dislikes." He labeled his opponents "a bunch of fuckers, I mean really incredible" because they fight his objective facts with political opinions.

Donald Falk attempts to draw similar bounds (albeit in a gentler way), since what people

> "need to be able to depend on in science is a lack of bias. Well, values and bias are not the same thing. They're very, very different. And when, in other words, Ed Wilson or Tom Lovejoy or Peter Raven make a statement about the loss of diversity in tropical rain forests, people count on them to be saying something that is factually true. That's what science owes to society, to say, to describe things as they are to the extent that

the scientific method can elucidate them. That has nothing to do with whether any one of those people might go on and say, 'I think it's great that we're destroying the forest.' That's a value judgment. But the accuracy and lack of bias is one dimension; the bias is separate."

Opponents of biodiversity (for example, the infamous Julian Simon) contest the notion that even something so unproblematic as estimates of biodiversity loss are factual. Ariel Lugo, a proponent of biodiversity, suggests that sometimes biologists' figures are biased by their emotions.[90] But even if these critics are wrong, the biologists they criticize do make value judgments, which may or may not stem inevitably from their data. Bias, values, data—how do you extricate them? Once you admit values into the picture, as Falk seems willing to do, bias seems to have to follow.

Noss takes an approach similar to Falk's, although he does not try to draw unenforceable bounds between values and bias:

"You know, it's a fine line between a bias and a moral principle, I think. Moral principles are often associated with certain sets of biases, so they're not always clearly distinguishable. But I guess we have to consider what kinds of bias are objectionable, and which are not. Which are the long-range consequences of different kinds of bias, and in various displays of bias. And I think we can do that through a rational analysis. A lot of what is motivating us is intuitive, but we can also as scientists be objective about the consequences of different moral positions. And I think we need to think about those things very carefully."

I would certainly agree with his last statement. Who can decide what counts as a bias and what counts as a moral principle? Noss wants it to seem that his deeply felt beliefs are not biases. Rather, they have the force of moral certainty behind them: his biases are the right ones, the factual ones. For example, "the proposition that biological diversity should be maintained is not a bias." Leopold's Land Ethic "is a statement of moral principle, not an expression of bias."[91]

For Noss, to be effective, "it is essential that we dispel the notion that conservation of biodiversity is a 'special interest' in the same league as the interests of the timber industry, commercial fisherman, Elk hunters, river-runners, or backpackers. Can anyone think of a more *general* interest, here on Earth, than maintenance of life? Any biologist who does not advocate this general interest has little credibility as an intelligent being."[92]

It does not matter whether you feel the way Noss feels about the importance of preserving the riches of life on Earth. I certainly do. But try putting Noss's proposition that biodiversity advocates represent the "general interest" to someone from any of the "special interest" groups he cites. What represents a special interest (and therefore a biased position) and what represents a general interest (and therefore a position working for all without bias) is contested ground. Noss stakes the general-interest claim, the moral-principle claim, to remove the taint of bias from biologists' policy and value prescriptions.

Peter Brussard is keenly aware of the perils conservation biologists face when they become advocates. When I asked him if people might stop listening to scientists if they were perceived as partisan, he replied,

> "Well, how can you be partisan about biodiversity? Preserving the biological machinery that makes the world work I don't think is a partisan issue, quite frankly. There's not another group of scientists that have any credibility, except in my mind, except some half-cocked economists, and I think it's arguable about whether they're scientists or not, that think we can do without the biological world. So from that standpoint, I think it's value-neutral. It's just like a physicist arguing about the utility of gravity: it's hard to argue against it."[93]

For Brussard, those who argue for biodiversity are not partisan, since biodiversity's values shine by the light of their own goodness. Note his boundary work. The only ones he judges might have any expertise are other scientists, and he then attempts to put economists, particularly those who doubt him, beyond the boundaries of acceptable science. He draws in the physicist, the paragon of scientists, as an ally, and cites ineluctable gravity as an analogy to biodiversity. To maintain his scientific stature, Brussard has

> "tried pretty hard not to be an advocate of any particular—let's see, how can I best phrase this? I mean, given that conserving natural eco-

systems, conserving species, conserving gene pools is a value-neutral thing, then there's also good scientific advice on how to do it and how not to do it. And it's a viable population or it isn't. Or this reserve is too small. Or this land-management policy will ultimately result in the extinction of this species or something. *That's* a scientific call, as well. And so what I've tried hard to do is stay away from advocacy outfits like the Sierra Club: 'Well, won't you come and testify about the importance of. . . .' Then I say, 'No.' I just say, 'Well, there are other people who can talk about that stuff, but I'd rather not.'"

His value-free premise is not a given to the opponents of biodiversity conservation, particularly in his home state of Nevada and its neighbors (opponents, he notes, call biodiversity "another green plot to take away one more freedom"), but he attempts to make what follows from that premise seem cut-and-dried. As I suggested in Chapter 3, however, so little is known about so much to do with biodiversity that "scientific calls" may well be influenced by one's passionate concern for what is being discussed. That Brussard and peers choose to work on endangered species and choose to call themselves conservation biologists makes them advocates. Staying away from the Sierra Club is boundary work. It may rid him of the veneer of advocacy, but not of its essence. And if biodiversity is an unambiguous good, why stay away from an organization that fights to defend it? For Brussard, "there is a confusion over what is science and what isn't science." And he tries to exploit those confusions to help with the definitions. For example,

"Now where it tilts one way or another is you say, 'Okay, we have three scenarios. If you graze this number of cattle on your land, then in the next ten years this little dickeyweed will go extinct.' Well, a good environmentalist would say, 'That's absolutely unacceptable; we've got to get rid of the goddamned cattle.' The rancher might say, 'To hell with the dickeyweed; the cattle are much more important and there are plenty more weeds where that one came from, so get off my back.' Those are value judgments that society has to make. And they don't have much to do with science. Science can make prescriptions and say that there's a prediction. But then where you come down on that is how you personally—What your own value system is."

But these value judgments have everything to do with science. Scientists named and classified the dickeyweed. Probably scientists identified it as being endangered, testified on behalf of the Endangered Species Act driven by their value judgment that no species should be eliminated, and drew on their scientific clout to appear as experts. Scientists develop the biology that goes into management plans for the dickeyweed, and scientists are promoting the dickeyweed's value to broader society. The kind of demarcation that Brussard and others believe is possible between what is and what is not scientific isn't possible. They continually cross boundaries that they claim are impenetrable.

In 1991, *Science* published a special "Perspectives on Biodiversity" issue. Terry Erwin asks in his contribution: "Should conservation strategy be scientifically or culturally based?" He goes with the former, since "scientific rationale may transcend cultural changes through time, whereas economic and political grounds certainly will not." Given this, "for scientists, the question is what can we provide from our science that will help generate a long-term, transcultural foundation on which conservation strategy can be based?" Finally, "acceptance of a nonhuman yardstick to measure environmental health—that is, evolutionary processes—and implementation of a scientific approach in conservation policies will provide a strategy to achieve a lasting stability for global environmental health because the basis for conservation will not be tied to the whims of human culture."[94]

Boundary work abounds here. Science is not culturally based. Scientific thinking does not change with the times as economic and political thinking do. Science can inform conservation in a transcultural way. Science can approach a value-free basis for conservation, above the fray of cultural relativism. Science has no "whims," which are in the realm of "human culture." The "yardstick" for conservation may be beyond science, even beyond humanity, if measured in evolutionary processes.

Erwin's specific transcultural policy recommendation is that we focus our efforts on conserving habitat that contains recently evolved hotbeds of speciation, as revealed in cladistic analyses (see fig. 4.1). By focusing conservation efforts where systematists have

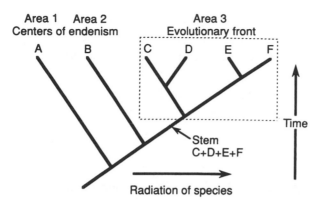

FIG. 4.1 Erwin's (1991) transcultural focus for conservation (*endenism* should read *endemism*). © 1991 American Association for the Advancement of Science; used with permission.

revealed such speciating fronts, we are most likely to allow maximum future evolution, and thus maximum future biodiversity.

I cannot agree that this way of thinking about conservation or about the world transcends culture in any way. It is firmly rooted in the culture of science, which is firmly rooted in a Western way of thinking; science has historical and cultural roots and is continuously culturally molded. What is held as truth changes continuously. Scientific truths and scientific methods are considerably different than they were a hundred or even twenty years ago. Erwin would disagree:

> "I think probably today we understand science better because we're much more rigorous with the basic hypothesis-testing scenario that everybody uses. Two, three, four hundred years ago, it was really—Science was a hobby for doctors and lawyers and Indian chiefs, really. And it wasn't until really the past couple of hundred years that scientists have become very rigorous in how they approach things. The scientific method is in concrete. And so I don't feel that it's going to change. Our methodology is going to improve—but not science."

Somehow "science" stays the same, even though methodology and factual content changes. "Science" is presented as not merely transcultural but transhuman.

Erwin's transcultural rationale reveals his interest in seeing the evolutionary process continue. While you or I may agree with this, may picture the world in these terms, valuing something as abstract as the evolutionary process is very much a cultural artifact. A Kayapo Indian in the Amazon, a rural herdsman in Senegal, or an Indian farmer at the foot of the Himalayas may value biodiversity as much as does Terry Erwin, but for different reasons. Their reasons will be as inextricable from their culture as Erwin's reason is from his culture. I asked Erwin if he really believed that science is in some way transcultural, that it doesn't root in our culture. He said:

"I guess what science does is work outside of genetic preservation. If you're dealing with an indigenous tribe in the Amazon, everything they do, they do because the tribe survives when they do it and the tribe disappears when they don't do it. And of course, that's a big gap there in this discussion, but obviously little changes, generation after generation, have adjusted those tribes that are surviving, in the evolutionary process of survival, to use the environment in a balanced way: not to overuse it, certainly not to underuse it. But use it just right. Set up all the taboos, the shaman getting the right plants for treating the right infirmities and so forth. So theirs is an evolutionary, practical thing that is almost built into their genes. We've lost all that. And in fact, most of the people living along the Amazon today have come from the mountains or from the city, and they don't have it either. And that's why exploitation is more than what the forest can bear. It's because people are using it in ways they don't know how to use it, whereas the indigenous people knew how to use it from generations, many generations past. From a scientific point of view, I think we can reach back and learn how to use it through scientific methods. So we're paralleling what goes on naturally. I think basically that's it."

An Amazonian tribe operates under the laws of natural selection in a way that Western scientists don't: so that makes their values cultural in a way that scientists' aren't? Might not the reverse be true, that the Amazonians operate under biological, and therefore transcultural laws, while the scientists are the ones with a firm foothold in a specific culture?

Erwin draws a sharp boundary around science to separate his prescriptions from the jostle and fray of competing value claims. Vickie Funk is a systematist (plants instead of beetles) who works in the same building at the Smithsonian. Erwin and Funk share a culture. Funk describes efforts by systematists to identify conservation priorities focusing on the bases of cladograms, on the endemic species prone to extinction that should be saved, not only because they are curiosities, but because they are crucial tools for understanding evolution. So not only is Erwin's prescription not transcultural; even within his own tight subculture of systematics, others disagree with him. Funk takes this one step further when she describes a visiting group of foreign scientists:

> "They don't have any concept of objectivity the way we [U.S. scientists] do. To them, the political aspect of what their decisions are going to have in science is just about as objective as they can get. Because moving up in [deleted country] really is a much more political thing than it is here. So that political to them is very objective. So we have a hard time—like, if we're trying to do cladograms, they don't want it to come out this way, they want it to come out that way because the expert in that group feels like it comes out this way. And they don't consciously realize they're skewing their data to make it come out that way."[95]

So even "objectivity," the flagship of Erwin's transcultural scientific method, may be rooted in culture. All this passes no judgment on whether Erwin's conservation prescription has merit. Rather, I am trying to show that all conservation prescriptions are value-laden, even if the prescription is something as innocuous as "biodiversity is good." Simultaneously, it is a political maneuver, a piece of boundary work for the biologist making the prescription to attempt to label his values or biases as truths, as ineffably scientific and therefore beyond criticism. It is a political act to disguise values or morals as part of the scientific process, and thus draw on that process's clout. Erwin's rationale (or any other biologist's for that matter) is merely one conservation priority among many, posed as if it were transcultural, objective, and therefore naturally entitled to override competing claims (I return to this point in Chapter 8).

ENVIRONMENTAL HISTORY, SCIENCE STUDIES,
AND THE ILLUMINATION OF ENVIRONMENTAL ACTIVISM
BY BIOLOGISTS

Michael Soulé urges us to break the shackles of multicultural relativism so that we may crusade against abhorrent practices: for example, when Chinese healers grind up the bones of some of the last tigers on Earth with medicinal intent, we must speak out in protest. But Soulé provides us with no guide as to how to discern universal right from wrong. If nature provides us with no external referent for the derivation of cultural or ethical imperatives, we must start by finding new ways to evaluate knowledge and value claims.[96]

This book takes a stab at that project, by using science studies and environmental history to evaluate fact and value claims made on behalf of biodiversity by biologists. We cannot merely accept that biologists are objective purveyors of knowledge, and that we should thus adhere to their conservation prescriptions. Those (like myself) who wish to see effective, just, enduring conservation need a more nuanced depiction than that. We should all understand who speaks for nature and why.

Humanities scholars—here, environmental historians and science studiers—can illuminate these murky waters. In the drive for biodiversity conservation, biologists redraw personal, professional, geographical, and biological boundaries. Science studies analysts can watch them do so, can highlight the work that goes into demarcating, defending, and changing those boundaries. They can look on as conservation biologists struggle to balance internalized norms, professional strictures, political goals, and environmentalist values.

By thick descriptions such as the one I provide here, rooted in specific events, times, places, and personalities, science studiers and environmental historians may teach more general lessons. William Cronon suggests that environmental history's unique strength lies in its use of narratives that engage the reader and present suggestive parables rather than conclusive proofs.[97] This strength is not unique to environmental history; science studies case histories also engage the reader by attention to specific episodes from which we might draw broader conclusions.

These stories, told in both science studies and environmental history, emphasize the ironic.[98] Science studies is a lot about revealing the ironies behind scientific facts and scientists' claims and behaviors. Things are seldom what they seem to be. Language and practice in science may pose as one thing but in fact be the opposite: objective may be subjective, professional may be personal, apolitical may be political, rational may be nonrational, and truth may be fiction.

Environmental history chronicles the ironies that result from nature's recalcitrant insistence on not cooperating in the ways people expect when they manipulate it. Both these disciplines contribute to the study of environmental problems by showing us that we are all preeminently fallible, scientists as much as anyone else. A take-home message from science studies and environmental history (and, of course, conservation biology), then, is that we should treat the Earth with less hubris and greater humility. And if, as I contend, ideas can act as forces of nature—if representations can have tangible ecological impact—then what effects (anticipated or not) might be felt when biologists propagate a new idea of nature? In the case of biodiversity, the lessons of ironies past lead us to question the claims of those who seek to speak for all of nature, and those lessons may also promote self-reflection in those who would make such claims.

In using environmental history and science studies to patrol these boundary crossings, I would have practitioners of these disciplines rethink their own boundaries as well. Environmental historians are not content to analyze the interactions between the natural world and human forces of production, behaviors, ideologies, and values; they also seek to manipulate those very factors. They hope their analyses of the boundaries between nature and culture will help teach lessons about the bounds of human activity; like conservation biologists, they wish us to reformulate acceptable behavioral and normative boundaries so that we form a more appropriate relationship with the Earth. Science studies, too, is concerned with how such bounds are constructed. But environmental historians follow in a recent tradition in which historians explore hitherto obscured factors such as race, gender, and sexual orientation with intent to influence the politics and culture they document. Like conservation biologists, these new historians seek to change scholarship so that the normative goal of influencing real lives—human and nonhuman —lies within the acceptable, even required purview of the scholar.

As I commingle these disciplines, the boundaries of environmental history expand to include as subject matter realistic portrayals of the institution of science. While environmental historians attempt to mediate the forces with which society and nature are molded, they require more sophisticated tools to understand their own and others' mediations. I would have environmental history gain from science studies a methodology and even a worldview for looking at science and society, a kind of analysis that looks at scientific and nonscientific claims reciprocally. Ideas—or useful tools—such as boundary work and Latourian resource webs, which always exist in environmental disputes, can be seen, and then analyzed, by incorporating a science studies perspective.

For science studies, to concentrate on environmental problems might arrest any impulse to deconstruct without attempting to put the pieces back together. To scrutinize case studies from the standpoint of environmental history might result in the epistemological admission that nature does constrain the facts and values scientists attribute to it. Engagement with environmental issues might result in greater commitment to affecting the worlds science studiers study, with concurrent commitment to accessibility in terms of style and jargon that would expand use of analyses to new audiences.

"People are constantly engaged in constructing maps of the world around them, in defining what a resource is, in determining which sorts of behavior may be environmentally degrading and ought to be prohibited, and generally in choosing the ends of their lives," Donald Worster observes.[99] In a postmodern, vertiginous, information-overloaded, contentious world, we may sometimes need help in steering our lives toward worthy ends. If we, along with millions of other species, are to survive, we need to become more sophisticated—about science, about nature, about culture, about the choices we make as individuals, as communities, as nations. The idea of biodiversity has been created and disseminated by biologists so that we rethink the choices we make in our relationships with nature. By synergistically pairing environmental history and science studies, I hope to help conservation biologists help a wider audience understand complex environment-society interactions, enabling us to make more informed decisions about how we shall lead our lives.

5

When promoting biodiversity, biologists are not attempting to shift our ideas to one image of nature. Rather, they appeal to the manifold images and meanings that we can find in the natural world; they exploit our predilection for seeing ourselves in nature's mirror. And the range of meanings we seem to find in biodiversity find neat parallels in the multiplicity of reasons biologists tout to preserve biodiversity. Biologists fervently try to convince us that biodiversity is important: it has *value*. This is a tough word, as elusive, in its way, as *biodiversity*.

"We all face, with alternating joy and repulsion, the idea that there is no absolute standard for social mores," Neil Evernden asserts. Instead of such absolutes, we now talk of *values*, a "contemporary term for the human-generated norms which we possess in place of absolute norms or 'instincts.' "[1] Langdon Winner points out that value used to inhere in things—things themselves were considered good or bad or beautiful or immoral—whereas now we each hold values of our own, which we assign to things, and often attempt to persuade others of their superiority.[2] Others rue this alleged lapse into moral relativism or value nihilism, as it is difficult to argue for the innate superiority of subjectively held values. Of course, even back in the "good old days," values that were said to exist outside of the human valuer were often placed there and defended by those in power. If, say, frugality or patriotism were good and laziness or adultery were immoral, at least part of these value attributions came from the elite, who knew their preferences would carry more weight if they were not seen as human ones. And the way values said to inhere in things change over time—nature the dangerous, nature the ugly becomes nature the spiritual, nature the sublime—belies the notion that values ever really came from anywhere different than they do now.

Furthermore, I am not sure who Evernden's "we all" are, for many of "us"—probably the majority of the West's and the world's citizens—still invoke absolute standards for social mores; one need only watch Sunday morning television to be convinced of that. And contra Winner, some people, even some humanists, do still suggest that value inheres in loci other than the human brain.

Still, we might think that scientists, paragons of humanism, would rush to abandon the notion of values inhering in extrahuman objects, or at least would avoid arguing that some subjectively based values are more valid than others. How can a scientist speak for what she cannot prove, what cannot be subjected to the method that allegedly defines her profession? It would seem a priori reasonable for scientists to treat values as beyond their expertise.

At the same time, if scientists wish to see their cherished values adopted and spread, they have reason to claim the right to speak for them. Confronted with ecological destruction, biologists are creating and defending an axiology that will serve to persuade others of the value of biodiversity. They are attempting to expand the boundaries of accepted scientific behavior to include the contested realm of values. As we saw in Chapter 4, they require considerable skill to capture new turf while simultaneously defending precious, hard-won territory.

With subjectively based values, my values have no greater claim to truth than do yours—unless I convincingly tie my values to something bigger than both of us. Biologists commonly attempt to tie the array of values they attach to biodiversity to that something bigger; since they attempt to convince us that their values are in some fundamental way better, truer, worthier of our allegiance, they seek to compel us to share these values.

Note that some of the references I discuss in this chapter predate the creation of the term *biodiversity*; for, as we have seen in Chapter 2 and elsewhere, prior to 1986, biologists espoused terms such as *nature*, *wilderness*, *natural variety*, *endangered species*, and *biological diversity*. Biologists now use *biodiversity* as a device, a strategy, a shibboleth around which they focus their values, as well as their advocacy on behalf of those values. With biodiversity comes serious activism by biologists on behalf of the natural world and aggressive boundary work that simultaneously seeks empowerment to speak for a host of values hitherto thought to be beyond the pale of scientific expertise.

We can mine nature for what we want. Some of nature's value is extractable; it can be brought under human control. Nature provides the raw material that we shape and mold, be the final product aspirin, Rainforest Crunch, or the biodiversity concept. Each of these products mixes natural reality with human labor; they are human constructions with a natural base. As we examine the values that biologists extract from nature, we should ask: how much nature, how much construct? The more successful the biologist is in persuading us that those values really do derive from nature and not from subjective preferences—or the more they really *do* derive from nature and not from subjective preferences—the more convinced we may be that we should respect or adopt them. A corollary question to keep in mind is, who benefits from the alleged value, and how?

The analysis of values assigned to nature has spawned a cottage industry of eco-philosophers. Arguments on the subject, and arguments about the arguments, come in an array of flavors and colors, alternately dazzling and deadening the reader. Countless books and articles have been and are being written in this vein; I have simplified many of their arguments here, focusing on those that biologists offer when they promote biodiversity. The journal *Environmental Ethics* gives over its pages to the ongoing debate about environmental values. Works by David Ehrenfeld, Warwick Fox, Bryan Norton, Michael Soulé, and Stephen Kellert offer broad, contentious (and therefore more interesting) overviews of reasons proposed for conservation. In addition, the writings of Mark Sagoff, Holmes Rolston, and John Tierney have provided me with much scope for contemplation of the thorny issues raised in this chapter.[3]

Two dichotomies are noted throughout: is the given value of biodiversity one rooted in humans—the human as the holder of values—or is the value one that inheres in biodiversity itself, apart from human valuation of it? And is the value selfish—in other words, can

humans benefit from it—or is it unselfish—which is to say, of benefit only to biodiversity? Various permutations of these are possible,
as we shall see. They have in common the attempt to make a claim
about a value that is bigger than the claimant, that moves beyond
"This is why biodiversity is important to me" to "This is something
that is much bigger than me, and you should assent to it too."[4]

SCIENTIFIC VALUE

The most obvious value biologists might promote for biodiversity—although one cited less often than others—is that it has value
for science. As the raw material for biological study, biodiversity
is essential for the scientific endeavor to continue unhindered. If
we recognize science and its goals as unquestionable, overarching
goods, then this value transcends mere subjective preference. "As
living organisms," asserts Thomas Lovejoy, "we have a vested interest in not limiting the growth of that branch of knowledge known
as the science of life. For that reason alone we must be concerned
with the survival of each and every species." We are not just people
with conflicting value systems: we are "living organisms" who depend on that science and the raw material it requires. Testifying
before Congress in 1991, Lovejoy declared that "the variety of life
on earth represents an extraordinary intellectual resource, and is
essentially the basic library on which the life sciences can build . . .
the kind of rapid loss that we are experiencing in the 20th century
is a form of book-burning and one of the greater anti-intellectual
acts of all time."[5] The book-burning metaphor scares us, and the ad-
jective *anti-intellectual* shames us into assenting to biodiversity's
scientific value.

Not only is biodiversity the "living library" of biology, it "can
inspire larger activities"—living things may supply "inspirational
value" for larger discoveries. Lovejoy notes that the discovery of how
the venom of the bushmaster viper works led, by analogy, to the development of the prescription drug Capoten for high blood pressure,
which nets $1.3 billion annually for its makers. An NSF panel advo-
cates saving the endangered desert pupfish, which survives in only
one western hot spring, as it might be a model for human understanding of heat tolerance. And biodiversity has scientific value in

inspiring a curiosity that is perhaps more finely honed in biologists, but that we are all said to have. Funk says, "Learning about the unknown and figuring things out and understanding what's going on is part and parcel of our nature." Edward O. Wilson also feels we have this drive to know; biodiversity is necessary to satisfy the need to explore in each of us, as well as for the formalized exploration called science.[6]

So biologists claim that biodiversity must be preserved as raw material for them to study, while simultaneously seeking support to study it. Biologists depict themselves as obligatory points of passage for a society that wants knowledge of biodiversity, wants to know what it is, how we can use it, how we can save it. The more they understand it, the greater the number of uses we find for it, the greater the demand for its preservation. Its scientific value increases concurrently with its other values.

ECOLOGICAL VALUE

As we might expect, ecologists and conservation biologists proclaim biodiversity's ecological value. These ecological arguments can be interpreted as human-value-centered and selfish or non-human-value-centered and unselfish, or some permutation of these. "Ecosystem services" may have value of and for themselves—in other words, it may be argued that keeping ecosystems healthy and functioning has value apart from any human valuer or any value humans may obtain from them. We may thus value biodiversity because we value the continued healthy functioning of ecosystems as such, regardless of any services biodiversity performs for us. More often, however, humans are said to benefit from such ecosystem services. Half a century ago, Aldo Leopold warned: "Recent discoveries in mineral and vitamin nutrition reveal unsuspected dependencies in the up-circuit: incredibly minute quantities of certain substances determine the value of soils to plants, of plants to animals. What of the down-circuit? What of the vanishing species, the preservation of which we now regard as an esthetic luxury. They helped build the soil; in what unsuspected ways may they be essential to its maintenance?"[7] More recently, Jane Lubchenco feels "very strongly that people are in fact much more dependent on ecosystem services that are provided by both managed and unmanaged ecosystems than is generally perceived to be the case. So I think it's sheer folly for us to act in ways that are undermining the ability of both managed and unmanaged ecosystems to provide these services that we're dependent on. And that we're doing that more and more as we pollute and destroy habitats, or alter habitats in one fashion or another. And I guess the bottom line is that we're changing the environment faster than our ability to understand the consequences of how we're changing it."

Most predictions of eco-doom are predicated on this argument, and many are stated in much more dramatic terms than those Lubchenco employs. As the argument runs, a myriad of organisms, especially "little things," comprise ecosystems that provide countless services that keep the Earth's biotic and abiotic processes up and running.[8] According to Soulé, "Many, if not all, ecological processes have thresholds below and above which they become discontinuous, chaotic, or suspended." Biodiversity may regulate these processes; among its many talents, biodiversity is said to create soil and maintain its fertility, control global climate, inhibit agricultural pests, maintain atmospheric gas balances, process organic wastes, pollinate crops and flowers, and recycle nutrients.[9]

Confusion in this line of argumentation ties back into why the concept of biodiversity has risen to prominence. Remember that biologists have scant understanding of the roles that species or populations play in maintaining ecosystems. In interviews, Lovejoy, Falk, and Ray confessed that you can strip away many species from an ecosystem without loss of ecosystem function. Ehrlich points out that by the time a species is endangered, it has probably stopped playing an important role in keeping the system functioning anyway.[10] Furthermore, it is not clear whether we should focus on species as functional cogs in the ecosystem wheel, or whether ecological services are emergent properties of ecosystems themselves.

With the biodiversity concept, these dilemmas become nearly moot. Biodiversity embraces lists of species, lists of ecosystems, the interactions of species within ecosystems, and the processes that species may maintain or control. When arguing on behalf of biodiversity, one need not focus on the specifics—specifically, the specifics of what we don't know. It is enough to explicate some of the functions that keep ecosystems running, or that ecosystems provide for us, and then extrapolate to the dangers associated with declining biodiversity. Peter Raven bases his thinking on Leopold's observation "To keep every cog and wheel is the first precaution of intelligent tinkering":

> "In every sense, in the sense of communities that will preserve soil, promote local climate, keep the atmosphere, preserve water, and everything else, the first rule of being able to put together communities well or have the world go on functioning well, or to keep climates as they are, or to retard disease, to produce products we want sustainably, because, after all, plants, algae, and photosynthetic bacteria are the only device we have to capture energy from the sun effectively—in all those senses, and in the sense that we're losing the parts so rapidly, I consider the loss of biological diversity to be the most serious problem that we have—far more serious than global climate change or stratospheric ozone depletion, or anything else."

"Habitat destruction and conversion are eliminating species at such a frightening pace that extinction of many contemporary species and the systems they live in and support . . . may lead to ecological disaster and severe alteration of the evolutionary process," Terry Erwin writes.[11] And E. O. Wilson notes: "The question I am asked most frequently about the diversity of life: if enough species are extinguished, will the ecosystem collapse, and will the extinction of most other species follow soon afterward? The only answer anyone can give is: possibly. By the time we find out, however, it might be too late. One planet, one experiment."[12]

So biodiversity keeps the world running. It has value in and for itself, as well as for us. Raven, Erwin, and Wilson oblige us to think about the value of biodiversity for our own lives. The Ehrlichs' rivet-popper trope makes this same point; by eliminating rivets, we play Russian roulette with global ecology and human futures: "It is likely that destruction of the rich complex of species in the Amazon basin could trigger rapid changes in global climate patterns. Agriculture remains heavily dependent on stable climate, and human beings remain heavily dependent on food. By the end of the century the extinction of perhaps a million species in the Amazon basin could have entrained famines in which a billion human beings perished. And if our species is very unlucky, the famines could lead to a thermonuclear war, which could extinguish civilization."[13] Elsewhere, Ehrlich uses different particulars with no less drama:

> What then will happen if the current decimation of organic diversity continues? Crop yields will be more difficult to maintain in the face of climatic change, soil erosion, loss of dependable water supplies, decline of pollinators, and ever more serious assaults by pests. Conversion of productive land to wasteland will accelerate; deserts will continue their seemingly inexorable expansion. Air pollution will increase, and local climates will become harsher. Humanity will have to forgo many of the direct economic benefits it might have withdrawn from Earth's well-stocked genetic library. It might, for example, miss out on a cure for cancer; but that will make little difference. As ecosystem services falter, mortality from respiratory and epidemic disease, natural disasters, and especially famine will lower life expectancies to the point where cancer (largely a disease of the elderly) will be unimportant. Humanity will bring upon itself consequences depressingly similar to those expected from a nuclear winter. Barring a nuclear conflict, it appears that civilization will disappear some time before the end of the next century— not with a bang but a whimper.[14]

Stephen Jay Gould presents an equally chilling picture. It is in our "enlightened self interest" to treat Mother Nature nicely: "We had better sign while she is still willing to make a deal. If we treat her nicely, she will keep us going for a while. If we scratch her, she will bleed, kick us out, bandage up, and go about her business at her planetary scale."[15] Nature is personified as a woman who cares not a whit about us; we, however, must value her supremely, as her biotic processes hold the key to our future.

David Pimentel expresses this somewhat more soberly: "We can't have agriculture without these species, we can't have forestry without these species, we can't live without these species, and that's the essential part." Walter Rosen, Jane Lubchenco, and Gordon Orians offer similar arguments as in Orians's statement, "I'm very much concerned about preserving the capacity of living systems to provide the resources upon which a quality human life depends." Bryan Norton points out that since humans reside at the end of food chains, we surpass most other organisms in our vulnerability to extinction.[16]

Not only does biodiversity sustain us; it provides an "early warning system" that alerts us when it—and therefore humanity—may be in peril. Similarly, biodiversity is a "barometer of environmental health." According to Falk, endangered species are "meaningful primarily because they tell us where there is trouble. And not just geographically. They are excellent ways of spotting problems." Raven calls biodiversity "the key to the world's stability, in terms of the fact that rich biodiversity provides an index, a canary in the coal mine kind of thing, to the stability and healthiness of the world." Ehrlich ties his study organism to this way of thinking: "Butterflies are key indicator organisms for the health of ecosystems, systems that provide *Homo sapiens* with indispensable services without which civilization cannot persist."[17]

These ecological-value arguments for biodiversity attempt to convey values much bigger than their spokespersons' individualistic preferences. Biodiversity keeps the world's ecology running, which in turn keeps human civilization running; or biodiversity *is* the ecological world in its entirety, which not only has immense value in itself, but also sustains humanity. Biodiversity's ecological value, therefore, looms inexpressibly large, virtually unknown, but incalculably important.

The ecological argument does have its detractors, who suspect the claim that diminished biodiversity means diminished prospects for human survival.[18] I would argue that since it's not necessarily untrue, why not err on the side of caution?

Nonetheless, ecological gospels used to promote conservation have sometimes turned out to have been preached by false prophets. Ecologists and their audiences may have trouble separating signal from noise, data from intuition. Allegedly precise scientific concepts have been revealed to be somewhat imprecise. For example, Daniel Simberloff reviews some of the "false leads" and "red herrings" offered by ecologists in attempts to render conservation prac-

tice more rigorous and enduring; theories and data on optimal reserve size and shape and the effective population sizes necessary for
long-term survival of species can be confusing and misapplied. Mills
et al. note that some policy makers have seized upon "keystone
species" as the most important foci for conservation efforts. These
species are thought to be most crucial to survival of many other species and the ecosystems they inhabit. Unfortunately, the notion of
a keystone species "is broadly applied, poorly defined, and non-
specific in meaning. Furthermore, the type of community structure
implied by the keystone-species concept is largely undemonstrated
in nature."[19]

To return to a familiar example, ecologists used to proclaim that
ecosystem stability stemmed from organismal diversity. The conservation corollary was: diminish the diversity, and the ecosystem
web will gradually unravel, leaving it unstable and nonfunctional.
Appealing as it was to common sense, how could this view of the
world be false? It is reflected in Leopold's now-classic conservation
dictum, "A thing is right when it tends to preserve the integrity, stability, and beauty of the biotic community. It is wrong when it tends
otherwise."[20] This mantra was easily conjoined to the desirable,
romantic portrait of a balanced, homeostatic nature. Unfortunately,
proponents of the diversity/stability hypothesis may have been projecting human ideals of harmony and stability onto the natural
world. As David Ehrenfeld puts it, "these ecologists were themselves
part of a human environment that instilled a strong, highly developed sense of a normative community, a balance."[21] And so they
reversed the naturalistic fallacy where the "is" of nature is taken as
the "ought" of humans by projecting the "ought" of humans as the
"is" of nature.

Thomas Dunlap shows that this view of complexly interconnected, elegantly balanced nature derived from ecological science
played a crucial role in changing public attitudes toward wolves and
other top carnivores. For the ecologically enlightened, these animals were no longer varmints; rather, they were positioned in an
intricate web that kept nature stable and balanced. Sagoff notes that
the diversity/stability hypothesis has been a cornerstone in arguments for virtually every piece of U.S. conservation legislation; he

contends that biologists continued to use it even after a series of blows crippled the theory in the mid 1970s.[22]

Although doubt continues to be shed on its correctness, all is not yet lost for the diversity/stability argument: a 1994 *Nature* study reports that the thesis holds for grassland ecosystems.[23] Who knows what tomorrow will bring? Some ecological theories are like articles of clothing: if you hold on to one long enough, it may eventually come back into style.

Furthermore, the natural world is so variegated that theories that hold true for some ecosystem types or geographical locations will not hold water elsewhere. In his book *The Balance of Nature?* conservation biologist Stuart Pimm illuminates the sources of misconceptions about diversity, stability, and homoestasis. Among other things, Pimm notes that under the label *stability*, ecologists group five distinct ideas; moreover, they conflate three definitions of "complexity" and focus on three levels of ecological organization. Thus, when generalizing about the relationships between diversity, complexity, and stability, ecologists may actually be talking about one of forty-five different relationships; and this is before they distinguish among particular systems with particular organisms in particular places.[24]

Perhaps those touting biodiversity's possible roles in maintaining ecosystem function tacitly recognize the fallibility of ecological science in supporting concrete, universal claims about biodiversity's ecological value. The "argument from ignorance" is built on the ground of what is not known, inasmuch as what is "known" may or may not turn out to be true a few years down the line. And if ecologists confuse and conflate the meanings of their own concepts, how is the layperson to make sense of these ideas? Little wonder that biologists do not rely exclusively on ecological values when promoting biodiversity's value to broader audiences.

The final ecological argument highlights biodiversity's value for future ecologies. Habitat destruction and global warming, the fruits of current and past careless valuations of biodiversity, are leading to biotic impoverishment.[25] Future enlightened humans will want (or need) to reconstruct ecologies, reconstitute diversity. Thus biodiversity has value for restoration ecology, the attempt to restore what has been lost. Wilson asserts that "the next century will, I believe, be the era of restoration in ecology." To Soulé, "it is apparent that the emphasis in conservation biology will gradually shift from the protection of quasinatural habitat fragments (there will be none left that aren't either protected or doomed) to the opportunistic construction of artificially diverse landscapes." Soulé calls this effort "recombinant ecology." He believes artificial nature will someday transcend oxymoron to become standard. Most of the restoration will be technologically sophisticated, relying on captive breeding and futuristic biotech fixes. Soulé suggests that "since we have no choice but to be swept along by this vast technological surge, we might as well learn to surf."[26]

Arguments for restoration ecology and the value biodiversity holds for this effort are predicated on the belief that people will come to recognize biodiversity's value more and more in the future. Biologists are banking on the belief that the construction of nature they are promoting will take root in people's hearts and minds. In fact, restoration efforts are under way all over the globe. In Costa Rica, Daniel Janzen's Guanacaste National Park project aims to restore a huge swatch of degraded cattle pasture into a biodiverse tropical dry forest. In the Dominican Republic, the government is taking things one step further. It has declared a huge area of abandoned agropastoral land as Los Haitises National Park, with the intention that one day in the distant future, it will have the resources to restore it; as James Wise puts it, the country "is scrambling to protect a forest that used to be."[27]

Restoration has its detractors.[28] Hugh Iltis, who thinks efforts should be concentrated on saving what is in good shape already, dismisses restoration ecology and its proponents: "As Ernest Hemingway said about writing, people must start to develop a built-in, shockproof crap detector." Nevertheless, if ecological arguments for biodiversity become more substantive, and if biodiversity advocates prove successful in promoting them, then restoration ecology is the ultimate proof that we accept biodiversity's ecological value. Biodiversity is said to have value because it will be needed for restoration efforts; and restoration efforts will symbolize the success of those who promote biodiversity's value.

ECONOMIC VALUE

Although many are unaware of it, the free ecosystem services provided by biodiversity save us billions of dollars annually. To lose them might bankrupt us, not only ecologically, but economically. This argument for the value of biodiversity is difficult to sell, because it is so hard to quantify in the only terms the audience for this kind of argument is interested in: dollar (or yen, pound, franc, peso) value.

When biologists and others assert that biodiversity has vast economic value, they mean we can extract from nature materials and services that directly augment human wealth and well-being. In this, the ultimate selfish argument, we are the valuers, and we stand to benefit. Of course, biodiversity may also benefit; this, in fact, motivates many biologists who promote its economic value. If economically driven people can be convinced that biodiversity is priceless, they will invest more time, care, and effort in its conservation. Biodiversity may be rent, cleaved, and plundered so that we shall value it enough to preserve it (as I will discuss in more detail in the next chapter). Dan Janzen asks, "What direct goods have the tropics provided?" He answers:

> For a start, chickens, eggs, elephants, turkeys, beef, pyrethrum, corn, rice, coffee, corsage orchids, tea, chocolate, morphine, tobacco, cocaine, dahlias, cotton, marijuana, aquarium fish, marigolds, strychnine, parrots, bamboo, macadamia nuts, rum, pepper, honey bees, vanilla, milk, peppers, cinnamon, dates, quinine, rubber, gardenias, bananas, avocados, mahogany, pineapples, impatiens, humans, sorghum, rosewood, coconuts, Brazil nuts, peanuts, potatoes, sweet potatoes, manihot (tapioca), squash, chimpanzees, pumpkins, beans, cane sugar, molasses, tomatoes, cats, guinea pigs, citrus, white rats, palm oil, rhesus monkeys. How many potential polio victims realize that their vaccine was grown in a chicken egg, and chickens are nothing more than tropical pheasants specialized at preying on bamboo seed crops (which an Illinois farmer mimics with his chicken feed).[29]

This message, which is promoted with great vigor by many biologists, is that tropical rain forests (ocean depths, agricultural soils, abandoned lots) harbor chemicals, fibers, flesh, resins, enzymes, genes, and whatnot that we can manipulate, extract, breed, purify, and pummel into products that will cure our diseases, feed our hungry, and line our pockets.

E. O. Wilson is among the foremost promoters of this point of view, pitching hard-edged economics with poetic prose: "Any number of rare local species are disappearing just beyond the edge of our attention. They enter oblivion like the dead of [Thomas] Gray's *Elegy*, leaving at most a name, a fading echo in a far corner of the world, their genius unused." "Biodiversity," he declares, "is our most valuable but least appreciated resource." But enlightenment is spreading:

> A revolution in conservation thinking during the past twenty years, a New Environmentalism, has led to this perception of the practical value of wild species. Except in pockets of ignorance and malice, there is no longer an ideological war between conservationists and developers. . . . If dwindling wildlands are mined for genetic material rather than destroyed for a few more boardfeet of lumber and acreage of farmland, their economic yield will be vastly greater over time. Salvaged species can help to revitalize timbering, agriculture, medicine, and other industries located elsewhere. The wildlands are like a magic well: the more that is drawn from them in knowledge and benefits, the more there will be to draw.

Many in both camps may be surprised to hear that the war between conservationists and developers has ended, "except in pockets of ignorance and malice." Those arguing for the economic value of biodiversity attempt to convince developers, be they megaindustrialists or starving peasants, to forgo short-term gain for long-term sustainability. They attempt to convince conservationists that putting fences around wild areas to keep out the masses no longer obtains in a world where the demand for increasing wealth, or just bare-bones sustenance, grows daily with no sign of abatement. And so, by Wilson's reckoning, "The race is on to develop methods, to draw more income from the wildlands without killing them, and so to give the invisible hand of free-market economics a green thumb."[30]

Tom Lovejoy has been equally pragmatic. In an unpublished brief written to educate President Bush on biodiversity in preparation for the Earth Summit, Lovejoy highlights biodiversity's importance as a "significant new source of wealth drawn from the variety of nature." Elsewhere, Lovejoy has extolled "the storehouse of biological properties provided by the wealth of wild species (collectively referred to as biological diversity) with which we share this planet. The ability to reach into those resources at the level of the molecule is creating a significant new source of wealth where biotechnology and biological diversity intersect." He also argues that biodiversity has a contribution to make to pollution cleanup, medical diagnoses, and nanotechnology.[31]

These economic arguments have great appeal. In 1979, Norman Myers published *The Sinking Ark*, still an indispensable reference for anyone interested in the gamut of reasons why we should value nature. In the 1983 sequel, *A Wealth of Wild Species*, Myers narrowed his scope to economic reasons. Why? Because everywhere he went, that was what people wanted to know about: these economic reasons pack clout. Myers is the epitome of a biologist who values biodiversity for a host of personal, spiritual, aesthetic reasons.[32] But to preserve nature for his spiritual contemplation, he sells it as dollars and sense. How does he justify this? "If species are enabled to survive through crass economics, should that detract from the pleasure of the purist who gazes at zebras and polar bears with a spirit that is not jaundiced with considerations of mere money?"[33]

Myers has argued that the loss of wild species "could set back the campaign against cancer for years." Biodiversity has chemical value; we can exploit it for a panoply of medical and industrial uses. These arguments are not new; in 1959, Hugh Iltis asked, "What month goes by when we do not read of a new antibiotic or a new drug that originated from plant sources?"[34] They are more widely touted today because successful examples are more numerous, the methods of extracting chemical value have become more sophisticated, and the line of reasoning has proven more successful in galvanizing conservation.

S. J. McNaughton notes, "Most people I don't think recognize that the most utilized over-the-counter drug, aspirin, was initially a botanical product that was derived from folk medicine people that chewed on willow twigs to get rid of headaches and aches and pains. Lots of that out there, I still believe." Aspirin's formal name, acetyl-salicylic acid, derives from the willow genus *Salix*. Peter Raven's research, which he explained in congressional testimony, shows that gamma-linolenic acid derived from evening primroses is found else-where only in human milk; this substance may prevent heart disease, eczema, and arthritis. Vincristine and vinblastine are alkaloids derived from the previously obscure rosy periwinkle of Madagascar; they aid victims of deadly Hodgkin's disease and acute lymphocytic leukemia and bring the manufacturer of derived medicines $180 million/year. Taxol, from the Pacific Yew, a logging "waste product," may fight ovarian and breast cancer. It is not only plants that are "natural biochemical factories," as Raven put it. Squalamine, found in all shark tissues, has been shown to kill bacteria, fungi, and parasites. Over three thousand antibiotics (tetracycline and penicillin among them) have been developed from microorganisms. In all, according to Wilson, over 40 percent of all prescriptions in the United States are organism-derived.[35]

Eisner travels the world, hawking chemical prospecting to all who will listen. Not only are the chemicals of an organism exploitable; in oft-quoted congressional testimony, Eisner called species storehouses of genetic information:

> In these days of genetic engineering, a species is to be viewed as a depository of genes that are potentially transferable. . . . The implications of this technology are tremendous and the subject of intense current discussion. The extinction of a species, in light of these advances, takes on new meaning. It does not simply mean the loss of one volume from the library of nature, but the loss of a loose leaf book whose individual pages, were the species to survive, would remain available in perpetuity for selective transfer and improvement of other species.[36]

As chair of the newly organized Endangered Species Coalition (created to lobby Congress to save the Endangered Species Act), Eisner has organized a "medicine bottle campaign"; he urges citizens to send empty medicine bottles to legislators to remind them of the medical importance of wild species. At one press conference, he brought together environmentalists and AIDS activists to stress the linkages between intractable diseases and wild species.[37] By screening such species for useful substances, we might cure diseases, boost economies, and place a higher value on biodiversity.

This hope underlies Costa Rica's National Biodiversity Institute (looked at in further detail in Chapter 6). The World Resources Institute has published a guide to *Biodiversity Prospecting*, a state-of-the-art manual on how biodiversity benefiting humans benefits biodiversity. While Gardner M. Brown, Jr., suggests that screening of wild sources has failed before—so much current effort may be for naught[38]—biologists and others are proving successful in tying their love of biodiversity to others' love of money and wishes for health. If monetary value is the ultimate "value" in society today, and if the biologists who promote biodiversity's chemical value deliver a fraction of what they promise, then more and more people will be convinced that they, too, should value biodiversity.

Humans have domesticated a fraction of potentially edible plant and animal species. "Among the millions of populations and species of plants now threatened with extinction there undoubtedly are many Cinderella plants potentially equivalent to the ancestors of wheat and barley, but they may be doomed to disappear without ever making it to the ball," the Ehrlichs tell us.[39] "I think we should all be concerned about the capacity of the world out there to produce the things that human life depends upon," warns Orians; "Food doesn't come from the grocery store." Janzen's list, quoted above, tantalizes us with what tropical biodiversity has yielded thus far; biodiversity veils both new food sources and wild relatives of current food sources that might improve existing strains.

Over 75,000 edible plants are thought to exist; humans have, at one time or another, made use of 10 percent of them.[40] Depending on who you listen to, either very few plants actually feed the world or quite a few actually do.[41] Either way, biodiversity must be conserved so that we can find new usable species or exploit the genetic diversity of existing species to improve and spread their uses. Norman Myers, Mark Plotkin, or E. O. Wilson will take you on mouthwatering tours of underappreciated gustatory sensations; consider, for example, the endangered maca, whose "swollen roots, resembling brown radishes and rich in sugar and starch, have a sweet, tangy flavor and are considered a delicacy by the handful of people still privileged to consume them."[42] (These authors also tell you what biodiversity can yield in the way of industrial oils, fibers, waxes, exudates, tannins, and dyes.) Bison, once on the brink of extinction, are returning to the western United States in great numbers; many are bred for our dinners. As Harold Danz, executive director of the American Bison Association, succinctly puts it: "Animals that people eat do not go extinct."[43]

New food sources await discovery, old food sources await rediscovery, and wild relatives of domesticated crops await exploitation. In the late 1970s, a wild species of maize, *Zea diploperennis*, was discovered on a Mexican hillside. This species is teaching biologists much about the evolution of corn, but more important from the biodiversity advocate's standpoint, it is resistant or immune to the seven most important viral diseases of corn, is adapted to high altitude, and as a perennial can grow not only from seeds but from rhizomes; for this alone it might be worth billions of dollars annually. Iltis tells how the genes of a "tangled, yellow-flowered, sticky-leaved, ratty-looking wild tomato" he collected in the Andes have improved domesticated tomatoes to the tune of $8 million a year.[44]

If you take new species of commercial trees and new or improved species of crops and combine them with the various ecosystem services that biodiversity performs in wild as well as managed ecosystems, you glimpse the enormous value of biodiversity to agroforestry worldwide. From biodiversity we can also derive new strategies of pest control, either from using beneficent organisms to destroy pest organisms or from new biodegradable pesticides.[45]

Biodiversity also bolsters economies in other ways. According to the U.S. Fish and Wildlife Service, in 1985, fish and wildlife tourism contributed $55.7 billion to the U.S. economy. By 1993, tourism had become Costa Rica's largest revenue generator, surpassing bananas and coffee. In Rwanda before its devastating civil war, ecotourism was number three, primarily because of the mountain gorilla, the nation's star attraction.[46]

In the above examples, biodiversity can be bought and sold as a commodity. Biodiversity also has amenity values—we are willing to pay so that it continues to exist and we can experience it or simply rest content knowing it is there.[47] Economists (or biologists) talk about opportunity costs or option values: we face grave economic penalties in the future from loss of biodiversity and, with it, the unknown economic value we might have derived from it. Nuances of economic analyses of biodiversity's worth are beyond the scope of this book,[48] but biologists have a multiplicity of arguments showing that the economic value of biodiversity is beyond calculation. In a world where decisions are too often made using economic cost and profit as the sole criteria, what could hold more sway than that?

Nonetheless, heavy emphasis on biodiversity's economic value remains problematic. Many species have no present or future economic value. Furthermore, tourism, industrial harvesting, and even supposedly low-impact exploitation of extractive reserves may destroy the very resources biologists hope to preserve. Janzen et al. note that if biodiversity is ever valued economically as it should be, it might cause political friction between neighboring nations with different bioeconomic policies, all wanting a piece of the biodiversity pie. "A war over biodiversity is not as far-fetched as it might first appear," they note.[49]

SOCIAL AMENITY VALUE

At our interview, Dan Janzen advocates broader consideration of biodiversity's economic value:

"If you take a bunch of people out of any tropical region, and they don't have that level of biological literacy, because they live in the middle of a sugar-cane field, or an oil-palm plantation, or in an urban ghetto where there's zero communication about—they don't know beans what their grandmother knew. . . . If you want those populations to manage their own natural resources, I don't know—water, pesticides, or air, or any of those decisions—we've got to get them back up to some basic understanding of what the natural history is of the organism that they are managing. . . . And you just go into how appalling, how appallingly biologically illiterate those communities are. Well, if you go then and you educate those people—and I say give them biocultural restoration, meaning give them biological literacy back—that is a very economic argument, a very pragmatic one. You're [producing a population who are] happier, healthier, saner, easier to manage, and can manage their own affairs better. . . . And a minister of finance gets interested . . . not a minister of culture, but a minister of finance gets interested in this."

Biodiversity has social amenity value. It can improve standards of living, make people proud, and help them lead more fulfilled lives.

In a way of thinking closely related to the economic arguments outlined above, biodiversity is said to have value as a resource for "sustainable development," another catchphrase of the late 1980s and 1990s. A World Resources Institute publication boldly asserts: "The conservation of biodiversity is the management of human interactions with the variety of life forms and ecosystems so as to maximize the benefits they provide today and maintain their potential to meet future generations' needs and aspirations." This philosophy also undergirds *Global Biodiversity Strategy*, partly sponsored by WRI: as biodiversity reinforces economic and social security, economic and social forces will in turn reinforce biodiversity.[50]

"Poverty in the Third World," declares Myers, "is a luxury we can no longer afford." Raven strikes a similar chord:

> Until a clear majority of us live lives of dignity, with the reasonable expectation that our needs and those of our children can be met through our own individual efforts, the world will continue to be a hostile and unstable place, and the biodiversity on which an enhanced quality of life for us all might be based will continue to disappear rapidly. . . . People everywhere would clearly become more interested in preserving biodiversity if they were convinced that it is an essential resource for sustainable development.[51]

By Raven's reckoning, biodiversity holds the key to respectable standards of living, and thus to reasonable levels of security for all the world's people. Orians, Wilson, Lovejoy, Janzen, and Brussard, among others, expressed similar views during our interviews.

David Western suspects that "the best hope for all species is linked to a single, uncompromisable human goal—the improvement of human welfare." Only when poverty is alleviated, people are healthy and have enough to eat, when the trade and income balance favoring the North is rectified, will people enjoy the luxury of being able to conserve biodiversity. These factors drive biodiversity loss, but biodiversity may be precisely the resource that can mitigate them. It can pay the debt, feed and cure the people, fuel the industry. The lopsided concentration of economic wealth in the North balances against a lopsided concentration of biodiversity wealth in the South; exploiting this wealth can bring Southerners closer to parity.[52]

Of course, some biodiversity advocates and many Southerners might have reason to doubt the sincerity of Northerners. In 1990, in a speech entitled "Diplomacy for the Environment," James Baker, President Bush's secretary of state, declared: "Finally, we know from our *own* experience in this interdependent world, that we must 'think local and act global.' We cannot serve America's environmental interests effectively unless we address *worldwide* environmental concerns. . . . In other developing countries around the world, we are fostering biodiversity. By so doing, we can increase the availability of natural products for commercial purposes."[53] Reversing the popular environmentalist credo "Think globally, act locally," Baker was clearly proclaiming that by addressing biodiversity issues in developing countries, we help ourselves get richer.

Many (but not all) biologists seem genuinely concerned about the overlapping problems of poverty in the tropics and tropical biodiversity. Janzen and Hallwachs ask: "Want to save tropical biodiversity? Then generate a populace with a developed world's standard of living and that views tropical wildland biodiversity as an important member of the domesticates in the human toolbox."[54] But this shift toward sustainable development as an inextricable companion to biodiversity conservation is not gospel among interested observers. "The early success of scientists in publicizing biodiversity and sustainability is manifest in the prominence of these concepts in science, policy, and public understanding," Redford and Sanderson note.[55] They believe these concepts have been linked without a clear picture of what either means, and without any notion of whether or how they may be compatible. Ray thinks the pendulum has swung too far toward "sustainable development," which he sees as an oxymoron, "a very fuzzy biopolitically driven paradigm."

Deriding the Brundtland Report to the United Nations, which calls for a jump start of the world economy with a dramatic increase in standards of living for the poor, the Ehrlichs assert, "If anything remotely resembling the Brundtland population-economic growth scenario is played out, we can kiss goodbye to most of the world's biodiversity, and perhaps civilization along with it."[56]

Iltis rails against the supporters of sustainable development, arguing that by putting people first, you ultimately destroy them: "It's a slippery slope from which you can't get up once you start sliding. Very, very dangerous." You cannot really help people unless they help themselves, he contends; this approach he calls "tough love," even though there doesn't seem to be much love involved— at least not for people. Iltis also dislikes the claim that indigenous groups have the knowledge to unlock new uses of biodiversity, and that they have been and remain its best managers, since biodiversity has value for them.[57] Iltis argues that if left alone, "if you give them shotguns and chainsaws, they are going to undercut their own environment."

Alluding to "the profane grail of Sustainable Development—the odd delusion of having your cake and eating it too," Soulé suggests that if indigenous peoples were indeed kinder to their environments, it was because population densities were low, or because they conserved isolated components of their environments on which they depended.[58] "I don't happen to believe there's any such thing as the noble savage," Ray notes. Asserting that "penicillin would never have been discovered by following indigenous leads," Eisner argues that the future for biodiversity prospecting is molecular, and that indigenous knowledge "is not going to have much value."

In short, although biodiversity may contribute to sustainable development, not all biologists believe that sustainable development contributes to preserving biodiversity. As with many other social arguments, it is not clear which biodiversity advocates believe more strongly: that biodiversity has value for people, or that people need to be enlisted so that they have value for biodiversity.

Biodiversity may also, however, be a source of national pride. Stephen Kellert announces: "In a world tending toward an increasing homogenization of values and customs, and a consequent erosion of national identity, the uniqueness of indigenous fauna and flora can be cited as one important distinguishing characteristic of a nation. . . . Employing this approach in Madagascar, local support for species protection was strengthened because the people became proudly aware that, despite their country's political and socioeconomic backwardness, the island was among the world's most outstanding biological areas." The erstwhile Mexican president Carlos Salinas de Gortari spoke proudly of biodiversity as his nation's *patrimonio natural*, and INBio's ambitious cataloguing effort is often advanced as a source of pride for Costa Ricans. Natural heritage can be a source of pride in developing as well as in developed nations. Once pride comes, responsibility in protecting biodiversity follows, and the nation that responsibly protects biodiversity can gain credibility in international diplomacy.[59]

Putting "farmers' skills at biomanipulation to work for the conservation of biodiversity," reviving ecological literacy, and restoring cultural appreciation of the natural world are among Janzen's goals in the ecological rehabilitation of Guanacaste National Park in Costa Rica.[60] By enrolling people to restore the park's biodiversity, he enrolls the park in restoring biodiversity to the people's psyches. "The single largest specific task facing tropical humanity today is reinventing the traditions that adjust population size and demands to the size and nature of the resource base," Janzen says.

> Focus on biodiversity for a moment. For millennia, it is just there—the huge gene pool from which the occasional Indian grandmother selects a new crop plant, the rope-maker selects a new fiber, the shaman selects a new heart-stopper. It is omnipresent, exuberant, hardly used by a local culture, mostly unknown. It almost always reseeds itself when trashed. . . . biodiversity must be lifted from the grave, planted back out, nursed a bit, and explicitly left in peace. Humanity is not about to give the earth back for another 4 billion years of evolution, and biodiversity has lost the greatest friend it ever had—agricultural and social inviability for most of the tropics. Perhaps one of the most delicate of all traditions— the incorporation of biodiversity's intellectual stimulation in the village social and mental fabric—is about to be taken from the three-quarters of humanity that live in, and always will live in, the tropics. A mil-

lion square kilometers of intensive agriculture can be highly sustainable agroecology, and the most deadly boring green cocoon in the universe. The mental health of the tractor driver, mayor, and mother is as much an ingredient in a sustainable agroecosystem as are integrated pest management, polyculture, and reforestation. The destiny of humanity is not glorious fields full of human draft animals picking beetles off bushes. Biodiversity conserved in wildlands has far more function than its very great value as a global gene bank and phytochemical cornucopia. . . . When humans have the option, it is pretty clear that they take the advantages of modern society and decorate the margins with biodiversity and unsullied water, air, views, and food. Someone challenged me on that once—he said that the oil palm plantation worker would rather have a TV set than a national park nearby. Who said either/or?[61]

In Janzen's view, natural selection historically kept human populations tied to the resource base; it gave us, not only biological diversity and human cultural diversity, but also human-biodiversity mutualism that sustained both. Humans enlisted biodiversity as a tool that increased their odds in the selection process; simultaneously, humans modified biodiversity and modified the evolutionary process, but not beyond recognition. Not only did biodiversity keep humans alive; it kept them enthralled. And humans, in turn, developed a psychological dependence on it: "Biodiversity provides a huge package of variety, which the human mind *needs* to be fully functional" Janzen observes.[62]

So we doubly tempt evolutionary fate. By destroying biodiversity, we endanger the resource base humans require. And by ignoring the evolutionarily adapted human-biodiversity mutualism, we endanger our psyches.

BIOPHILIC VALUE

According to E. O. Wilson, "We really can't afford to lose any species; they are a crucible of future human creative effort."[63] He takes Janzen's biocultural restoration effort a step further: by reason of our deeply rooted "biophilia," reintroducing rural—or urban—people to the secrets of the surrounding biodiversity provides not only intellectual fulfillment but genetic fulfillment as well.

Wilson's 1984 book *Biophilia* brought to the public the idea that love of nature may have been hardwired into our genes by natural selection. Wilson first used the term in 1979 in a Harvard University Press advertisement column in the *New York Times Book Review*, saying, "Our biophilic descendents will regard species extermination as the greatest sin of the twentieth century." He defines *biophilia* as "the innate tendency to focus on life and lifelike processes."[64]

The idea, however, was not new. Hugh Iltis had been pursuing the same theme for several years, albeit without using the neologism. A paper by Iltis, Loucks, and Andrews (rejected in 1967 by *Science* and published in 1970) declares: "Nature in our daily lives must be thought of, not as a luxury to be made available if possible, but as part of our inherent indispensable biological needs. . . . a monotonous environment produces wave patterns contributing to fatigue. . . . Biotic as well as cultural diversity, from the neurological point of view, may well be fundamental to the general health that figures prominently in the discussions of environmental quality."[65]

In a nutshell, Wilson's biophilia hypothesis presents love of nature as a universal biological adaptation of humans, selected during the course of evolution. Contact with biodiversity can therefore awaken passions encoded in our genes and can rekindle human appreciation of and reverence for the Earth's biotic riches. Conversely, by ignoring our own biophilia, we simultaneously endanger our psyches and imperil the Earth.

Wilson and others cite such evidence as humans' universal and deep fear of snakes, love of scenic vistas, affiliation with chimpanzees (our close genetic relatives), decoration of homes and environs with plants and pets, and reliance on animals for metaphorical expression. They note the prevalence of certain natural patterns in landscape paintings, the more rapid recovery rates of postoperative patients with views of parks than of those with views of brick walls, and the preponderance of leisure activities that involve nature (including the tidbit that in Canada and the United States, more people visit aquaria and zoos than attend all professional athletic events combined) as evidence of our biophilic impulses.[66] Iltis and Orians, who has also done work on what he calls "environmental aesthetics," cite similar examples. For example, Heerwagen and Orians published a study demonstrating that office workers without windows decorated their offices more often with reminders of the natural world. Orians writes, too:

> Human beings, like all other species, select habitats from an array of options and, like other species, display strong emotional responses to landscape configurations and plant shapes. An evolutionary perspective suggests that strong emotions generated by objects or situations signal the action of long-term natural selection for those responses. . . . If the arguments advanced here have merit, only those environments emotionally associated with high resource levels for people should be able to evoke strong positive responses.[67]

As Orians, Iltis, and Wilson see it, we "naturally" develop strong ties to what sustained us in the past.

BIOLOGISTS ON BIOPHILIA

I asked the biologists I interviewed whether they were familiar with the concept of biophilia, and, if so, whether or not they thought the notion had merit. They were, and most did. Here are some of their responses:

PETER BRUSSARD: "I don't know. I hate to say I don't agree with it. But I find it a little far-fetched. I mean, if you look at other cultures, some of them have what you can interpret as a deep appreciation of biodiversity. And other cultures have none whatsoever. So I strongly doubt that it's genetic." [Later, after reminiscences about his childhood and reference to his enthusiasm for biodiversity despite lack of encouragement by his parents:] "Maybe there is some kind of genetic predilection. I don't know."

PAUL EHRLICH: "When I saw the book and heard the term the first time, I thought, 'Jeez, this is something that's probably illegal in most U.S. jurisdictions.' It reminded me of the old story of the guy who went to Cornell to study animal husbandry until they caught him at it. . . . I think there is some pretty good evidence that human beings need to relate to nature. . . . I don't think it's been put to a rigorous test. But I think it's an interesting idea that's worth pushing around. . . . [It] makes a lot of sense, subjective sense to evolutionists, a just-so story that if you come from a couple of billion years of evolution, hundreds of millions in a green world, for example, this probably has some impact on you. The test could not be run ethically, let's put it that way [laughs]."

THOMAS EISNER: "Before Wilson termed a very good term, namely *biophilia*, I used to talk about the Camp David Syndrome. . . . No matter how rotten the president is feeling, he takes a helicopter and communes with nature out in Camp David. Whether the president happens to be sympathetic to nature or whether he is someone like Reagan who appoints [Secretary of the Interior James G.] Watt to basically oversee the rape of nature, personally, these guys escape to their little refuges. You know, in the Middle Ages, and so forth, the British kings went to their hunting preserves, which are the big tracts of land now, thank goodness, that are available for conservation in England. In the United States we see this as a rather fundamental need. You know, if you take a train out of Grand Central and you go up to Connecticut, you go through Harlem and you see the flowerpots and these fairly poverty-stricken apartment buildings. I mean, it's the same phenomenon. People like natural things. I mean, we obviously don't pollinate any flowers, there's no reason we should like flowers; that's the insect's job. And for years and years and years that used to be the angle I used to pick as a conservationist."

TERRY ERWIN: "Exactly. Yeah. And—If we're dependent on it, and we're destroying it, then we're destroying ourselves, which might not be such a bad idea, actually, for the rest of the world [laughs]." "We have a lot of subsumed instincts that are evolutionarily derived from our long process coming from savanna chimps. And I don't think those things have really left us. I think they've just been buried deep maybe, wherever: subconscious, spirit, or somewhere *deep*, deep inside. And that's what I was mentioning: that even people who live in Manhattan have plants in their house. Regardless of what they say, there's some contact there. And by raising one's awareness, I think that old contact comes out. . . . It rekindles the coals of evolution."

DONALD FALK: "I'm saying that ecosystem services and evolution aside, it's important to what it means to be human to be connected to the nonhuman elements of the living world." [Asked whether he agreed with the concept of biophilia:] "Yeah. Yeah. Very much along that line."

JERRY FRANKLIN: "I don't know if I can relate to it in exactly that form. I think as humans, we relate to living things. We're very conscious of our mortality. I think it becomes entwined in our whole unconscious and conscious view of the world, and our concern with our own mortality. Now whether it's a product of our consciousness or our genes, you could argue the point." [But then in answer to a question on aesthetic appreciation:] "Oh, I don't know. That's kind of a tough question. I would say that a lot of it, in fact the core of it, is probably innate. It isn't learned at all. It's felt. It's intuitive."

VICKIE FUNK: "I don't know whether it's genetic. . . . The color green has always been considered a soothing color. Psychologists will set up green-colored rooms, and you know, stuff like that, for restful experiences. And going out and sitting in a forest, where it's quiet, you know, and everything is green and leafy and the birds are singing has always had a restful effect on everybody. Whether that's genetic or . . . the color green is [just] restful on our eyes, and the quiet, not having the city noises, is very restful, I don't know. . . . I'd come down on the side of genetics probably 70 percent of the time for most of what we do. I think that genetics probably has a lot more to do with everything we do than we think it does."

HUGH ILTIS: "They're in love with it. It stimulates them. It stimulates them because as a species of mammal, we are genetically programmed to be stimulated by biodiversity!" "There are *thousands of ways* you can see that this is really what happens. People *love* flowers. They love pets. And they love what's healthy for them, if you give them a chance to get it." "Just like we produce a language, we produce a mental picture of the world by very clearly genetically programmed ways to get this into us, because this is the basic survival strategy of all mammals, of all animals. So we have pattern-acquisition devices. The point is, if you don't give the child a chance to utilize his genetic propensities, and as Muriel Beadle points out and as Paul Shepard points out, *in the proper sequence*, you're going to get a screwed-up adult." "I mean, we are genetically programmed to be full of wonder at the diversity of it all. All of our antennae on our head, all of our antennae and our intellect, all of the things school gives you. Here you see a gorgeous flower: it just makes everything bristle in terms of reception of the aesthetics." "We're genetically programmed to explore the environment. I mean, just watch a cat go into the woods. Watch a rat or a mouse go sniffing around in the woods. We're the same way; there's no difference. That's how we made our living. And we get a great charge out of it. And we are in the cities being deprived of this." "Our color senses are clearly biologically programmed. For example, chlorophyll and blood, red on green, is the most highly contrasted color, and it's very easily explained: we lived in an environment in which blood on grass—We were partly carnivores, and green is everywhere, you know. . . . Steve Gould made *fun* of some Germans who said that we have a desire to climb mountains, and they called it something or other philia: mountain-climbing love. I don't think that's funny. I don't think it's crazy either. When you sit on top of the mountain—The chimpanzees do it! They climb to the top, and then when they are on the top, they go ten feet down so their silhouette wouldn't show, and then they watch the landscape. Of course. Or take color blindness: that's a genetic adaptation, that's not an abnormality. That's an adaptation. Because all animals that hunt are colorblind because colorblind makes you escape the camouflage." [Iltis read from a 1974 manuscript, on an earlier visit to a remote corner of the Great Plains:] " 'The vast, endless treeless horizons empowered my imagination like nothing I can remember before, or in a way since. Behind us in the bottom of the valley, a small ranch with a windmill, some haystacks, a field and a pond with ducks, and a few cattle. As low-key and fragile an ecosystem as man could devise. Before us, not a house in sight, not an engine within earshot, only the undulating yellow-green vastness of the great plains landscape. I wondered, then, if this is not what Moses must have felt when, leading his people out of crowded, oppressive Egypt, he stood on the steps of the promised land of milk and honey. If this is the force which drove the Mormons' elders to the Great Salt Lake, the Norwegian pioneers to Minnesota, empowered by a vision that was not their own. A yearning [for] a landscape free of people, no competition. For a landscape to live in without strife and stress. The grandeur, the majesty, the peace of it all. The sense of timelessness, of infinite distance. Is this the vision which the last slow and majestic [movement] of many a great symphony depicts? Be it Brahms's first symphony, Beethoven's sixth, or Sibelius's second.'. . . Now let me reread one sentence here: I talk about these symphonies. By the way, these symphonies are very interesting. If you look at these last movements, it's a very slow heartbeat. It's the kind of a heartbeat you get when you see something that's absolutely astonishing. And the rhythm in this is the same, in practically every major symphony. Take, for example, Tchaikovsky's fifth [sings]." "There's no accident in this: music and psyche are very closely related. And I've listened to many many a symphony since.

And I'm quite convinced: it is the realization of making a major opening for yourself and for your people. It is a majestic type of movement. And you can see this, by the way, in movies like *The Emigrants*. When they go with their wagons, for three hours they have hardship. And in the very last scene, the old man in the first wagon climbs up to the hill and looks in front of him and there's this enormous plain of central Minnesota. And all he can see is a few animals in one corner and nobody there. And he nods to his fellow travelers in that covered wagon and says: 'Here is where we are going to stay.' And there's a lot of biology and a lot of human culture. Alright, I talk about Sibelius's second: 'A psychic benediction dedicated to tranquility, to the feeling of a nature humanly uncrowded and free of competition seen in all its peaceful splendor from a high hill. Whatever explanation, whether an appreciation of landscape beauty culturally determined, due to traditions handed down for tens of thousands of years, or whether a visual satisfaction hormonally induced, due to powerful adaptive and innate feelings selected by natural selection over *millions* of years, the vision of an ecologically open habitat of perceiving ecological goodness in a vast humanly unfilled land has ever stirred the minds of men.' And I might add the souls."

DANIEL JANZEN: "Biodiversity provides a huge package of variety which the human mind *needs* to be fully functional." "Hey, you're hardwired for color vision. You're hardwired for all the complexity to hear complex music. You're also hardwired to respond to all the little creepy-crawlies and the flowers of the morning and all that kind of stuff."

K. C. KIM: "If you have an enrichment of biophilia in a system, that reduces—there's no scientific data, I'm just hypothesizing—it reduces the crime and everything else." "So biophilia I define as innate need of nature, not the need of a species or biodiversity per se. It is nature. In other words, synergistic, synergism of all different living organisms involved."

THOMAS LOVEJOY: "You bet. I start my [Amazonian] field camp with it. . . . I think there is if people give themselves half the chance. I think it's suppressed in most people. Because as social animals, we have increasingly turned inward. As we've been successful with agriculture and urbanization and everything else, we dwell more and more upon ourselves rather than the larger context we're in. And so I think a lot of those people have suppressed it. But I think it's inherent."

JANE LUBCHENCO: "I guess I wouldn't go as far as Ed does in that. I think I'm more skeptical about the extent to which those kinds of feelings or behaviors have been selected for."

s. j. MCNAUGHTON: "I'm convinced of it."

REED NOSS: "I think that Ed Wilson is correct, that people have an inherent biophilia, but some people more than others. And I think probably the vast majority of biologists have an inherent biophilia that's stronger than much of the rest of the population." "And my impression was that the vast majority of children have demonstrable biophilia. They have a 'sense of wonder,' as Rachel Carson called it. They're just fascinated by life. But probably due to factors of safety, and it's probably natural selection or a *cultural* selection, has led to a situation where contact with nature is discouraged by parents and teachers because of some inherent danger out there."

GORDON ORIANS: "There's also biophobia. So what one wants is I think a more neutral term that simply says we have strong emotional responses, affiliations with the living world, which can include negative and positive things. . . . you get incredible snake phobias with people who have never seen a snake, if they see a picture. Just like that! And they persist. . . . So there's a body of research coming that says that, that begins to give us some understanding of the nature of the wiring. And the big agenda is not, is there such a thing as biophilia, or bioattachment, or whatever the word we want for this. The real issue is, what form does that take? What are the specifics of it? It's some of that that we're trying to get in some of this research in environmental aesthetics, is to try to probe that.

PETER RAVEN: "Yes, of course I do. . . . Look at the role of biodiversity in art. Look at the role of biodiversity in anything. It's tremendous. Fundamental."

G. CARLETON RAY: "Yeah, I totally agree. I totally agree. Again, it is controversial. You can't prove a thing like that. Proving love of animals by looking at genes—it's a pretty difficult experiment. And a lot of people will say it's pretty naive to assume that."

MICHAEL SOULÉ: "Yeah. There is something. But it's not very strong. Prima facie evidence for its weakness is that generations of people grow up, are born, and die in cities, never seeing nature. Never even seeing a plant in a pot, in some places. They see rats and cockroaches. Also some of the biophilic responses are phobias. So there's some negative aspects of the so-called biophilic responses."

DAVID WOODRUFF: "Yes, I subscribe to I suppose Wilson's biophilia-type thing. Yes, somewhere in my hard-wiring are pleasure centers that I recognize when surrounded by nature or even half nature."

Support for biophilia seems widespread among these biologists. While some expressed doubts about the depth or universality of biophilia, only Lubchenco outright doubts its existence.

For all this, how do biophilia proponents explain why humans are so bent on trashing their environment? If biophilia is so fundamental, why is it so easily suppressed? And for all the lawn paraphernalia and house plants and ecoescapes, we are also fatally attracted to booze and cigarettes and guns; while these items may be biologically derived proximately or ultimately, they do not suggest biophilia. Furthermore, city dwellers may crave green space, but that does not explain why many choose to dwell in the city in the first place.

For something so ingrained, why do we have to work so hard to inculcate, or release, or arouse it? Wilson responded to this question:

> "They're not running away from it. Have you overlooked the prominently displayed fact that we're now a suburban nation? . . . Given a choice, people in Europe and the United States, especially the United States, the wealthiest country, did not all rush together into some idealistic arcosanti, or whatever they call that thing, that termite nest in Arizona. [I take it he means Biosphere 2.] They spread themselves out in true biophilic fashion with their own private domain which they proceeded to invest huge sums in to recreate small savannas around the house. And put substantial emphasis on living organisms, greenery, and life. . . . In other words, it's no coincidence that they're crowding the national parks in the United States in record numbers. I don't think people have run away from nature. I think they're desperate to get back to nature. If they had their choice of the ideal place to live, they would be surrounded by greenery, by diversity of organisms—take delight in them."

Some of these responses beg the question. Why suburban and not rural? Americans may create mini-savannas around their homes, but these are usually artificial monocultures, in which great effort and expense are exerted to exterminate diversity. While we may be trampling the national parks in record numbers, many still choose to live from day to day in distinctly unparklike environs. I am not presenting these as reasons I think biophilia does not exist; but the evidence in its favor seems selective—just-so stories with alternative interpretations, and not, of themselves, convincing. For example, in *The Biophilia Hypothesis*, Stephen Kellert cites American and Japanese attitudes toward nature that show "an aloofness from the biological matrix of life, restricting their interest to a narrow segment of the biotic and natural community." But rather than taking this as evidence against biophilia, Kellert uses it as a warning that people should start paying attention to what other, tendentious evidence suggests should be there.[68]

It also remains to be explained how much of biophilia is in response to bio*diversity* per se, as opposed to landscape patterns, for example. Wilson admits, "They're only loosely linked; it's true. I don't see anything necessarily in the type of biophilic responses described in the book—the ideal place to live, preoccupation with serpents, and so on—as naturally producing a special value for maximizing biological diversity." Orians agrees that

> "I know of no particular research that looks at the aesthetic responses where the number of species is a variable, let's say. So suppose you were to get people to rate pictures of big flocks of birds. And in fact the psychologists and I are planning to do something like this. We have a big flock and it's all snow geese. Then you have a comparable scene, and it's several species of geese in the flock, let's say. Or geese and cranes, whatever. And to find out what extent does the pattern of richness of species in a scene or whatever you're testing augment or not the strength of the affiliative response. We do know that there's an enormous amount of listing and ticking, you know, trying to see a lot of different bird species, and keeping life lists. There's clearly some sort of affiliation with this."

Orians does offer a convincing just-so story to answer my query that if we evolved on the savanna, and the savanna is "the highest valence environment for people," why do we have such strong aesthetic and emotional feelings about the closed, lush, incredibly diverse tropical rain forest?

"Well, it's a compound question. One is the closed forest itself, and the other is diversity. Let me take the diversity one first. . . . Even if one is living in an environment that is not especially species-rich, like a savanna, nonetheless if you start looking at the resources people use in those, richness does make a difference. Particular resources may be important at particular times of the year. That knowing about the richness of the resources, knowing where things are, knowing what's available at different times of the year is important to survival. And if you look at traditional cultures, hunter-gatherers, they eat a lot of things. And they use a lot of plants for fiber and construction and all these sorts of things. So they make tremendous use of the biological diversity they have. So attention to it, and the variety of species and what you might get out of them, I have very little problem explaining that. And that might very well be the basis for why you're responding to diversity itself, because knowing what's out there has had throughout our history benefits. And people were aware of the richness of species and the potential ways you could use them. . . . Now the issue is closed forest. And so you're asking in a sense of why should we be so turned on by closed forest if, according, to the arguments I've made in papers that we're fundamentally a savanna animal. And I think even people that are raised in areas of closed forest, in terms of general living, etc., prefer it open. We don't make any parks, 'This is closed forest land.' You don't find any parks in Seattle that are constructed that way. I've known nobody, if we go hiking in the Cascades, and you're going through the really dense conifer belt and you hike on up and get up to the open subalpine part, and I have yet to meet a person who doesn't have a real catharsis when you open up into that. It's just absolutely universal in people who are in the rain-forest belt here. And you see that in the tropics, too. And I think most people enjoy going in and experiencing a rain forest because it's so incredibly rich. But they're not going to want to live there. So I think, in thinking about responses to environments, there are different scales of time and space and use that make the whole response thing a very complicated one. And to say that anybody is always going to prefer a savanna scene under any circumstances is nothing that I would ever claim. Nor would I expect, even though I think that fundamentally that for long-term usage that's probably the highest valence environment for people."

Orians's explanation is as compelling as it is untestable. While he could, perhaps, prove that humans (all of them?) have aesthetic preferences for savannalike environs, that would still get us no closer to why that should be so. It would not justify the biological determinism that kicks in to explain the phenomenon.

It is worth perusing Iltis's lengthy meditations on biophilia quoted above. Iltis's thoroughgoing biodeterminism finds convincing genetic bases, not only for our love of nature, but even for how our responses to certain music stem from our distant and recent ancestors' searches for resources. Iltis has a just-so story for every occasion. This does not make his explanations false. It just makes them open to question, unproven and unprovable by standard scientific procedures. To my question about whether biophilia is tied to biodiversity per se, Iltis responded: "Well, I don't think we need biodiversity in particular, except as highly intelligent animals, the diversity that you find, let's say in the tropics, is so utterly astonishing, so utterly marvelous. And the point is: it makes you happy!" Who is "you"? I mean, it makes me happy, deliriously so at times. But is what makes Hugh Iltis, his peers, and me happy what makes everyone—what makes "you"—happy? Isn't it suspicious to take one's passion and declare it a universal genetic need, especially in light of evidence to the contrary? Again, this is not to say I don't believe in biophilia; but it also may be that biologists are pasting a scientific rationale over what may simply be an acquired or coached delight. In Chapter 7, we shall explore the role biophilia plays in E. O. Wilson's overarching strategy for conservation. For now, suffice it to say that if love of nature is genetically determined, that brings its investigation into the realm of biology, and also makes it more reasonable—indeed, evolutionarily ordained—that biologists should speak about love of nature. For who would have this naturally selected trait more highly developed than biologists, who elect to spend their time surrounded by the diversity of life?

Reed Noss concurred with my statement that biologists' biophilic sense calls them to biology, as a priest would be called to religion. A self-consciousness that seems dangerous pervades the promotion of the concept of biophilia. We may well have a psychological need for nature; indeed, I would like to believe that the things I feel are universal, genetic, and thus easily tapped into to motivate conservation. Yet it seems too much of a coincidence that those promoting this rationale for biodiversity's value are also those who may have this trait most finely developed and are in a position to manipulate it. The circumstantial evidence for biophilia consists largely of stories that sound convincing on the face of it, but they can certainly be countered with contrary stories, which may also compel you, depending on what you think of biophilia in the first place. Biologists' studies of the biophilia phenomenon bear watching. Biologists promoting biophilia as a reason why biodiversity should be preserved—and, in turn, as a resource to be drawn on if we want biodiversity to be preserved—bear watching, too.

TRANSFORMATIVE VALUE

In his 1987 book, *Why Preserve Natural Variety?* the philosopher Bryan Norton asserts that the most compelling basis for an enduring conservation ethic is that natural variety has transformative value:[69] contact with biodiversity can be the occasion for us to reconsider our shallow, consumptive preferences and make us adopt values that are, in some way, objectively better. By "objectively better," Norton means values that sustain ecological and evolutionary processes— the very processes that effected this transformation. In this transformation, we simultaneously consume less and value biodiversity more. And the more contact we have with biodiversity, the further and deeper the transformation will go.

Where does the locus of this value lie? Are we the valuers, and does biodiversity merely stimulate something in our nervous system? Below I discuss the idea that biodiversity has "intrinsic value." One meaning of this may be that something inheres in biodiversity that ineluctably transforms all who come into contact with it. In one interpretation of Norton's theory, by having the power to transform, biodiversity is empowered to save itself.

I asked Norton to clarify his notion of transformation.[70] He believes that ideas about what we now call biodiversity have gone through three stages in the recent past. First, a focus on endangered species, or what he calls "elements," informed conservation; we see evidence of this in legislation such as the Endangered Species

Act, which atomized nature into distinct parts. Around the time of the National Forum on BioDiversity in 1986, ideas were in flux. The "element" concept persisted, but hints of something broader informed the discourse, and, I'd add, were subsumed under the neologism. This view also influenced Norton in *Why Preserve Natural Variety?* Now, he regards biodiversity as a process. The focus is not on the elements but on biological complexity at all levels and on the interactions between levels. Norton does not see "transformation" as something external nature effects in a human observer. Rather, it occurs when, amid biodiversity, you find yourself surrounded by a process of great majesty and antiquity and feel yourself to be part of that process and responsible for its continuance.

Norton urges that we no longer rely on unproven axioms about biodiversity's intrinsic value or right to exist when defending it; instead, the locus of value should be the human valuer. But if biodiversity is a complex historical process, then we are part of that process: a human valuer apart from the source of value does not completely make sense. Soulé's ideas about transformation (see below) somewhat clarify Norton's. When surrounded by biodiversity, the very dualism that suggests a phrase like "surrounded by biodiversity" vanishes, and one identifies with the natural world; one is inextricably part of it. The transformation of values occurs partly because if you are inextricable from the grand process of nature, by consuming it or altering it, you irrevocably hurt yourself.

Note that this idea of transformative value dovetails nicely with Janzen's biocultural restoration value or Wilson et al.'s biophilia value. In Janzen's view, this process of transformation has historical precedent; it has been taken for granted during eons of human evolution, and must be resuscitated today for our survival. For those who believe in biophilia, transformation would work by reactivating what our genes encode. For either, evolution ordains transformation.

No matter how we look at it, if biodiversity has the power to transform, then it puts biologists in a better position to talk about its values. Biologists spend their time focused on biodiversity; they immerse themselves in biodiversity and are transformed and renewed by it. Whether this power of biodiversity to transform is biological fact or complex social construction, biologists are among those most likely to be in biodiversity's thrall. This transformation, even if not universal, does occur. And the result, at least by my value standards and those of the biologists I profile here, seems beneficial: who can argue with life-changing experiences that make one consume less and preserve the natural world more? If we view humans as biological organisms adjusting to their changing environment, values can be seen as adaptations governed by selection. We are part of evolution that is self-conscious; transformation is when we realize this and select more adaptive values, where "adaptive" means those values that will allow ourselves and our offspring to survive.[71]

Linda Graber describes how devotees of wilderness believe they commune with a sacred power that illuminates their lives and their relationship to the universe. She borrows the term *hierophany* from religious studies to describe experiences in which something sacred shows itself to us. Wilderness becomes a center of purification for the sullied, overwhelmed, modern Westerner. Thomas Dunlap says turn-of-the-century Romantics shared a similar view: wild animals and places provided an antidote to the artificial, petty life of civilization, and experiences in nature were said to transform the experiencer.[72]

Stephen Fox suggests that such a hierophany struck John Muir in Yosemite; Muir felt played upon by an overwhelming spiritual force and became a zealous conservationist as a result. Norton assigns a name to what happened to John Muir, and to many of the biologists I portray here. Biodiversity has transformed people. Transformation can work through the beauty of an organism: think of advertising campaigns using pandas or golden lion tamarins as icons of worship.[73] Transformative hierophanies may come from understanding the beauty of the process of evolution, from experiencing the mysterious and unelucidated impossible intricacy of ecosystem relations, or simply from appreciating the riches of an undisturbed forest.

Arne Naess writes of "highly dedicated persons who cannot help but work for conservation and for whom it is a vital need to live with nature."[74] Some among them may have experienced this transformative moment. The biologists I interviewed, many of whom "cannot help but work for conservation," seem to agree, albeit not unanimously, that biodiversity has transformative value.

PETER BRUSSARD: "I think that lack of support for biodiversity often springs from ignorance of biodiversity. And if you can get people to zoos, if you can get people to go to natural history museums here, they're not bored but actually entranced by what they see there, that's certainly going to help. So in that case, yeah, I think I would agree with that. But it's not magic by any means."

DAVID EHRENFELD: "I think that's, to a large extent, true. I think I said something very much like that in my recent lectures to the New York Botanical Garden in Manhattan. . . . I talked about the danger of people growing up with no contact with nature. That is deadly. . . . That's why to me, I'm more and more convinced that the most important place and way in which we can deal with this whole problem of environmental degradation is in education at the level of kindergarten, first grade, second grade, third grade. That is *the* critical thing."

PAUL EHRLICH: "It's true that it has that possibility."

THOMAS EISNER: "I think he's right. I just don't know, I don't know what the cultural, genetic predisposition is of people as a group and individually to absorb that type of philosophy."

TERRY ERWIN: "That's a really anthropomorphic way of looking at it. I don't think there's any transformation. I think maybe it's a recognition of something that's deep inside."

DONALD FALK: "Yes. I know Bryan's work and I think he's very articulate. It's probably not necessary to resort to that complex a formulation. If the . . . millions and millions of people who go to Disneyland every year and experience culture and nature in that sort of setting spent an equal amount of time experiencing nature, and culture for that matter, in a less controlled environment, I think the message that would be communicated would be vastly different. . . . I find in order to keep myself oriented to what's important in life, a periodic experience of something other [than society's predominant values] . . . is very important, and nature is one place I can do that."

JERRY FRANKLIN: "I agree with that as something that happens, and it is a value."

VICKIE FUNK: "I think that's true, but I think that could happen to somebody sitting out in a eucalyptus forest with sparrows hopping around. The intricate nature of a tropical cloud forest or paramo or rain forest, wherever you are—you know, the fact that you've got 380 species of tree in a hectare instead of 10 is not going to be appreciated by most people. I mean, they could sit in an oak-hickory forest here and get the same experience as if I took them to a tropical rain forest. . . . I think that, yes, that says something about preserving the out-of-doors maybe, but, no, it doesn't help us with our arguments of why the biodiversity levels, you know, why the species diversity in tropical forests, are so important."

HUGH ILTIS: [Agrees.] "He doesn't ask the why because he's not trained or schooled in the kind of basic, gut feeling of evolutionary understanding. The gut feeling is missing."

DANIEL JANZEN: "Yeah, that's another way of saying perhaps what I was saying very clumsily at the beginning."

K. C. KIM: "I really don't know beyond the fact that I consider biophilia is [a] necessary component of human life. And from that perspective, if you can relate human life in terms of biophilia, then naturally to preserve biodiversity for the sake of biophilia is natural to people."

THOMAS LOVEJOY: "Oh, I think that's perfectly true, that's perfectly true. There is a magic in living things. And, I mean, that's partly because they're the closest things to us that we know of in the universe. That's why, you know, zoos and aquaria have a greater annual attendance than all major sports events. Yeah, sure, there's tremendous value there."

JANE LUBCHENCO: "I think I would phrase it differently, but with the same result. I don't think it's biodiversity per se that's transforming. I think it is being in beautiful habitats that are natural."

S. J. MCNAUGHTON: "I think that is an attractive idea, and I think that anybody that supports that idea ought to be contributing to programs that get kids out of the city and into farms and the countryside and the woods that they wouldn't see otherwise, because I think it is a transformative experience. And I think that the encapsulated life history I gave you indicates that it transformed *me*. I mean, I grew up experiencing this in a way that transformed me. . . . I think *anybody* can. I've known people who were confirmed urbanites, who had never really been out of the city in their life, including people who are very near and dear to me, who came to the Serengeti, and while they did not necessarily enjoy the experience at the time it was happening, it was transformative in the way they looked at the world—retrospectively. They found it uncomfortable, they didn't like the bugs, they didn't like the dust, but afterwards it was without doubt one of the high points of their lives. I think it can happen to anybody. It doesn't happen to everybody—I've known people . . . it doesn't happen to."

REED NOSS: "Intuitive knowledge depends on an emotional bonding and is renewed and strengthened by regular, direct contact with wild nature."[75] [Interview:] "Yeah, I think that's true, in general."

GORDON ORIANS: "That's one that I resonate with. Because you can't value something you've never experienced. If you haven't walked in an old-growth forest here and seen it, . . . it's not going to mean much to you. If you haven't been on the Skagit Flats in the winter and seen the thousands of snow geese wheeling around—if you don't even know they're there, you can't possibly be concerned about their going. That may be an overstatement, but nevertheless, this sense of value comes with contact and experience. Which is why Tom Lovejoy takes the senators down and shows them the Brazilian Amazon. Which is why the Organization for Tropical Studies runs courses for staffers, congressional staffers from Washington, D.C., to come down for an intensive week down there, and they see the tropics, and they see where things go wrong. And they internalize it and they come back motivated and concerned about it. So yeah, that's so obviously true."

DAVID PIMENTEL: "If these people get an appreciation of nature by hunting, by fishing, by birdwatching and that leads them to a desire to learn more about their environment and the ecosystem, I think this is great. And I think that's part of the argument for the people who are pushing the appreciation of the mammals—the cheetah, the lion, the elephant, and so forth—if they get to be interested in those, it may lead them—I'm not saying everyone, but some—to be interested in nature, and this is quite a legitimate approach, I think."

PETER RAVEN: "Couldn't agree more. I mean, that goes to the moral and aesthetic kind of arguments I was making in the beginning. No, I couldn't agree more. And I also—a lot of people have pointed out a related thing, which is [that] our minds are built up entirely in relation to biodiversity."

G. CARLETON RAY: "I think it's right. . . . I just think it's generally true."

WALTER ROSEN: "I don't know whether it happens. I'd like to assume that it does. It's almost essential to assume that it has that potential for us. I'm extremely interested in restoration ecology, and a former student of mine has become a leading spokesman; his name is Bill Jordan and he's at the Arboretum at the University of Wisconsin at Madison. . . . And one of the things that Jordan does is to talk about the spiritual component of restoration, that it provides the means for participants to get reconnected with nature, that we're all so alienated from the natural world. Restoration is a way back. . . . I find the personal contact so enriching and so enjoyable that I like to assume that, in that regard, I'm not unique, and others would, too."

MICHAEL SOULÉ: "I could quibble like any academic about the word *transformative* and what he means by it. We could spend a lot of time trying to figure out what he means by it. But I think I intuitively understand what he's saying. . . . When you talk about transformation, there's sort of an object and a subject. There's the transformer and the transformee. That's a dualism which I have a hard time accepting. . . . I prefer to talk in terms of identification, as Arne Naess does. And not that nature is acting on me and transforming me, but there's some mutual interaction or a mental process of identification."

E. O. WILSON: [Not in direct response to a question posed on this topic:] "I believe that the more that you understand organisms, each species in turn, its natural history, its evolutionary history, behavior, the more involved people become with them." [In response to the question:] "Yes . . . I don't think Norton, neither Norton nor [Peter] Singer, nor for that matter any other philosophers, environmental philosophers, have yet fully grasped the significance of a genetically based human nature. . . . The potential is there. The cognitive direction, the prepared learning rules, you know, what we can respond to most quickly and most deeply are there. And we don't know just their exact form. . . . But these ethical philosophers like Norton should start paying a great deal more attention to cognitive psychology and evolutionary biology as they sharpen their arguments about transformation and valuation."

DAVID WOODRUFF: "My personal experience is, yes. . . . I do find that personally surrounded by nature, my attitudes toward everything are different from and improved over the way I viewed the world when sitting in this traffic jam this morning. So urban life divorces me from part of reality that I miss. . . . I also have trouble with this argument in the Third World. And the argument there is, this is just irrelevant. What we're really dealing with here is poverty, economics. So I find the argument doesn't sell. Even though I accept Norton's philosophical argument. And I accept Wilson's hard-wiring argument, or primeval past, the 'whispering within' argument. I find them of little utility, little operational utility."

Many of these biologists (Ehrenfeld, Ehrlich, Falk, Franklin, Lovejoy, McNaughton, Noss, Orians, Raven, and Ray) strongly support the notion of biodiversity's transformative value. Lovejoy even banks on it when bringing the Powers That Be to the Amazon; he assumes that the rain forest will transform them, just as Norton might predict it would. Funk and Lubchenco tend to agree with the notion, although they both assert that it's not necessarily biodiversity per se that effects the transformation. Brussard, Eisner, Rosen, and Woodruff tend to agree—perhaps they see it operating in their own lives, but question the universality of the principle. Woodruff's questioning of how this idea would work in other cultures points to the Western-centeredness of this supposedly universal phenomenon; others, too, have noted that nature's restorative powers are not necessarily appreciated in poorer quarters.[76] Quite a few (Erwin, Iltis, Kim, Raven, Wilson, and Woodruff) immediately tied transformation to biophilia: they thought biodiversity transforms us by tripping neurons evolution selected to be activated in just this way. Of course, some may have this trait more finely honed than others, but that would still not explain a North/South dichotomy in the transformative power of biophilia.

Perhaps the ground must be prepared for some people to experience transformation. In Chapter 4, I discuss how Funk takes people on tours of the National Worm Collection; how Ehrenfeld wrote a book to turn people on to nature so that they would get back into nature to be transformed firsthand; how Lovejoy founded the *Nature* series on public television years before he started schlepping congresspeople to the Amazon, because if you can't bring people to nature to have them transformed, you bring nature to them in whatever diluted form. Stephen Fox suggests that when the Sierra Club's president, David Brower, published gorgeous books of natural photos and prose, "instead of bringing people to the wilderness, Brower's publishing program brought the wilderness to the people— with much the same conversion effect—through books that were hard to put down."[77]

Conservationists seek to encourage this "conversion effect" by putting people in direct contact with biodiversity. Biologists may feel such conversion is possible because they themselves went through precisely this kind of transformation, usually in childhood. John Muir is quoted as saying, "Merely present a boy to nature and nature does the rest. It is like simply pressing a button!" Fox analyzes the "radical amateurs" who have driven the U.S. conservation movement: "At an early point, usually in childhood, often during a period of relative isolation from human contacts, they began to love some aspect of the natural world. Nurtured into maturity, that love was then expressed in conservation work. . . . Generally conservationists were neither born nor made but reborn in a riveting instant of instant conversion."[78]

E. O. Wilson exemplifies Fox's thesis in a charming autobiographical sketch worth reading for anyone interested either in this eminent biologist or in how people become hooked on biodiversity. Wilson's family relocated frequently, and he attended sixteen schools in eleven years. Thus, "because of the difficulty in social adjustment that resulted from being a perpetual newcomer, without siblings, and younger than most of my classmates, I took to the woods and fields. Natural history came like salvation at a very early age. It absorbed my energies and provided unlimited adventure. In time, it came to offer deeper emotional and aesthetic pleasure. I found a surrogate companionship in the organisms whose qualities I studied as intently as the faces and personalities of boyhood friends."[79]

Hugh Iltis fits the pattern of the peripatetic boy who fell under the thrall of living creatures. Like Wilson, he has sought in later life to codify this affiliation with nature into something natural that we all share, and deeply so. "Why do I get an irrepressible urge to defend what I love, the beauty and diversity of nature, and especially the disarming loveliness of flowers?" he asks rhetorically, and answers: "One of my earliest recollections is joyfully picking huge and wildly unorganized bouquets of flowers on a Moravian mountain meadow, scabiosas, bluebells and daisies, and then lying on my back in a 'nest' surrounded by tall, tall grass watching the bees and the clouds. Ever since then, I have been an addicted botanist, 'half plant,' as an old friend of mine used to describe me." As he became older, his early botanical transformation melded with his growing awareness of threats to the natural world, so that "slowly, slowly but surely, my childhood love and aesthetic awareness of nature matured into a passionate desire to save it."[80] Just as Iltis was transformed (or, as he might put it, his biophilic hard-wiring was tripped) by crucial contact with biodiversity during his formative years, many of the other biologists I profile here had analogous experiences:

PETER BRUSSARD: "I think that I've always been interested in birds as a kid. I mean, some kids were always interested in baseball and I was always interested in jars of spiders and things. It certainly wasn't encouraged by my parents by any means. Maybe there is some kind of genetic predilection, I don't know."

DAVID EHRENFELD: "I grew up in the suburbs. Densely populated North Jersey. . . . We had a vacant lot next door. I liked that. I was sent to camp in Maine, and I think that was very important, because I was not at all sports-oriented. So I hated that and was afraid of it and that sort of thing. But I did like our canoeing trips. When we got away from the camp, got into the woods, I enjoyed that a good deal. That was an exciting kind of thing: that was formative. . . . I maintained that kind of contact on an occasional way, basis, until I went to college. I didn't have any kind of conservation consciousness or awareness when I was in college. I didn't really develop it until I became a graduate student of Archie Carr's after medical school."

PAUL EHRLICH: "Oh, my mother always talked to me about nature and so on. I was sent to camp when I was a kid, introduced to butterfly collecting and things like that. It goes back so far I can't really—I was collecting butterflies when I was maybe ten. And before that watching butterflies and birds and snakes and whatever."

THOMAS EISNER: "I started collecting insects when I was six. And then my parents built, when they first started saving a little bit of money in Uruguay, they built a little summer house that we had. And that was in a fairly wild area. It wasn't jungly or anything, but it was an area where we started going when I was ten years old, twelve years old—by that time already totally enamored in biology—not at all exposed to the kind of biology I liked in class—very much a loner. I didn't care whether anybody knew what I was doing; in fact, I was always weird. . . . And I was sort of exposed to the smell of woods, to looking under rocks and looking under logs. And there was just an overwhelming feeling. The first person who's ever been able to put that in writing is somebody who writes so well, and that's Wilson. And so the opening thing in *Biophilia*. And there it is; it's all there. . . . [Tells charming anecdotes of kids with whom he corresponds and who are completely into insects.] "These people suffer from that kind of disease; and there are not many. But I've long since given up trying to tell people what it feels like to do that. I still get that. I mean, I can spot people who catch that disease."

TERRY ERWIN: "I don't really know, to tell you the truth. When I was five years old, I was hiking in the Sierra—I'm from California—I was hiking in the Sierra with my grandfather fishing every summer—summer after summer after summer. And at five I caught my first limit of rainbow trout or brownies, or brookies, or whatever they were. And I continued that experience until I was about eleven or twelve or thirteen, when I first started recognizing girls. . . . But deeply embedded was really the experience from my grandparents. . . . [Relates college discovery of entomology.] It was a very gradual thing. But its roots were probably back there fishing with my grandfather."

DONALD FALK: "Certainly early childhood experiences. My family tended to spend a lot of time outdoors in the summer and we'd go skiing in the winter. In other words, our mode of relaxing and traveling, whether on weekend[s] or for larger blocks of time, tended somehow to be in a natural setting. And that's why I say that I think we're in a very dangerous moment now with fewer people in society having that serious, sustained relationship with nature on its own terms. And you know, there's contact with nature in a very controlled, mediated setting, and that's better than nothing. But I think that sends us a message, it teaches us a lesson about nature that we're only willing to experience it in a sort of Disneyland setting."

JERRY FRANKLIN: "I had a sense when I was as young as nine years old that I didn't like to see trees cut down. Particularly big trees. And I didn't like to see areas developed. And I didn't like to see organisms extirpated. Somehow that was not *right*. . . . It obviously came from my parents, in part, and the kinds of environments my parents put me into. For all I know, some of it might have come from Walt Disney movies that I saw as a kid, too. But you know, it goes way back into my roots. I remember when I was sixteen, kicking around deer hunting with a friend of mine, I ran across a natural area on Forest Service lands. There was a sign talking about how it was set aside for science and educational purposes. And I thought, 'Gee, what a neat idea. It's a really great idea!' And so once I began to get into a position where I could influence those things, I did. And I've always had an appreciation. As a kid, I camped in national parks and that whole idea."

VICKIE FUNK: "Well, I was a late bloomer. . . . I went through high school and college interested in absolutely everything. . . . And it really wasn't until—I mean, I'd always enjoyed camping and hiking—but the idea, the appreciation for things out there being interesting to really investigate came when I took a field biology course after I'd already gotten my bachelor's degree. And just for fun, I went to a field station for summer and actually got hands-on research experience. And that's what converted me. I mean, I went off like a rocket. I went from being interested in biology as a general concept to just absolutely enthralled with the idea that I could think of something interesting and go out and figure it out."

HUGH ILTIS: "I was fortunate. I grew up in a big house with a grandmother and a grandfather downstairs. He was an old grouch, but she was wonderful. My parents were very busy. Weekends, however, were spent out in the countryside. My father was a biologist. And the garden was full of flowers, full of fruits, with a huge sandbox. . . . And by the time I was six or seven years old, I had my own herbarium and my own botanical garden. I had a hill in which my father encouraged me in no uncertain terms to dig out plants of certain types and plant them: and they grew. So I grew up from early stage on with animals and plants. We had white mice. We had aquarium fish. We had hamsters. . . . The point is, we then became refugees, which usually implies a great deal of loneliness, because, what I said earlier, we are cut off from our social contacts. But at the same time, we landed in Virginia . . . a beautiful estate on top of a hill overlooking Fredericksburg, Virginia, called Fall Hill. And in Fall Hill, we got there in early February 1939, and I spent the whole time just running around in the woods. Oh, my gosh, discoveries you know. Everything new. I knew a lot of biology and natural history by then, but everything was new. Similar enough to Europe, because the flora is basically the same. But the species were all different. And many things were totally new like possums and raccoons and things like that. And all these butterflies, and all these birds, and all these flowers: oh, it was just incredible! So the loneliness didn't really hit me particularly. I mean, by the time I was sixteen or seventeen, I had a herbarium of 2,000 specimens in addition to what I brought over from Europe—I brought about 300 or 400 specimens with me. And I had my life sort of spelled out for me. And the psychological high, the eureka feeling, one can even get when one is ten years old, to find something really weird: there is no substitute for that."

DANIEL JANZEN: "Well, I'm that old school of people who started out collecting butterflies and shells and all that kind of stuff. And I just kept up with it. Just as long as I can remember. I suppose, I don't really know, six years old, seven years old, something like that. My father was the director of the U.S. Fish and Wildlife Service. All his lineage is farmers, people who play with living things. . . . Frontier farmers clearly are not quite the same as some kind of corporate farming today. . . . So, German farmer on one side, Irish farmer on the other side. Both of them had the response to my butterflies and shells, 'That's fine. Keep doing it, keep doing it.' So there was no resistance."

K. C. KIM: "You know, I have been sort of a natural historian since seventh grade. I was motivated by one of the biology teachers in seventh grade that got me going up to this point. I haven't changed my interest or my commitment ever since seventh grade. . . . I spent most of the spare time in the woods, chasing could be snakes, could be frogs, could be butterflies, whatever. . . . My father hated it. Because he never figured out how ever in the world I'm going to make a living. And so he detested it until as a matter of fact, five years before he died, we patched [things] up. He finally realized, this guy is not only making a living, he's pretty well known. . . . Until then, he and I were [at odds], because . . . that kid's considered bright, goes through the best schools there were, and all he's doing is chasing the bugs, so to speak. And I guess that through this process I began to appreciate what [biodiversity] means, not just to a biologist, but to [the] whole human species at large."

THOMAS LOVEJOY: "Well, you know, that's very hard to explain. And everybody, or most people when they're a kid, has a certain amount of fascination with ants or whatever it may be. And I certainly did all that. I think it's perfectly possible that if life had taken a different turn or two, I might never have gotten turned on. But what happened was I went away to this school that had a zoo in Millbrook, New York. And I chose it because it had a zoo. And I really didn't know that it was going to be such a big deal for me. It wasn't a huge, overwhelming conscious thing. Although I

did choose it because it had a zoo. I said, I want to go here and I don't want
to look anymore. They had a fabulous biology teacher, one of these great
characters, and he was bright as hell. . . . Every student in that school was
required to take his biology course, which I think was an extremely far-
sighted decision of the headmaster. It was either at ninth or tenth grade. . . .
It was an old-fashioned biology course, because you basically went through
the kingdoms in the course of the year. It was not old-fashioned in terms of
the kinds of things you learned about. But you in fact were exposed to the
diversity of life on earth. But it was three weeks was all it took. And I was
hooked. You know it was all these incredibly fascinating, different living
things. You know, whether it was *Volvox* or *Stentor* or, you know marine
annelid worms, or you know, whatever it was. There was just the richness
of fascination. And I was hooked."

JANE LUBCHENCO: "I'm not sure that there was anything having to do
with biodiversity directly. I grew up spending a lot of time in the out-of-
doors. I did a lot of backpacking—I grew up in Colorado. I spent a lot of time
in the mountains. And have always appreciated beautiful habitats, beautiful surroundings."

S. J. MCNAUGHTON: "My interest in biodiversity was before I even had
a career, it was when I was a young person. I lived, for a while, on a farm
in northwestern Missouri. I grew up in a big extended family of aunts and
uncles, and part of that time was spent on a farm. . . . Well, I hiked, you
know, and I engaged in outdoor sports. I hunted, I fished, I trapped, I did all
that stuff. And it was probably my interest in that that turned me into an
ecologist, rather than the fact that I was an ecologist that got me interested
in biodiversity. And I lived with two uncles for a time who were bachelors,
and they had a lot of land. . . . And part of the policy on that land was to
leave the wild animals alone. . . . They didn't shoot natural predators. They
didn't shoot foxes, they didn't shoot coyotes and things like that. As I say,
there were no coyotes there when I first got there, but in the five or six years
I was around. . . . coyotes were coming back, the predators were all coming
back. In addition to that, their policy was to buy old farmland and seed it
down into grass. By the time they died, the native prairie was coming back.
So it was all of that childhood experience."

REED NOSS: "When I was a kid, I lived, I was fortunate to have some wild
areas—not extremely wild, not wilderness, but wild areas near my home—
where I ended up spending most of my time, you know catching snakes and

climbing trees and playing in the creek and all that kind of stuff. And as I grew up, I saw those very areas I played in being destroyed. I grew up in a relatively rapidly developing area near Dayton, Ohio. And my favorite playgrounds were being destroyed right before my eyes. It was a personal thing for me, no longer any place to play. But also it always just struck me with horror and sorrow to see living things killed. And like most people I suppose, at first it was individual suffering of animals that troubled me a lot. You know, seeing people torture snakes or cats or any other living thing. But also just seeing trees, you know, cut down, just really made me sad, as a kid. So, God, by the time I was—and I don't care if this is on the tape— by the time I was, let's see, around third grade, my friends and I would be actively vandalizing bulldozers and construction equipment that would be tearing down these woods that we loved. . . . You know, it was just a natural response. Here were these goddamned people with their machines tearing down this beautiful woods: what else would you do? I don't think we ever stopped any development. I don't think we ever stopped a house from being put up in one of our playgrounds. But who knows, we might have slowed it down a little bit. But we weren't thinking strategy. We were just thinking gut-level emotional response, a defense, a kind of extended self-defense that a lot of people talk about. I think many people do have that feeling, that there is a larger self. And when they're defending nature, they're defending that larger self. So that's what—I was drawn to it from that defensive position early on in life."

GORDON ORIANS: "I started birdwatching when I was seven. That's why I got into science. I had little notebooks of my observations, bird lists, going back to when I was nine. When I was about thirteen, I discovered people got paid for doing this, and I knew I was going to major in zoology and I was going to become a biologist. So I *love* living organisms."

DAVID PIMENTEL: "I guess it really was being intrigued with nature right from the start. And it was primarily insects that I was intrigued with. Way, way back when."

PETER RAVEN: "I can just say briefly that I was interested in these things from the time I was six or eight years old. I really liked butterflies and plants and nature and everything else. I grew up in San Francisco and I always liked it. It was just wonderful. I read little books about it and I just got into it. It's always been an important part of my life."

G. CARLETON RAY: "Turning over rocks when I was barely able to walk and picking up worms and sticking them in my ear and all that kind of stuff. You know, I have a different approach. People say, 'How did you get interested in animals?' I say, 'How'd you lose your interest?' I mean *everybody*—there isn't a kid alive who doesn't like animals. And you don't have to do *anything* to keep that alive. What we've done in our culture—I remember when I was working at the New York Aquarium, part of the New York Zoological Society, I was a curator right out of college, when I told you I was doing a park and all that. And zoos and aquaria were listed in *New York* magazine—all of them—under childrens' entertainment. All the art museums were adult entertainment. That was literally true. . . . Well, city kids don't so much have the opportunity. But most kids have a pretty good opportunity to understand animals and diversity and all that. It comes totally natural."

WALTER ROSEN: "I don't know that that's answerable. I was a nature lover when I was a little kid. I grew up in Forest Hills, New York—not exactly the bare streets of downtown tenement area, but a very urban environment. And I appreciated growing vegetables in our little tiny backyard. And I appreciated as a boy scout getting on the subway and riding for an hour in order to be able to hike over the George Washington Bridge and go bopping around the Palisades State Park. So my earliest encounters with nature I remember fondly. So I guess one piece of my concern came from my early childhood. I used to spend summers as a kid at summer camps out in the country, and I loved that."

MICHAEL SOULÉ: "I've always been an outdoor naturalist, since I was a kid."

E. O. WILSON: "As I told you, I've always been congenitally predisposed, I mean congenitally in the sense of probably personality traits and early childhood experience and so on, toward biodiversity studies, broad natural history and synthetic studies. I take immense pleasure in doing what I'm doing right now, for example, which is drawing, making multiple drawings—ten each, approximately—of each of 600 species of the ant genus *Pheidole*. I just completed last night species 386. And a lot of people would consider that mind-numbing in terms of trivial detail. But when I pick up another species of *Pheidole*, for me it's like seeing the creation: I'm the first to see this species. I'm the first to see why it's different, what's unique about it, and so on. And that gives me immense pleasure in just dealing with the fine detail. God is in the details: that's really what it comes down to. So

I've always had that feeling. When I was in my teens, I used to—my love of nature and doing this sort of thing goes back to when I was about nine years old—when I was in my teens I was enthusiastically rolling around in the fields and forests of Alabama. And even then, so many of the streams were toxic—so many of the hills were denuded, you know—and running in deep red-clay gullies. And I was distressed by this, when I saw that something had gone wrong where I grew up."

DAVID WOODRUFF: "I'm very lucky. I'm one of those people who probably made my life career decision really early—like around age five or something like that. I was playing in a pond and a frog jumped out of my bucket. And I got real excited by that, and just got turned on to nature."

Few of the respondents reported, or seem to remember, epiphanic experiences, single events that transformed them. The frog jumping out of Woodruff's bucket is one exception. McNaughton supplies another: "I can remember a specific example, being out just walking in the fields in Missouri and it was snow, it was probably December–January. It was a pretty cold day, rolling hills, hedgerows, fields, a large pond on the left, with snow cover, fresh snow, and it was snowing in fact. And I saw a fox, a red fox. And I watched it for probably 15–20 minutes; I just stopped and watched it. That was an epiphany to me. And I would be sadder if there were no foxes in the world."

Most, though, were always enchanted with biodiversity, or arrived at such enchantment gradually. Some (Ehrenfeld, Funk) came to it late in life. Some (Brussard, Kim) explored nature despite family desires. Raven has a theory "that all little kids are interested in biology and natural things. The only question is whether they stay interested or not. If I had to pick one thing that makes the difference, it would be encouragement."[81] Raven and others (Ehrlich, Falk, Franklin, Janzen) report such encouragement from family members; for Kim and Lovejoy, teachers helped effect the conversion.

Earlier, I noted that when asked about biodiversity's power to transform, several biologists immediately invoked biophilia: we are genetically coded to respond to biodiversity in this way. Noss's response above (echoed by Iltis and perhaps Wilson) takes this a bit further. For Noss, this biophilic urge leads one not merely to respond with love but to defend, passionately, the object of one's love.

Of course, like Soulé, Noss would reject the dualism of my last statement. He and his friends were part of the natural world; they identified with it. And as such, their early attempts at eco-sabotage were self-defense.

I am sympathetic to this view—as I am to the idea of transformation in general—while at the same time remaining skeptical. I, too, took to nature as refuge from a solitary youth. Next to our house in suburbia was a small parcel, an acre maybe, of scrub woods. It was my kingdom. Even if I didn't know their names, I knew the plants and animals; I knew where each tree was. Then one gentle, third-grade day, I got off the schoolbus and walked over the hill to find it was all gone. A bulldozer had wiped it out in the blink of a morning. And when I got over my mourning, I had been transformed into a lifelong environmentalist. I had been twice transformed, once into a lover of nature, and then into a fighter for nature. I remember feeling that part of me had been destroyed; as backhoes annihilated my neighborhood and the rest of Long Island, I felt as though a series of wounds had been inflicted on me. My paltry attempts at eco-sabotage felt to me like self-defense, just as they did for Noss.

But much as I'd like, I still can't express this in universal terms. What happened to me just doesn't happen to everyone. I can't translate my personal experiences into biologically determined and therefore inevitable predilections. On the other hand, so many people are fascinated—are able to be transformed—by contact with biodiversity.[82] And I applaud the results when these people become born-again conservationists, no matter at what age the conversion occurs. If the transformative value of biodiversity were taken seriously—and I believe the actions of biologists show they do take it seriously—renewed effort would be spent on getting people out into nature to effect this transformation.

In the Third World (and also, of course, in the United States), parks are often fenced off, defended from the local population who might infringe on the integrity of the area, and thus on its potential for sucking in the almighty dollar of the (usually) Northern tourist. Even if not officially barred, people are often excluded by lack of transportation or inability to pay entrance fees. I am sure that their level of resentment toward this use of their land is inversely propor-

tional to the amount of time spent there. You simply cannot love
what you do not know. Whether or not the transformative value
of biodiversity is a universal phenomenon is not really important.
It *does* work, at least, with many people. Here and abroad, a lasting conservation ethic—one that will endure when the government
changes hands or when the tourists migrate elsewhere—can only
take root when people care, deeply, about the natural world. This
love may still matter little when people are hungry, when they need
this land or its creatures to satisfy their most basic needs. But after
the basic demands of life have been met, you must have been transformed at some point and in some way by biodiversity if you are to
value it at all. If biodiversity has transformative value, then familiarity with biodiversity breeds affection for it, breeds a desire to
consume less and salvage more of it.

INTRINSIC VALUE

Going beyond the testable assertion that biodiversity has transformative value, some biologists proclaim its intrinsic value in and
of itself, apart from any human valuer. Humans thus have no right
wantonly to destroy biodiversity. Such assertions may be justifiable
from certain religious standpoints. If God or some other deity or
sacred process created the natural world alongside humans, then all
creatures are imbued with sacredness: all have intrinsic value.

Yet most biologists have no such religious views. In asserting
biodiversity's intrinsic value as one of the reasons why we should
conserve it, they move well beyond the realm of what we might expect scientists to acknowledge and defend.[83]

How the notion of intrinsic value is received will, of course, depend on what is understood by *biodiversity*. If biodiversity is a list
of species, then one must try to understand what it would mean for
a species to have intrinsic value. But if biodiversity is more inclusive—and I have suggested that many of the biologists promoting
its value hold this view—then the question is, rather: Do all the
Earth's creatures, their interactions, and the processes that gave rise
to them and to us have value in and of themselves? This conception
makes it more difficult to say no to the concept of intrinsic value,
particularly if humans are part of biodiversity. But it still doesn't

help much in daily life, when we may be discussing the wisdom of an activity that will eliminate an element of biodiversity—a species, a vacant lot, an ecosystem.

We have seen that assertions of intrinsic value are not original to modern-day biodiversity proponents. Aldo Leopold, Charles Elton, and Rachel Carson all believed something of the sort. David Ehrenfeld's *The Arrogance of Humanism* contains a passionate defense of the idea. More recently, Ehrenfeld has elucidated some of the qualities inhering in nature that might comprise intrinsic value: "For the people of the next millenium, the qualities of nature—honesty, reliability, durability, beauty, even humor—will be necessary landmarks for survival, there for the finding, unless the damage we are doing now proves too great."[84] The Ehrlichs, who believe in nature's intrinsic value, assert, "This is fundamentally a religious argument. There is no scientific way to 'prove' that nonhuman organisms (or for that matter, human organisms) have a right to exist."[85] For deep ecologist Arne Naess, this lack of proof presents no problems. In a conservation biology textbook, he asks: "Is it my privilege as a philosopher to announce what is of intrinsic value, whereas scientists, as such, must stick to theories and observations? No, it is not—because you are not scientists as such; you are autonomous, unique persons, with obligations to *announce* what has intrinsic value without any cowardly subclass saying that it is just your subjective opinion or feeling."[86]

Naess just says no to value subjectivity; he urges biologists to do the same. Of course, many biologists, even if they are inclined to agree with Naess on intrinsic-value theory, will hold back from such pronouncements. We can understand why. Such pronouncements might be considered the antithesis of traditional scientific expertise. To proclaim intrinsic value without standard scientific proof for one's proclamation begs others to question your authority.

Nonetheless, some biologists make such pronouncements. In 1985, Soulé wrote "What Is Conservation Biology?" to introduce the fledgling discipline to nonadherents and to stake out common ground for believers. Among these common precepts are normative ones, which "are shared, I believe, by most conservationists and many biologists, although ideological purity is not my reason for proposing them." As I have mentioned, these norms are to be taken as part and parcel of the discipline. They include the notions that diversity of organisms, ecological complexity, and evolution are normative goods. The "most fundamental" postulate of all is that *"biotic diversity has intrinsic value,* irrespective of its instrumental or utilitarian value."[87] Soulé builds his science on this unimpeachable yet untestable assertion.

The biologists I interviewed ran the gamut on biodiversity's intrinsic value:

PETER BRUSSARD: "Well, I mean the whole concept of value and use is a human construct. So I mean, who values things? Well, people put a value on things. So if people were to disappear, that would mean value would disappear, right?"

DAVID EHRENFELD: [From *BioDiversity*:] "For biological diversity, value *is.* Nothing more and nothing less."[88] [Interview:] "Well, I couldn't prove it, I guess. I just believe it."

PAUL EHRLICH: "Well, that's part of my quasi-religious view. I mean, that's part of the quasi-religious end of it. The question is, would there be no value at all if there were no human beings to assign value? Or do these things have some intrinsic worth on their own? And I guess having had the experience, I guess, of being with a lot of organisms in the wild, I just have the feeling, first of all not only do they have more than worth, I think they are almost capable, some of them, of assigning worth. If you watch chimpanzees or gorillas in the wild—I've had the privilege of doing both—chimps really extensively—I just can't have the feeling that the only value they might have is what they might mean to us. But you can't possibly defend that scientifically. Would the universe be a worse place with no life? I think it would be a worse place, but that's a very 'lifeocentric' view."

THOMAS EISNER: "I find myself not being able to answer it because I need to have a definition of what you mean by value there." [Segues to biophilia and why things have value to him.]

TERRY ERWIN: "*Intrinsic value* is a humanistic term. I mean, we're trying to say something else, but we can't say something else. It's impossible. These things are because they're passing through the universe: we are, they are, it is, everything is. And so, you start having to invoke God or something if you're going to start talking about *value*. It's just: these things exist.

We exist, they exist. And it's all interesting. It's just hard to grapple with an intrinsic value. I don't think anything has an intrinsic value."

DONALD FALK: "Yes, absolutely." [Does he use this argument?] "I do. It really depends, in a sense, who you're talking to. If you're talking to business executives in a field that depends on plant diversity economically, then I'm likely to brush by that argument fairly quickly, and I'm not going to try to engage them on a moral level. . . . [I have] a deeply personal belief that all that out there . . . has a right to exist and we are an integral part of it, no more and no less."

JERRY FRANKLIN: "Oh, I basically think so, yes. But I haven't given a whole lot of thought to it. Fundamentally, I think things have a right to exist, regardless of their use to human beings."

VICKIE FUNK: "Oh, I think that's true. I think it may be more apparent to people who are more fully able to appreciate it. But I think that other things have a right to live on this earth instead of just us and cockroaches. And denying the right of other life forms to exist just because we want more space is not necessarily justifiable."

HUGH ILTIS: "Well, there is intrinsic value in the sense that the web of life, it all hangs together. And we are part of the web of life and we cannot predict what is going to be useful and what isn't going to be useful. [But he does not believe something has intrinsic value] just because it's out there. . . . That's just sort of nonsense. By all means, I'll defend it and I'll fight it harder than Norton does! In favor of it."

DANIEL JANZEN: "The word *value* is anthropocentric. . . . That's a contradiction in terms."

K. C. KIM: "Sure, sure. Yeah, I agree with that. [Each species] is a result of evolution which is unique. No other species like that, however closely related. . . . Personally, I believe that a species has intrinsic value by itself."

THOMAS LOVEJOY: "Yeah, I think I agree with that. I mean, I think it's exciting that there are other forms of life. And that's value in itself."

JANE LUBCHENCO: "I think that that's an argument that's a very personal one. And that some people would buy that and some people wouldn't. I do think that ecosystems function best when they have sort of the natural contingent of species present. And that that is one kind of value apart from any value that we might place on them."

S. J. MCNAUGHTON: "I don't see how anything can have value outside of a value that human beings place on it, because value is really something that is uniquely human, isn't it? I mean, unless there is a Divine Creator that has a value system, then it has value through that Creator. But aside from that, how can something have value in the absence of humans, which so far as I know are the only organisms that have a concept of value."

REED NOSS: "And so it gets back to the fundamental argument that if we're important, as almost every religious tradition in history recognizes that human beings have intrinsic value, that we're important as individuals and as a species. If we're important, the only logical corollary is that *all* species are important. There's no basis for saying that we're more important than other species. . . . Well, there's a lot of rational, philosophical reasons for questioning any kind of value independent from an observer. There's a lot of good scientific reasons for questioning that. But if indeed it is an intuitive appreciation that many of us have, I think that speaks strongly in favor of recognizing intuitive value. To me, it's not a rational insight at all. It's an emotional, direct experience that things have intrinsic value. But the fact that that experience is so widely shared—I think it really is widely shared—again, certainly with regard to fellow human beings, all ethical traditions have accepted intrinsic value. As far as intrinsic value in other living things, I'm not sure where the majority stands on that. But to me, it's something I can't deny, that sense of intrinsic value in nature. . . . My intuitive, you know, spontaneous experience is that things are valuable. And I can't bring myself to believe otherwise. But I think the value that's there is probably in the interaction between whatever it is out there and whatever I'm experiencing. And so to place the locus either in myself or in nonhuman objects would be missing the point, that it's the interaction between nature and me that causes me to recognize intrinsic value. It's somewhere in that interaction."

GORDON ORIANS: [He has written: "Each living species is a result of unique evolution and thus has a limited claim to value in itself."[89] I asked, why "limited"? And what did he mean by "value in itself"?] "I think all values are limited. For example, let's say, we may claim that the value of a human life is unlimited. But we don't really believe that. Or at least we don't act as if we believe that. . . . I suppose for me, it goes back to a quasi-Schweizerian respect for life. The beauty of it, and the complexity, and the processes that created it. These are awesome to me. And I think in part it's this sense of awe out of which comes my sense of 'intrinsic.' Insofar as I can clearly define it. I find this a murky arena that I have a lot of trouble with, knowing what I feel."

DAVID PIMENTEL: He believes in some notion of intrinsic value, but "in trying to protect or conserve nature, to use the argument of intrinsic value gets you—well, I just don't think it sells very well."

PETER RAVEN: "Yeah. Well, I believe it. It follows from what I've been saying. If you just repackage it slightly. The world is a place that's filled with biodiversity. And I look at it holistically. And I look on us as one species in it. And I say that does have intrinsic value. It has value that is absolutely intrinsic because that is what it is. That's all it is. That's fundamentally it. That is the world. That's almost beyond making a decision or even thinking about it. That's the planet that we live on. You know, and in that sense, I think what they say is just *profoundly* correct."

G. CARLETON RAY: "Most of the time when people think of the intrinsic value of biodiversity, they mean exactly what it says: the intrinsic aesthetic or spiritual value of biodiversity. . . . I would just say, yes, there is intrinsic value from all sorts of standpoints. . . . It's a very complicated topic. . . . The statements that are made . . . sound great. But they're so *vague*. And a lot of people are, I think, 'Oh, my gosh, biodiversity has intrinsic value.' But you walk away and ask, 'What was that man saying?' And those are the people who haven't the foggiest idea what it's all about. I don't really know how it translates to most people. . . . A lawyer would probably have you for lunch over that, tie you all in knots."

MICHAEL SOULÉ: He strongly agrees with this assertion. See his comments earlier in this chapter and in the last section of Chapter 3. As we have already seen, Soulé vigorously asserts that "*biotic diversity has intrinsic value*, irrespective of its instrumental or utilitarian value."

E. O. WILSON: [I asked him to interpret his observation that "each species is unique and intrinsically valuable."] "By that I mean that—I don't mean metaphysically intrinsic, I mean that because it is so unique. It has such a vast and ancient history. And because it has so much potential value in many dimensions to humanity, that each species is intrinsically valuable. That is to say, it has certain value for humanity. . . . It's a modest definition of intrinsic value."

DAVID WOODRUFF: "I am anthropocentric, I would suppose. . . . I think there are other arguments that are easier to use and are more persuasive. So

I don't want to deny the merits of these values. . . . I think our philosophical understanding of biodiversity's merits is interesting to some people, more to others than to myself. But it doesn't help us deal with the day-to-day approach of nature that is where the action is."

Eisner and Woodruff were circumspect in their responses. Lubchenco respects the argument, but did not commit to it. Iltis doesn't necessarily believe it, but would use it if it proved effective in convincing people to conserve biodiversity. Brussard, Erwin, Janzen, and McNaughton believe that "value" is a human construct, and they therefore cannot conceive of value inhering in nonhumans.

The majority of biologists I interviewed do believe, in some way, that biodiversity has intrinsic value, even if they may hold different conceptions of what this means. So, for example, Wilson's idea of intrinsic value does not seem intrinsic at all; he is still discussing the value of biodiversity for humans. Franklin, Funk, Kim, Lovejoy, Pimentel, and Ray may personally hold this belief, but they do not necessarily publicly offer intrinsic value as a reason why we should conserve biodiversity.

But Ehrenfeld, Ehrlich, Falk, Noss, Orians, Raven, and Soulé would or do use this argument. Raven's rationale for accepting intrinsic value sounds somewhat like Erwin's rationale for rejecting it. Read Noss's response, and see how he is grappling with something difficult to put into words. So it is with intrinsic value: it's a feeling many biologists and others have about the world, a feeling disconnected from English vocabulary. When I read eco-philosophical treatises, I often find them tortured: their authors are grappling to express the ineffable.

This is the problem with the notion that biodiversity has intrinsic value, and that therefore we should save it. As Ray points out, the idea comes off as fuzzy nonsense to those not inclined to believe it in the first place. Intrinsic value appeals to those with whom you don't need to argue that biodiversity has intrinsic value: they just agree with you. They share that feeling, and that feeling leads them to behave in more protective, less hubristic ways toward the natural world. If the proponents of biodiversity aim to tie their value claims to something bigger than themselves, positing value that inheres outside of us could be effective. When humans are not the center of the moral universe, when all value is not human-centered, we may act with humility toward the rest of the Earth. But I question whether this argument will convince anyone not already convinced by it. Perhaps part of biodiversity's transformative value is to instill into some the feeling that nonhuman entities and processes have intrinsic value.

SPIRITUAL VALUE

If it seems a priori odd that some scientists believe and preach a concept like intrinsic value that cannot be proven scientifically—indeed, it can barely be expressed at all—it may seem totally bizarre that scientists talk about biodiversity's spiritual value. Yet, for example, at the National Forum on BioDiversity, Ehrlich noted: "Curiously, scientific analysis points toward the need for a quasi-religious transformation of contemporary cultures." He means that "a quasi-religious transformation leading to the appreciation of diversity for its own sake, apart from the obvious direct benefits to humanity, may be required to save other organisms and ourselves." Soulé believes that if biophilia is to become a real player in conservation, "then it must become a religion-like movement. Only a new religion of nature, similar but even more powerful than the animal rights movement, can create the political momentum to overcome the greed that gives rise to discord and strife and the anthropocentrism that underlies the intentional abuse of nature."

Ehrenfeld declares: "Clearly, we have to take this valuing out of the purely intellectual sphere, at least until the present phase of the scientific revolution has run its course. . . . What is left, if we eschew the cost-benefit approach, are the realms of religion and emotion, which fallible scientists should not despise. Within the purview of religion are several very different ways to celebrate diversity—some invoking God and some not."[90] Ehrenfeld has remonstrated eloquently against our overweening faith in human reason, or "humanism." Science stands at the apex of the humanistic enterprise as the most finely honed use of reason to understand and control the natural world. Could biologists, these avatars of humanism, really be preaching anti-humanist apostasy?

The answer is mixed. While some biologists profess a brand of spirituality that does not obtain in our traditional notions of science, others account for their own and others' spiritual feelings about biodiversity in mechanistic ways compatible with our images of the typical scientific worldview. I develop this argument further in Chapter 7. Here I start to show how some biologists are shaping history by fostering what may be an unprecedented merger between modern science and religion.

Biologists find spiritual value in biodiversity precisely because of, not despite, their science. People turn to spirituality, or profess spiritual feelings, when confronted with vast unknowns that defy logical explanation. In an analog to traditional religions, biodiversity's spiritual power is linked to our lack of knowledge of it. Since some biologists spend their professional lives surrounded by biodiversity, its unfathomable complexity and its sublime beauty combine with feelings of humiliating ignorance to infuse intense spiritual feelings. The more they learn, the more awe they feel; and the unknowns, the gaps the sacred world of science can't fill, leave further room for values and spirituality and aesthetics to rush in. Biologists have few axiomatic laws for their claims about biodiversity. As a result, the spiritual may become axiomatic, the scientific problematic.

Some historians have revealed a tradition of ascribing sacrality to wild places. Max Oelschlaeger traces a strand of such spiritual devotion back tens of thousands of years. Linda Graber's *Wilderness as Sacred Space* shows how American aficionados promoted wild places as valuable because they were holy sites, potential altars for human spiritual nourishment. J. Ronald Engel's *Sacred Sands* reveals how those who wished to preserve the Indiana Dunes in the early twentieth century used these tactics, and Susan Schrepfer contends that partisans of California's redwoods held and promoted similar feelings.

A large swatch of the U.S. conservation movement is woven from this fabric. Stephen Fox believes the movement is tantamount to a religious backlash against modernity. Historian Richard White suggests that environmentalism originally stemmed from science, then turned against science to become quasi-religious.[91] I believe biodiversity proponents bring this phenomenon full circle. They lend the imprimatur of science to spiritual arguments that have a history of resonating with the public. If the value of biodiversity were felt not merely in the pocket or in the brain but in the *soul*, then the most effective, permanent conservation ethic imaginable might result.

Engel asserts that "the attempt to convert other persons to a new vision of the world, so that their attitude toward others is transformed because of it, and they are motivated to form a more perfect community on its behalf, is the kind of activity characteristic of all missionary religious movements."[92] This description of those who fought to conserve the Indiana Dunes also fits those who fight to preserve biodiversity. A group of zealous partisans, many of whom happen to be scientists, proselytize so that we may incorporate into our minds, hearts, and souls a new vision of what is of value in the world, along with a set of rules we should observe to make sure this value is preserved. They would convert us to this new paradigm so that we become more spiritually fulfilled, and, simultaneously, so that the source of our fulfillment will endure eternally.

We should not conflate two distinct arguments here. In one argument, biodiversity is sacred: it has value in and of itself. This takes intrinsic value arguments one step further: the value inhering in biodiversity makes it sacred. Myers proclaims that "every species, as a manifestation of creation's life force on earth, deserves to have its own chance to live out its life span." Ehrenfeld avows that knowledge exists beyond the realm of the rational and that value cannot be measured in economic terms alone. His "Noah Principle" epitomizes one such nonhumanistic value. The Ehrlichs urge us all to embrace the Noah Principle, although they concede: "One cannot assert this ethical responsibility on scientific grounds. It clearly arises from essentially religious feelings: we believe that our only known living companions in the universe have a right to exist."[93]

In this argument, the locus of value is external to humans. That differs from the argument posed by those who value nature because they find spiritual nourishment there. They suggest that biodiversity has spiritual value for humans, and that we may become fulfilled by sharing the numinous experience biodiversity can prompt.

Myers warns that as biodiversity disappears, our "spiritual security is at stake." Kim and colleagues urge that "(n)either should we ignore the psychological and spiritual benefits to be derived from living together with other species with which we share a common evolutionary heritage." McNaughton notes that nature is "a spiritual as well as a material resource." The preface to a book about extinction warns that decisions about how we treat the current mass extinction spasm "will affect the face of this Earth, including the economic and spiritual welfare of your children."[94]

No matter what their philosophical predispositions, many biologists share a tactic: having found the spiritual in nature, they make arguments that are spiritual in nature. While some discuss feelings of biodiversity's spiritual value in private, they do not use this as a public rationale. Others seek to convert us so that we may view biodiversity with religious devotion, feel it with spiritual awe. They seek something akin to Norton's notion of biodiversity's transformative value, but stronger. It is not just that they would have our preferences change; rather, they would have us born again. When John Muir discovered Yosemite, he declared, "I feel like preaching these mountains like an apostle."[95] Some biologists have been born again, turned spiritual by biodiversity. As prophets who have witnessed revelations, they now must preach the gospel of biodiversity with evangelical fervor in a diversity of pulpits.

In Chapter 4, we saw where and how they preach. Let us now look more closely at some prominent biologists' views of biodiversity's spiritual value. During my interviews, I asked if each biologist was spiritual in any way, and then asked him or her to elaborate.

PETER BRUSSARD: "Mike [Soulé] really means that we need to convert to some sort of a religious experience having to do with biodiversity. And I just don't have it. But I certainly am happy to explain to people what I consider to be the very practical values of biodiversity. And if pushed, I will explain what my own personal values of biodiversity are. But I think they're rather personal things."

DAVID EHRENFELD: I did not ask him directly. He has written eloquently against humanism and has affirmed many tenets of Judaism. His advocacy of the Noah Principle, above, makes clear at least part of his spiritual leanings.

PAUL EHRLICH: "Well, I use the term *quasi-religious* because my own personal view is that science doesn't explain everything. Science does not explain, easily explain, a lot of our attitudes toward different things. On the other hand, in my view it's preposterous to think that there's somebody with a white beard up there on a throne. In other words, I am totally areligious. . . . But it seems to me that this is a matter of our basic feelings towards each other and towards the planet, and that's not something that falls within the realm of science. I can't convince you scientifically that it would be a good thing for your soul to enjoy nature and feel responsible for it and so on." [I asked him if he would call himself spiritual in any way.] "Yeah, I think all people are spiritual to one degree or another. I'm not the 'link hands and stand around the campfire vibrating' sort of spiritual. But I think the most important things in life are things like friends and sex and so on . . . that I can enjoy analyzing the evolutionary roots of. . . . When people say, 'Are you religious?' I always say, 'No.' And then they say, 'That means you think science explains everything?' And I don't see that as a dichotomy. That's why it's so difficult to use that term. The reason I use it, and the reason I put *quasi-* on it, is I think it transmits something to people. I try to get the message over that this is something—maybe spiritual would be better. But spiritual has taken on an artsy-fartsy connotation, and I just don't like it."

THOMAS EISNER: "I find it devastating that any of the Western religions are so anthropocentric and so totally, totally useless when it comes to getting involved in the conservation movement." [Is he religious or spiritual in any way?] "No. Spiritual? God, I don't know what you call spiritual. I mean feelings I have not the slightest way of interpreting scientifically? . . . That's my definition of spiritual. And feelings that I very much need, sure." [Could he elaborate on any of those in the context of biodiversity—does he have such feelings?] "Sure. Absolutely. As I mentioned, I mean I can be in a really high-tension situation from overwork or something—I'm fundamentally such a happy person that it's very rare that I find that I need to resort to some remedy—and I can go for a walk through nature and I get the kind of feeling where I suddenly understand other people's experiences, which they then verbalize through some kind of religious language. . . . And you know, I know it's all endorphins or whatever, and sooner or later somebody—I mean, I don't have to define it as spiritual because I figure sooner or later—I won't live long enough, but sooner or later, we're really going to understand enough about the central nervous system that we're going to be able to [inaudible] on which neurosecretory substances are going out, but which are going down to make that experience possible. And I can even give you a biological reason for why these biological phenomena occur. I mean, it's a way of physiologically programming a high in a context which the body politic tries to manipulate so that the highs express themselves into support for social cohesion and political support. . . . Am I religious? Absolutely not. In fact, I'm very tolerant of religion because I've been on the nontolerant receiving end of that. [He and his family were Jewish refugees from Nazi Germany.] But I'm suspicious of clergy if it represents a majority church, for the simple reason I see it as simply a political force, and to the extent that it is not aligned with any of the fundamental things I believe in. I often see it as rather counterproductive."

TERRY ERWIN: "Not at all, no. Zero. I'm just a traveler in time, that's it." "But as a scientist you can't be an atheist and you can't be a believer because you can't test the hypothesis. So your only recourse is to be an agnostic. There is no other possibility if you're a real scientist."

DONALD FALK: [I asked about his personal motivation.] He has "a spiritual, or a deeply personal, belief that all that out there exists and has a right to exist and we are an integral part of it, no more and no less."

JERRY FRANKLIN: "Well, I consider myself to be spiritual, but not religious. And I don't even know what that means. [Laughs.] . . . Again, I don't know what it means other than I feel a connection with other life. I feel, you know, that all of us have an impact throughout the stream of time. And yet I fundamentally don't believe in life after death. So I can't do any more with it than that." [I asked him if this connection could be explained through science, or whether it was metaphysical.] "It's some of both. And it's probably a little more rational. That is, you can provide a strong rational basis for that, as much rational as it is emotional. But I *experience* it emotionally, even though you can explain a lot on the basis of rationality."

VICKIE FUNK: "I'm not religious at all. That doesn't mean I'm not spiritual. You know what I mean? It doesn't mean that I don't have an appreciation for the wonder of things, for the gee-whiz kind of approach to things." [And you would call that a spiritual thing?] "I would call it a spiritual thing because it's something I can't define as a scientist, or put my finger on. . . . [Arthur Cronquist] "used to say to me that he could tell right away whether somebody had this fire-in-the-belly kind of thing that you needed in order to be a good scientist. And it's this burning desire to figure things out, you know, and to—But with it comes I think a wonder that these things even exist, that this intricate nature of how everything is interconnected and how evolution actually works and speciation and, you know, all that kind of thing. And I think the *wonder* of all that and the desire to figure it out, and maybe because you don't believe in God and that somebody designed the whole thing this way, one then wants to figure out, well, okay, given that that's not an acceptable explanation, how did this all get to be the way it is? So I think there is a spiritual aspect, if you want to use that word."

HUGH ILTIS: [I asked him what he meant in his papers when he says that love of biodiversity has to be codified into organized religion.] "It has to be codified into a religious ethic of some sort. Meaning by religious ethic something we teach our young people; it has to be respected. But then you should be flexible nevertheless. And I don't mean by this belief in an almighty being or anything like that. No, no, that's not what I mean at all. But the point is, religion evolved as a way to transmit knowledge that is absolutely basic for human survival and that they didn't have any other better way of doing it than to say, 'You learn the ABC, you learn the Ten Commandments.'" [Would you describe yourself as a spiritual person in any way?] "In a way, sure." [To illustrate this, he read to me the beautiful passage he wrote many years before about visiting a remote region of the Great Plains, quoted above in the section on biophilia.] "And spirituality, in itself, is a way for the human body to reward the mammalian brain with a hormonal high in saying, 'You're a good fellow; here is a shot of hormone, make you feel good.'. . . to get a high out of seeing a beautiful flower, a beautiful fertile landscape, in a sense what the body is doing, the hormonal system, is telling the brain, 'You're doing the right thing.'. . . It's another adaptation. . . . Spirituality in itself is related to evolution, even though we may not be able to untangle all of it."

DANIEL JANZEN: "At least not in the traditional sense. . . . The trouble is, if you use the word *spiritual*, it gets a little fuzzy . . . [if] by *religious* or *spiritual*, you mean living by a set of precepts and being fairly consistent by living by them, in that sense I'm religious. But in terms of the mystique part, I have zero—I'm a total pragmatist. I think I understand what the world is, even if I don't know the details. Miracles, religious experiences to me are very pragmatic things that you can explain." "*Spiritual* can mean six different things to different people. You know, some guy is on a marijuana trip, and says, 'I feel spiritual.' Well, he does. And that's his definition of the word *spiritual*. I say, well, he's on a marijuana trip and his physiology has been affected. . . . It's a very straightforward physiological response."

K. C. KIM: "It's simply when you are in a spectacle of nature some-where. . . . Everybody has a complete fulfillment of satisfaction. That is what E. O. Wilson's book called *Biophilia* is all about. You know, sitting in the jungle, I think anyone will go through that. . . . music is basically trying to bring you emotionally, psychologically into that state. . . . And eventually you bring yourself into a state of fulfillment, a state of the spirits or whatever you want to call it."

THOMAS LOVEJOY: "I really don't know how to answer that. I don't know, I think it's pretty astounding and miraculous stuff. And I think, you know, there's a strong ethical dimension, which can be a religious dimension. I mean we're talking, if you want to talk in religious terms, there's the state of the creation. If you go scraping around, you'll find that I said that a long time ago, much to the horror of a lot of people." ["By 'creation,' do you mean. . . ?"] "Capital C." ["A God-based creation, not sort of. . . ?"] "Yes. Which you can either look at it that way or not." ["And which way do you personally look at it?"] "I'm not a religious person. I believe I'm an ethical person. But I'm certainly not part of any formal religion. I don't disrespect people who are. I mean, I really don't. And there's something really interesting about ourselves as a species. And I think it's inherent. . . . I think it's inherent in the societies that there is a seeking for a larger force. I mean, just start looking around the world. Whether it's Mayans or Aztecs or ancient Egyptians or whatever it is, it's there. I don't know what that means. . . . What I'm trying to say is that people always have value systems. It's like we are hungry for value systems. And how that quite relates to this universal search for a larger force, I don't know. But there is some kind of connection. And I think I probably was heading in the direction of saying that a large part of the solution here has to be an alteration of our values to the point where we *do* think about these things with the same kind of appreciation we have for works of art or our own works."

JANE LUBCHENCO: "Religious, no. Spiritual, perhaps, in the sense that I derive tranquility from nature. I find it aesthetically very pleasing. And there is some spirituality that's associated with that, but it's not in any particular religious context. I'm not—I don't believe in any god. I don't participate in any formal religion, or in any informal one."

S. J. MCNAUGHTON: "Science can only aid in an understanding of some sort of objective reality—that is, of what the world is and how it functions. That's all it can do. There are other realities that are equally valid, that are subjective realities that involve art, for example. . . . I used to think that natural history was of no value whatsoever. If you could not attach a number to it, . . . it wasn't even worth considering. But working in the Serengeti made me feel that there are things about this place that I can know that is subjective knowledge that I'll never—that I can't attach a number to it. It's—you know, see, here we are, this is where we've made this transition into talking about subjective reality, and it's something that I can't properly state because of what it is—So, is this a religious experience? Yeah, probably, I guess so. . . . I can evoke [certain feelings], just as I said, when somebody brings in grass species that smells in a certain way . . . it creates an emotional state in me that I would have to characterize, as best I could, as a spiritual experience. . . . This photograph of me in front of the tent [on a wet Serengeti plain], however, is a very experiential thing because of the way it happened and, you know, there were no wildebeests there at all. It was a hard afternoon rain. By about 6:30, I could hear the wildebeests way off in the distance; I knew they were going to be there. They came that night. The soil was all wet. I was cooking breakfast when the picture was taken. I sat there after breakfast, wildebeest all around, soil all wet, and I just had this feeling that like here is this huge, operating *thing* surrounding me. . . . You could feel the soil sort of *pulsating* with this life that had been activated by this rain, and here were all these wildebeest and the grass was growing, and so on. That's a spiritual experience to me. And there's no way that I will ever attach a number to that, but it's very real. There's no way I can ever tell anybody about it that makes sense." [After the interview was over, McNaughton noted that he was a Christian, but declined to discuss this any further.]

REED NOSS: "I actually don't. But it may be a semantic point. You know, I don't feel the need for any ritualistic type of experience with nature or with any other being in the cosmos, or whatever, in order to fulfill myself. . . . Whatever spirituality I might have is more probably of the Eastern kind, of a direct awareness, like a Zen or a Taoist type of experience of nature. . . . [Early Zen and Taoism] were simply a way of looking at things, a way of living and a way of looking at things. And in that sense, I do feel a kind of spiritual or at least a nonrational connection to nature. But I wouldn't call it religious, because . . . [that] usually means a prescribed set of standards, more standards and beliefs which I can't say I really have. . . . even though I certainly could not claim that my philosophy or whatever about nature is not informed by my scientific training—I'm sure it has been. . . . The very root of it is this direct spontaneous experience of connection, and joy, and just awe that I feel. So I think that's underneath it for probably most biologists that I know. And when I've talked with others, often in very different scientific training and other backgrounds, they still have mentioned that kind of direct experience as being very important. Of course, many won't admit it, particularly the scientists who worry more about their status as objective observers and so forth, probably wouldn't publicly admit that kind of feeling."

GORDON ORIANS: "I suppose for me, it goes back to a quasi-Schweizerian respect for life. The beauty of it, and the complexity, and the processes that created it. These are awesome to me." [Do you consider yourself to be religious or spiritual in any way?] "Not in the narrower definitions of the word. I consider myself to be an atheist. And I don't believe—So in that sense, is there some entity to which one could communicate and hence cause events to change—in that sense, I'm not at all religious." [And spiritual?] "If what I have described to you about my sense of awe and appreciation is spiritual, then yes."

DAVID PIMENTEL: "I believe that almost anything is possible in nature. Not in what religious people believe is being religious. I'm awed by nature, and I guess that's in part why I've always been interested in it is that my lack of understanding and desire to know is what has fascinated me by nature. And when you see and read about all this stuff in books about all these different types of organisms, what they do and how they earn their livelihood, it's fascinating."

PETER RAVEN: "As far as we know, we're the only living things in the universe and I think we have a responsibility based on that fact, that spiritual fact, to guard and keep biodiversity." [Do you consider yourself spiritual or religious in any way?] "I think I must be intrinsically but not formally." [Intrinsically, that is to say, with respect to your relationship to nature or to biodiversity?] "To the world. To other organisms, to other people. I consider myself part of a system, a temporary part of a system in which what I do will have some limited effect on what other people do. And I think that means that I'm spiritual in a descriptive sense."

G. CARLETON RAY: [When describing his experiences studying walruses as part of our discussion of transformation] "There's no question that being out on the ice alone with these animals and listening to them underneath the ice making all their sounds, you know you're never going to understand what they're doing. So forget trying to do the ultimate experiment. You're listening to sounds that may reverberate over three hundred square miles and all this stuff going on. You can *describe* what's happening. But try to figure it out exactly. . . . You do definitely come away different. There's no question in my mind. If you *don't* you've got to be a stone." [Do you think of yourself as being religious or spiritual in any way?] "Spiritual yes, but religious absolutely not. Religion is a corruption of theology."

WALTER ROSEN: [Do you have any feelings about biodiversity, or toward biodiversity or when you're in contact with biodiversity that you would call spiritual?] "Yeah. Yeah. I think all the time. . . . It's become a fixture in the forefront of my consciousness, the diversity of life. And I think I appreciate it more now than I ever did as a student of biology, even as a graduate student."

MICHAEL SOULÉ: [Do you feel that your spiritual beliefs inform your biology, or vice versa?] "I think there's a little bit of reciprocity. I don't have many spiritual beliefs, actually. It's just, I mean I was impressed when I was studying Buddhism that it just seemed to make a great deal of sense from what I knew about science and psychology, and evolution. . . . they weren't right about everything. I still don't believe in reincarnation, for example. But I could be wrong [laughs]." [Do you identify yourself now as a Buddhist?] "I don't know. It's a fair question. Then we'd have to define what a Buddhist is. But I have affinities towards a lot of Buddhist ideas and ways of seeing things. For example, coming back to the concept of intuition, one of the Zen Buddhist practices is . . . what's called koan training, where you're given puzzles, statements, questions, and you're supposed to answer them. Well, the answers cannot be cognitive. You try all of them first, and you finally learn that that's not ever going to get you anywhere. At least with your Zen master, it's not. So then you're forced to—Some people find it very easy because they're naturally intuitive and aren't trained to think as scientists. But for scientists it's very hard, I think, at first to approach problems from an intuitive perspective. So what I did, it helped me to trust my intuition."

E. O. WILSON: [Do you consider yourself to be religious or spiritual in any way?] "Intensely." [And could you explain what you mean by that?] "In the sense not of subscribing to any particular creation story. Not in the sense of accepting a patriarchal supernatural being, or matriarchal. Not in the sense of going through that exquisitely pleasurable experience of surrendering to the tribe known as religious conversion, or being born again, or giving yourself to Jesus, or whatever. Not in any of those senses. But in the sense of recognizing that at the core of it all is the set of deep, almost mystical motivations to rise above ordinary human experience. To find meaning in life that transcends individual mortality. And it is the burden and also the extraordinary opportunity of the humanist to try and travel that route in a way that honestly embraces what we can know with certainty about the human condition. In that sense of participating in that search, which is at once highly public in being based upon scientific, objective information, and highly personal, in being very reflective and emotional and sometimes nonverbal, that experience that I find exhilarating, and in that sense I'm quite spiritual. . . . Spirit for me is that part of combined emotional and intellectual experience which is most profound, exhilarating, and also mysterious. In other words, that part which we feel we can refer our rational thoughts to with the hope that there is always far more to existence than our personal lives and to our normal day-to-day pedestrian thoughts."

DAVID WOODRUFF: "I was brought up Christian. I was fully processed by one Christian church. My family then moved to another part of the world, and I found I had to be reprocessed by another Christian church to become a human being. And I began to question the merits of this system where you couldn't carry your credentials around with you. I became areligious as a teenager. I am very sympathetic to people who need religious support. I regard religion as—the major religions to me could be understood historically and psychologically. They have no, they play no major role in my personal life. Am I spiritual? That's different from being religious, okay? Is that what you're getting at? . . . . Spiritualism. I don't know the answer to that, because I'm not a very introspective person. But do I have, am I spiritual in that I recognized when the frog jumped out of my bucket as a five year old, that was an exciting moment that shaped my whole life? Yeah. If that's spiritualism, yes."

With the exception of Kim, McNaughton, and perhaps Ehrenfeld, these biologists reject organized Western religion, sometimes quite forcefully. Noss and Soulé affiliate with Eastern religions. What is odd is that, despite their resolute rejection of religion, so many of them seem willing, even eager, to call themselves "spiritual." Even if that means different things to different respondents, we still might expect more to follow Erwin's dictum that "real scientists" must be agnostic, as no one can conduct the critical experiments to prove or disprove the existence of God or any other metaphysical force. Biologists might be expected to eschew what seems to be science's antithesis.

Janzen, Eisner, and Iltis, while professing feelings they describe as "spiritual," have thoroughly mechanistic explanations to explain them. Both Eisner and Iltis attribute what we think of as spirituality to naturally selected rewards for proper biophilic responses. Iltis calls nature "sacred to humanity" and opined at the National Forum that the value of biodiversity must become codified into organized religion.[96] Like Wilson (as we shall see in Chapter 7), he recognizes that religion is an ineradicable evolutionary adaptation. Given its ubiquity, some conservation biologists hitch their wagon—the preservation of biodiversity—to this force. If you can't beat 'em, join 'em, or at least adopt their language and even their strategies for attracting converts. Their message: don't think about it, just feel it and worship it for your own survival.

But Iltis's poetic musings on his epiphanic visit to a remote corner of the Great Plains, for example, reveal a kind of joy, awe, and wonder, to which both he and others affix the label *spiritual*. Eisner, Franklin, Lubchenco, and Ray discuss this. Rosen stands in awe at the diversity of it all, and Pimentel stands in awe at what is possible in nature. The wonder of it all makes Funk feel spiritual, and this spiritual feeling also encompasses her drive to make sense of it. Franklin's spiritual feelings stem from a sense of connection with other forms of life, and Raven's spirituality stems from, and is part of, the interconnectedness of all people and things. He feels a sense of identification, the same kind of identification emphasized by Soulé and Noss.

Such feelings run deep, infusing their bearers with sentiment. At a loss for language adequate to express this sentiment, they resort to the word one resorts to when one can't explain something: *spiritual*. For these biologists and for many others, being in nature—surrounding oneself with biodiversity—can almost not help but bring about experiences to leave the senses reeling, the mouth agape. The incomprehensible complexity of it all: we can't handle it. Our brains go numb when faced with such richness out there, so much bigger than ourselves. How can we help but feel awed? And biologists spend their lives digging deeper into the intricacies, developing profound awarenesses of both the mindblowing intricacies they have unearthed and the complicated skein they haven't begun to untangle.

Even if they can reduce their spirituality to physiological terms, many biologists are still moved by beauty, their appreciation informed by their scientific understanding of biodiversity's intricacies. They find great and deep fulfillment from being in nature. They may identify with the other organisms on Earth and the processes that gave rise to them, seeing themselves as inextricably part of those processes, with concomitant responsibilities toward enabling them to continue. They accept it as axiomatic that this beauty and the diversity that comprises it should be nurtured and allowed to endure.

This biocentric spirituality closely parallels the deep ecology movement. Based on the work in the 1970s of the Norwegian philosopher Arne Naess, adherents of deep ecology believe that the Earth is sacred, that other species of organisms have as much right to exist as do humans, that nature is imbued with value. Biodiversity is their icon, and they take their sacrament in wilderness areas where biodiversity is found. They will often go to heroic lengths to put their convictions into practice, to protect what they consider sacred. The radical tactics of Earth First!—tree spiking, monkeywrenching—are inspired by the deep ecological philosophy.[97]

Perhaps because they deem these tactics undignified, perhaps because "ecology" has already been co-opted by an ideological movement, perhaps because they do not wish to be associated with a spiritual movement that they have not defined and led, most biologists I interviewed were loath to associate themselves with deep ecology. Actually, many (Eisner, Funk, Lovejoy, Kim, Erwin, Franklin, Pimentel, and Woodruff) had not heard of it, although Eisner would promote it if he thought it would work. Ray and Wilson were reluctant to reject the idea outright, while Orians, Janzen, and Brussard ("I think it's a bunch of crap, but that doesn't mean it's a bunch of crap for other people") showed no such reluctance. Iltis rejects it because deep ecologists are "shallow in their knowledge of evolution": they need to get into the *why* underlying the mysticism, and this why, of course, is biophilia. Ehrenfeld says, "God knows what deep ecology is," although he recognizes that, on the basis of *The Arrogance of Humanism*, he's "sort of a hero to some of the deep ecologists." McNaughton, when I described what it was all about, agreed with the philosophy, but wished they wouldn't call it ecology.

Not surprisingly, Ehrlich, Noss, and Soulé willingly associate themselves with deep ecology. We have already seen how Noss took precociously to monkeywrenching; it was a *natural* response when destruction was wrought on nature, which he saw as part of himself even as a child. Biologists, he has urged, are biophiles, and therefore must be warriors to preserve what they study and love, what, in fact, they *need*. Other biologists, he is sure, share his joy and awe and reverence, but do not speak out publicly because they feel they must preserve the boundaries between rational and intuitive, mind and body, science and emotion. Noss's science is interpenetrated by his spirituality, his spirituality is imbued with his science: by his own admission, the bounds blur and dissolve.

Soulé, too, rejects this kind of dualism. He adheres to deep-ecological views, especially the teachings of Arne Naess, because Naess advocates identification rather than dichotomization, and because "it enriches my life to have that level of awareness" that allows for value to inhere in nature. He believes other biologists reject deep ecology because: "These are concepts which have nothing to do with science as it's routinely practiced in the minds of most scientists in their laboratories. And they consider it to be subjective, emotional, mystical. But see that's the difference between a movement, like deep ecology, and a discipline, like conservation biology or most sciences. A movement requires emotion; a discipline, we *think*, in many cases, simply requires objectivity and knowledge." Soulé wants to blur the boundaries between the movement and the discipline. The values are there already: why not be honest, making conservation biologists' work more accurate and holistic? Simultaneously, they'd be laying their values bare for others to emulate.

Still, Soulé slips into the kind of dualism he claims to avoid. He talks about shifting between cognitive levels, of knowing when one is trying to be objective, of not committing completely to being a disciple of Zen or being a deep ecologist, of appreciating the rational parts of Zen training. He is conscious of the boundaries he violates, or perhaps he struggles inside to balance the worldview "required" of a scientist and that required of a Buddhist or deep ecologist. Perhaps he has not resolved this dissonance.

Like Soulé, Ehrlich is caught between worlds. While at some level he knows science can probably explain what he feels, science still can't explain everything. Although an atheist, he feels spiritual; he calls for a "quasi-religious" transformation of cultures because the word *religious* conveys something meaningful to people, something he feels too, but may choose other words to describe. Deep ecology, or something akin to it, is part of the solution:

> The main hope for changing humanity's present course may be less with politics, however, than in the development of a world view drawn partly from ecological principles—in the so-called deep ecology movement. . . . Most of its adherents favor a much less anthropocentric, more egalitarian world, with greater emphasis on empathy and less on scientific rationality.
>
> I am convinced that such a quasi-religious movement, one concerned with the need to change the values that now govern much of human activity, is essential to the persistence of our civilization.[98]

Ehrlich feels biologists must play a pivotal role in creating this feeling for organisms, in effecting the quasi-religious transformation epitomized by deep ecology, through

> "talking about it. Telling people how good it feels to get out in nature, how marvelous—you know, one of the things religion does for a lot of people apparently is gives them a sense of wonder. I don't think biologists, I mean my experience is I know almost no biologists, certainly not population or evolutionary biologists that are deeply religious. They exist, you can find them, but I think our sense of wonder tends to be taken care of quite nicely when you put a stingless bee under the microscope and it looks like it's carved out of solid gold. Or you know, you

look at the intricacy of how a cell works or how populations perform, or when I was a kid I was into butterflies and birds and so on, and I think it's sort of a sense of awe and wonder. I think we have a society that's really pulled back from the natural world."

Returning to the natural world, immersing themselves in bio-diversity, provides biologists with the awe, the wonder, the feelings of spirituality people require:

> "The best adjusted and happiest people that I know as a group are prob-ably field biologists. They live long, they almost all enjoy their work, they often work Saturdays and Sundays. Their idea of a great time is to go out and bury themselves in a tropical forest doing fieldwork and so on. And they all tend to share this kind of view [deep ecology], whether they express it or not, they really value it. I can't defend to you scientifi-cally or rigorously something which is just not scientific or rigorous. If it's our substitute for religion, maybe human beings really need—when you face the horror of your own longevity and so on—need something. And I think scientists as a discipline often just go for that."

Some biologists have found their own brand of religion, and it's based on biodiversity. The biologists portrayed here attach the label *spiritual* to deep, driving feelings they can't understand, but that give their lives meaning, impel their professional activities, and make them ardent conservationists. Getting to know biodiversity better takes the place of getting to know God better. By trying to get others to share biologists' quasi-religious experiences, Ehrlich hopes people will also share biologists' quasi-religious feelings about bio-diversity's value. Ehrlich claims that these feelings fall outside the realm of science when in fact they are a feature of science and the motivation that drives it. Biodiversity has spiritual value in the sa-crality attached to it and in the numinous experiences derived from it. In order to preserve what gives their lives deep meaning, biolo-gists strive to convert currently destructive others to a more nurtur-ing spirituality by preaching on behalf of biodiversity.

AESTHETIC VALUE

Writing around the time of the National Forum on BioDiver-sity, environmental ethicist Eugene Hargrove asserted that "since the aesthetic tradition linking the natural history scientist with the artist and poet no longer plays a significant role in the professional life of biologists and other environmental scientists in this century, value issues in conservation seem mysterious, if not obscure."[99] In advocacy on behalf of biodiversity, biologists contradict Hargrove's assertion; they continue and revitalize this tradition. Biodiversity has aesthetic and emotional value. Biologists find it beautiful, and this beauty moves them, sometimes profoundly.

I fear I draw some arbitrary distinctions throughout this value taxonomy where none may be called for. Those who say biodiversity has intrinsic value may also posit that biodiversity's beauty is inherent, and not in the eyes of the beholder. Part of Janzen's cultural-restoration argument holds that biodiversity provides aesthetic fulfillment for rural people. The beauty and intricacy of individual organisms, landscapes, and organic processes in large part create biodiversity's transformative influence. Those who promote biophilia take biodiversity's aesthetic value for us as a given, and add the twist that we respond this way because we are genetically programmed to do so: according to Orians, " 'beautiful' landscapes are probably highly functional ones in that they potentially provide rich combinations of resources for human existence."[100]

Biodiversity's spiritual value is similarly inseparable from its aesthetic value. When I asked Soulé to elaborate on his aesthetic appreciation of biodiversity, he replied: "What you're really asking is: what tickles my pleasure center? Or my spiritual center, although the word *spiritual* can be misinterpreted easily. That part of you that makes shivers run up and down your spine, or makes tears come to your eyes or whatever, however you want to define it physiologically or operationally." What gives you pleasure is beautiful. What makes you feel transcendent you find beautiful. The more beautiful you find it, the more it moves you spiritually, the more you appreciate it, the more beautiful you find it.

Again, not despite their science but because of their science, biologists feel compelled to comment on biodiversity's aesthetic value. They spend countless hours pondering it, engrossed by it. Their scientific understanding uniquely informs their aesthetic. They have a distinctive, and if it were to be spread, possibly deep aesthetic that leads them to even greater appreciation of biodiversity's value.

Like assertions of intrinsic value, the argument from aesthetics is not new. Leopold, Elton, and Carson also talked about the value to humankind of the natural world's beauty. In a 1982 State Department Conference on Biological Diversity, David Pimentel noted that "the natural biota are of great aesthetic value to society," and Archie Carr felt that aesthetics "in the long run . . . is perhaps the greatest motivation for public support."[101] Biodiversity's aesthetic value is frequently cited in current literature as one of the major reasons to preserve it.

This motive also emerged in my interviews. The loss of its beauty was "the main thing . . . that bothered me about vanishing nature," according to Eisner. "I think we should all be concerned about diminishing the aesthetic richness of life," Orians warns. Brussard says that "aesthetic considerations are big for me. And the aesthetics go probably beyond simply visual aesthetics, but also the aesthetics of listening to things, aesthetics of place." Woodruff frankly admits to me that he picks his study organisms "*because* of their aesthetic appeal"; one of his main motivations is to unlock the secrets of how evolution has worked to produce this beauty, so he may as well pick the most beautiful organisms he can find.

Paul and Anne Ehrlich delineate different kinds of aesthetic value. For example, whooping cranes have "conventional beauty." All organisms have what the Ehrlichs call "beauty of interest"— they live out their lives in fascinating ways, finely adapted to the most extreme conditions: each insect "dwarfs in interest and intricacy works like the Mona Lisa which are valued at tens of millions of dollars, yet humanity exterminates them without a qualm."[102] The more we know about these adaptations—the more we understand about evolution—the more beautiful we find organisms, the more we value them. Hence we see one way biologists can claim a more finely developed sense of aesthetics and therefore a privileged position to discuss such "unscientific" things as aesthetics.

Soulé stresses that "one's attitude about nature is influenced by what you know about it. . . . I think the more you know the more you appreciate . . . knowing about the history [of an area—e.g., whether it has been cleared or restored] actually enhances my appreciation; it adds another dimension, an orthogonal dimension to my appreciation, enriching my appreciation in another way." Franklin is *constantly* delighted by the way in which nature has enriched itself, and in the incredible ways in which it's evolved to get jobs done . . . so there's a case where the learning constantly enriches appreciation for how incredibly creative nature is." Erwin expresses a similar idea:

"And as you look at some of these smaller creatures that most people never see, you see all kinds of different symmetry, etc. And any of these things could have been created by our greatest artists. But they weren't; they were created by the evolutionary process of selection. And just exactly how that process came about is so complex. What drives the evolution along this lineage, along that lineage? What kind of complex environmental situation could actually produce something like this? That's the mindblowing thing about it. You see it and it's just, God, it's just *beautiful*, absolutely beautiful. How did it come about? The process behind it must be even more beautiful, more intricate, more complex, more sophisticated, whatever. And it's a challenge to the human mind to figure that out."

Evolutionary understanding informs Erwin's sense of biodiversity's aesthetics. And his aesthetic appreciation drives him to understand evolution further. The insect is intricate and beautiful; the process that gave rise to it is similarly intricate and beautiful. And the occasional elegance that can arise from the attempts to figure it all out adds another level of aesthetic value to biodiversity.

Funk's sense of wonder drives her science, too, and her scientific findings drive her sense of wonder, and make her sense of aesthetics more acute. But this has its drawbacks:

"Go to Hawaii. Everybody thinks it's beautiful. Go to Hawaii as a botanist, it's *horrible*. There's something like two native orchids in Hawaii. And they're terrestrial. And they're green. And there's now 20,000 cultivars of orchids on the Hawaiian islands. All the plants that everybody thinks are beautiful, the birds of paradise, the gingers: all that shit's introduced. I go there, I *cry*. I hate it. You know, I have to hike six hours to get to an endemic plant. . . . And Hawaii, I think, is the world of the future. . . . And it makes me, as a botanist, really upset."

One's biodiversity aesthetic can be tied up with one's knowledge of whether an area is "natural." Of course, perhaps hardly any landscape is natural—that is, untouched by human hands. "How we deal with and how we respond to 'artificial' landscapes depends on the definition of *artificial*, and more particularly, our knowledge about whether [a landscape] is artificial," Soulé says. One's attitude toward nature is influenced by what one knows about it. Soulé's predicted world of artificially diverse landscapes could have unanticipated effects on biodiversity's future aesthetic value.

And there is an aesthetic in diversity qua diversity, in the variety of stimuli diversity offers in a landscape. People are drawn to tropical rain forests, not only because we can glimpse an occasional spectacularly beautiful organism there, but also because we can see so many different kinds of beautiful organisms; and we can be overwhelmed by aesthetic appreciation of the terrifying complexity of it all. With astounding individual variety, however, comes chaos. Donald Worster warns that as ecologists have replaced a portrait of a balanced world at equilibrium with an ecology of chaos, the natural world grows harder to love. Daniel Botkin takes up this theme, noting, "Nature that is inherently risky may seem less beautiful than nature that is completely deterministic."[103] He suggests we must change our ideas and aesthetics of nature to match our newly required view of nature as constant flux. But this complexity, this variety, this indeterminacy is a key part of biodiversity's aesthetic value for some biologists; they find the infinitely complex infinitely beautiful and infinitely challenging.

Janzen chooses to work in the tropics, he told me, because "I have a deep-set emotional response obviously, which is very fond of that variety." Maximum aesthetic appreciation for him

"requires two things—it requires that the biodiversity be there and that you are exposed to new aspects of it, either by being able to look inside a pine needle and being able to see what's inside as well as outside, or by, I don't know, going to Australia and seeing what pine trees look like in Australia. It needs to be a new thing all the time. Now the thing about biodiversity is it's very complex. It's like a very, very, very complex body of literature or music: it's virtually impossible to get to the bottom of it. You can't. . . . A tropical habitat is complex enough so that—I've been working in Santa Rosa [Costa Rica] for twenty years, pushing twenty years—and I can walk down the same trail I've worked down *thousands* of times and there's something new every single time. So in that sense, it's a perpetual renewing machine."

Novelty and variety renew Janzen's intellectual and aesthetic appreciation for biodiversity. For, according to Orians, "familiarity breeds aesthetic contempt."[104] Orians explains this via biophilia theory: our nervous system receives so much stimulus, we heed only the novel, or else we would go insane. By extension, endangered species become more aesthetically beautiful to us precisely because they are endangered. Like his aesthetic appreciation, Orians's appreciation of aesthetics is inextricable from his scientific identity.

McNaughton, too, cherishes both biodiversity's conventional beauty and the information content stored within. His aesthetics derives from

"just the desire to have different experience, and part of having different experience is seeing rhinos that looked different when there were rhinos, or seeing grasses of the same species that grow differently, that are morphologically or genetically different, seeing trees that are different. There's a satisfaction, I think, to new experiences, and part of those new experiences, particularly if you're trained to recognize it, is the differences between individual biological organisms. Individual trees are as different as individual human beings, if you're trained to recognize them."

Difference, variety, complexity, heterogeneity, intricacy of individual organisms, organismal interactions, ecological and evolutionary processes: from these spring the enormous aesthetic value biologists derive from biodiversity. This strong aesthetic-spiritual appeal, then, is a powerful motivator of biologists' conservation activities. Funk's conservation spirit was kindled when in Ecuador she discovered that "they're making charcoal out of these beautiful rain forests or cloud forest trees. Old, old forests that are just gorgeous, and they're being cut down for charcoal. I mean, for nothing. I mean, 'Ship them some briquettes!' you keep thinking." Noss's "strongest feelings about nature are still just that direct joy. I mean, I guess it's an aesthetic appreciation where I'm literally just brought sometimes to tears just by looking at a piece of moss or some other thing in nature . . . when alone is when I get that strong, almost overwhelming sense of joy. . . . And again that's what motivated the defensive kind of actions that I've felt towards the woods I played in as a kid, and which now motivates me, has motivated me to become a conservationist."

CHOOSING VALUES

The skeptical analyst—one who, perhaps, does not share these feelings—may find some of the above remarks dubious. Professions of emotion and spirituality can be read either as revealing personal comments or as tools for getting what the speaker wants, be that conservation of biodiversity or even personal gain. And it cannot be denied that biologists' advocacy on behalf of biodiversity—and advocacy of their role to speak for its values—has reaped financial rewards. Self-sought publicity has padded the research coffers, and publishing royalties and speaking fees yield considerable lucre. Then there is the power that comes with speaking for nature—a rather formidable constituency to represent.

Still, when reading these biologists' words or when talking with them, it would take a hardened cynic to feel that cash or power were their overriding motivations. Jared Diamond writes that "it is almost unbearably painful to have to watch the destruction of biological communities that one knows and loves."[105] McNaughton proclaims that biodiversity is "the glory of existence, it's the beauty and aesthetics of the world we live in. And without that, life is impoverished." Watching biodiversity, particularly large vertebrates, "not only enriches our life in terms of beauty, in terms of thrill, but also in terms of insights into ourselves. . . . We learn, we emote, we're inspired, all of these things by these critters," declares Franklin. For Erwin, "every time I see a beetle it blows my mind. . . . It's a high." According to Iltis, "Every time you see biodiversity, every time you see these wonderful things, it makes life worthwhile. It's a wonderful thing, the beauty of it all." Ray feels "that biodiversity is important emotionally to us. . . . It's emotional to the point of loving. . . . You're right down close to your core as you're able to recognize. . . . All these things are marvelous."

To save what Dan Janzen calls "the very things that give meaning to our lives," biologists seek to make those things more meaningful to everyone else's lives.[106] And so they attribute so many values to biodiversity. Given so little time to reverse our destructive course and so small a public attention span: on which values should promotion of biodiversity conservation ride? Are some values more important than others? Are some more morally correct to use? Are some more effective than others?

Let's start to answer this by looking at what's important to biologists. In a preliminary project in Costa Rica in 1990, I asked twenty-four U.S. and Costa Rican life scientists, ranging from graduate students to long-established researchers, why society should care about biodiversity conservation, and why they personally cared about it. Human economic and health needs were the most frequently expressed reasons why society should care, but they ranked far down the list of reasons why the biologists themselves cared. They were most motivated by biodiversity's aesthetic value for their lives, followed by a keen sense of ethical responsibility; this category included those who said biodiversity should be conserved because of its intrinsic value. Ecosystem arguments followed closely behind.

I attempted something similar in my interviews for this book. The results do not lend themselves to such easy tabulation.

WHY CARE ABOUT BIODIVERSITY?

I asked the biologists I interviewed: *A.* Why should members of society be concerned about biodiversity conservation? *B.* Why are you personally concerned? Some of their answers were as follows:

PETER BRUSSARD: *A.* Ecosystem services. *B.* Aesthetic reasons.

DAVID EHRENFELD: I didn't get to ask. His writings make it clear that both he and everyone else should appreciate biodiversity's intrinsic value, and that we have an ethical responsibility to do so.

PAUL EHRLICH: *A.* In his writings, he has consistently stressed the four reasons why society should care: economic, ethical, aesthetic, and ecosystem. *B*. He is personally motivated by "ethical obligation" and "the ecosystem services argument."

THOMAS EISNER: *A*. Economics (as in chemical prospecting) and aesthetics. *B*. Aesthetics.

TERRY ERWIN: *A*. In Isaac Asimov's *Foundation* series, the planet "Trantor was this glob made out of cement and I beams. At one little end of it, which was something like the southern tip of New Zealand, there was a garden. . . . I would hate to see my species be the one that drove this planet to become Trantor." *B*. "Every time I see a beetle it blows my mind. And working on these tropical species, I just get personally *excited* when I see the tremendous variety that's living with us on the planet."

DONALD FALK: *A*. Economic benefits; hubris—that is, "By the time that we start thinking that we're the only significant life form on the planet, we've guaranteed our own downfall." *B*. "There is a survival issue at stake here. . . . My personal choice just arises out of the conviction that as soon as we allow ourselves to think of human society as in any way, in any respect separated from the biosphere, we're doomed."

JERRY FRANKLIN: *A*. "Biological diversity is the basis for sustainability. And it seems obvious that we could go about a process of depleting the organismal base of this planet to the point where it simply doesn't work as well anymore." *B*. "What personally motivates me is fundamentally the love of life, and I don't want to see us responsible for the destruction of life forms, for the loss of processes. So it's a very ethical issue with me. . . . For me, it's the organisms themselves and our responsibility to them."

VICKIE FUNK: *A*. "There's a lot of standard, pat answers that come out of the books all the time: oh, we eat these things, they may have medicine. And I can parrot those, but I'm sure you've heard them from everybody else. . . . Those are compelling, but they're not the central core." *B*. "I think this desire to know what we don't know is at the base of everything. And I think that all these other things, for scientists, are just reasons that we think up that are *true*, but nevertheless they're just things that we can put out in front of the general public as why we do what we do." Her conservation motivation is to save the things that give her wonder; it comes from seeing her old research sites razed and realizing she has to be "a better world citizen."

HUGH ILTIS: *A*. "We are living on this Earth, only this Earth, no other

Earth. There won't be any other one. This is what we've got." We rely on it for physical and emotional sustenance: ecosystem services + biophilic need = powerful rationale. *B.* "Look: I love it! . . . Every time you see biodiversity, every time you see these wonderful things, it makes life worthwhile. It's a wonderful thing, the beauty of it all."

DANIEL JANZEN: *A.* "It and its variance, its variation, is a major piece of the stimulation that is required to set off major chunks of brain function." *B.* "I have a deep-set emotional response, obviously, which is very fond of that variety."

K. C. KIM: *A.* Resources, ecosystem functions, ethical obligation, biodiversity destruction as "symbolism of the fallacies of the technological society." *B.* He did not specify which are most important for him.

THOMAS LOVEJOY: *A.* Indicators of environmental change, a library for resources (both chemicals from nature and also natural examples of inspiring new ideas), ecosystem services. *B.* He is driven by how bad the situation is. We did not get more specific than that.

JANE LUBCHENCO: *A* and *B.* Ecosystem services.

S. J. MCNAUGHTON: Pragmatic reasons aside, the most important thing "is because it's the glory of existence, it's the beauty and aesthetics of the world that we live in. And without that, life is impoverished. By that, I mean, intellectual life, aesthetic life, the whole content of life is impoverished in the absence of biodiversity."

REED NOSS: *A.* Recognizes the economic arguments, although finds them problematic. *B.* Biophilia and intrinsic value.

GORDON ORIANS: Does not distinguish between personal and societal. Orians spoke of providing future generations with the "capacity for a rich and rewarding life" and respect for conserving the process of evolution. Moreover, "I tremendously enjoy living organisms, and interacting with them." He also mentioned ecosystem services, food, and medicines.

DAVID PIMENTEL: *A* and *B.*: Ecosystem services.

PETER RAVEN: *A.* Moral responsibility, sources of sustainable productivity, the basis for human welfare. *B.* "The thing that personally I find most compelling is the need to preserve, the need to understand and preserve biodiversity in order to be able to make the world as good as it can be when and if we do reach that kind of relative stability later on."

G. CARLETON RAY: Emotional, pragmatic, aesthetic. Ray said that he did not distinguish his own concerns from those of society.

WALTER ROSEN: Rosen said he wanted the human support system uncompromised and human aesthetic enjoyment undiminished.

MICHAEL SOULÉ: A. "Practically speaking, the most important thing about nature is it provides us with oxygen and calories and shelter. And we evolved—we're part of it—we evolved in it and are of it and can't be separated from it." B. Love of nature.

E. O. WILSON: A. Irreversible loss for people of both utilitarian and biophilic benefits. B. "I'm congenitally attracted to biodiversity."

DAVID WOODRUFF: A. "It affects [people's] health and their financial well-being." B. Woodruff said he had loved natural history since childhood and felt that working for conservation was a way not to have lived life in vain.

These complex responses are not easily classifiable. Suffice it to say that the economic bounty of biodiversity so often touted by them does not compel them personally. Rather, they are moved by biodiversity's beauty, its power to keep Earth systems running and human civilization functioning, and the ethical responsibility to assure these functions in perpetuity.

In his search for "unity among environmentalists," Bryan Norton argues that the dichotomous worldviews of "moralists" and "economic aggregators" are ideals that seldom do or should come into play in actual disputes. He declares that "moralists, as conscientious objectors, shut themselves out of the decision process."[107] Norton believes that multiple rhetorical strategies can and should be used when arguing for environmental goals, as these strategies are effective and reflect multiple principles and ideals held by those who argue them. While this may be true in some cases, I believe that often rationales—particularly those of an economic, pragmatic stripe—are being pitched to different audiences even by those who may not believe their own words. Norton also focuses too much on winning local eco-skirmishes, whereas how we ultimately conceive of the Earth will dictate how the Earth is ultimately treated. Furthermore, the moral position has carried and continues to carry great ideological and political weight in environmental battles. The staunch moral stand that refuses to yield all ground to economic expediency has been responsible for many conservation victories in the United States.

"Iatrogenic" illnesses are those caused by medical treatment. Some biologists' conservation prescriptions may induce their own iatrogenic side effects. Arguments for biodiversity's economic value carry perils, some of which I have already enumerated: much biodiversity has no utilitarian value; biodiversity as an economic crop may not outperform other uses of the land; attempts to put monetary value on biodiversity may result in "crackpot rigor," where economic models with poor performance records in more conventional situations are then seized upon to put a value on life;[108] where biodiversity proves to be economically valuable, it may be worse exploited than when it was seen to have no economic value at all; we have every reason to believe that biodiversity's exploitation will be at the expense of, rather than in the service of, local people whose livelihoods have depended or could depend on it; and "sustainable development," or the use of biodiversity to better people's lives and generate income from biodiversity, no matter how benign, still usually results in biodiversity's degradation.

Furthermore, Ehrenfeld argues that heavy reliance on economic justifications merely legitimizes the same economically rich but morally bankrupt system that is destroying biodiversity in the first place. We live in a society that values things and makes decisions based chiefly on economic worth. Beauty, ethics, responsibility, community. We seldom trade in these currencies when making decisions as nations or as smaller political or corporate entities. Many of us have lost our individual ethical compasses or are driven by desperation to make personal decisions based on how the outcomes will impinge upon our incomes.

I am not talking here about those living on the edge, who make pasture from nature because they must to meet their most basic needs. I am talking about the decisions of the more wealthy, decisions that add to our economic wealth at the cost of our biotic wealth. This world is dominated by a system that grants multinational corporations virtual carte blanche in the name of "free market economics." We do not question the morality or justice of an economic system that allows so few to profit while leaving so many hungry or desperate, that garners so much from resources the poor desperately need, that agglomerates economic wealth while incurring biotic and moral bankruptcy. Deriving economic wealth from biodiversity fuels, rather than curbs global greed.

The biodiversity-as-cure-for-cancer argument circumvents the basic problems: avarice, overconsumption, lack of public control of corporate activities, ignorance aggressively promoted by the powers that be and aggressively acquiesced to by citizens who are too lazy or too stunned to think. Touting biodiversity as an economic and medical panacea misses more radical solutions—political, social, and personal solutions that might get at the roots of these problems. Biologists feed into the very system that is destroying biodiversity by harnessing the forces of international business, labeling biodiversity another "resource," while leaving buried the causes of its destruction.

These economic arguments allow the hubristic, violent way we treat the Earth to continue unexamined and unabated. We sometimes profess outrage at the ravages of war or at the carnage of single acts of killing; we shake our heads in sadness and horror at such lunacy. Yet we seem to feel no such compunction when we plunder and pillage with hideous destructive fury all that the Earth has invented in its riot of creation. Economic valuations of biodiversity perpetuate the arrogant ideas that the Earth exists for our plundering, and that we have no corresponding moral responsibility to steward the amazing gift we have been proffered, but of which we seem so undeserving, so ungrateful.

For biologists making claims about biodiversity's economic value, reputations and credibility are at stake. As Langdon Winner expresses it, utilitarian arguments "create a disingenuous tone in environmental advocacy. It is as if those who had come to worship at the temple had decided to change a little money on the side."[109] Economic arguments may be a smokescreen providing cover for the real attack.

Several of the biologists I interviewed were uneasy at expressing this argument, which did not really move them personally. Eisner, a zealous promoter of chemical prospecting, said: "I sometimes ask myself, 'Why do I feel the way I do and do what I do and become politically involved?' I mean, my finding that emphasizing the commercial value of nature is such a timely argument, I have to say in the same breath that I would much rather be addressing the intrinsic rights of organisms, because I feel more comfortable with that argument than the market-economics argument."

Soulé has proclaimed: "I think we will have reached cultural adolescence when we can admit in public that conservation is not only for people, something most of us admit already in private."[110] If economic concerns are not their primary motivation for valuing biodiversity—and they do not seem to be for any of the biologists profiled here—then it is dangerous, disingenuous, and dishonest to pitch this argument too feverishly.

As I have pointed out, biologists know that different people will be swayed by different arguments; the gravity of the situation is such that biologists feel the need to do what they can where and how they can. But overreliance on economic rationales may underestimate the human capacity for complex understanding of problems with concomitant ethical action for a shared goal. So many arguments have an implicit, simplistic "tragedy of the commons" rationale, viewing humans as materialistic automatons who make decisions based only on self or immediate family and nothing else—as animals acting only to maximize their evolutionary fitness.[111] While not ignoring the harsh realities of the search for subsistence or the pursuit of greed, this argument still bears the whiff of hubris. Are poorer or less literate people really lacking deeper affiliative needs, really not able to understand or generate arguments that ask for sacrifices to the community, really not willing to appreciate and preserve beauty around them?[112]

Armed with cash and political clout, how often do we intervene, not for the health and well-being of people here and abroad, but sometimes to their detriment as we seek to realize fulfillment of our conservation values without being completely forthright about our motives? It is only fair to be candid about the values so often obscured in our defense of biodiversity. I'm not suggesting that Northern biologists cram their values down everyone else's throats. But certainly nonpragmatic reasons for appreciating nature rest not solely within the intellectual and emotional ambit of biologists, of Northerners. Much as local or indigenous knowledge of practical uses of biodiversity has been ignored in the past, so have local aesthetic and ethical considerations of the natural world. Surely all parties can learn from explicit discussion of biodiversity's values as long as the dialogue is reciprocal among peers.

The point is to acknowledge and share *all* the reasons why we devote ourselves to biodiversity, rather than condescending with simplistic rationales that mask our true feelings. Not only does everyone stand to learn quite a bit, but this exchange might curtail Northern arrogance and Southern resentment. I bet we'd find that people in Southern nations would love to preserve biodiversity if they had the economic margin to do so. We might even learn about different ways of seeing the world, ways that could abet scientific and nonscientific enterprises. This in turn might lead us to redirect some of our conservation efforts so that we attempt to mitigate the adverse socioeconomic conditions at the root of biodiversity loss. And it might turn our attention inward to our own society, to attempt to rectify a life way where people rampantly act not out of desperation but out of greed.

If the values motivating our drives to preserve biodiversity were on the table, we'd move beyond attempts to lure wealthy eco-tourists whose trickle-down cash is seen as an incentive for neighbors of biodiversity playgrounds to keep their hands off them. Rather, we'd make it a high priority to found programs to help any interested citizen spend time exploring his or her own national biodiversity treasures. You cannot love what you do not know. Enduring conservation efforts require that people in whose backyards biodiversity is maintained value biodiversity as much as do the scientists studying it and working to save it. A sense of stewardship for the Earth, empathy for nonhuman forms of life, a basic ecological literacy, and a reverence for the biotic world do not necessarily follow from chemical prospecting, ecotourism receipts, or sustainable development crusades.

I do not mean to denigrate these efforts, and neither biologists nor anyone else can expect to overthrow the dominant global value paradigm overnight. And for those who desperately require new sources of wealth, it is almost immoral *not* to put biodiversity to work for them. But biologists are driven by a set of values deeper than those they often profess, values that spring from their professional activities as ineluctably as do facts. These values are necessary for a lasting conservation ethic. As such, it is a matter of honesty, fairness, and prudence for biologists to accompany any advocacy of biodiversity's economic value with explications of the values they really care about.

Ehrenfeld writes: "Non-humanistic arguments will carry full and deserved weight only after prevailing cultural attitudes have changed."[113] Many conservation biologists are trying to change these cultural attitudes. Perhaps we are entering an era when such personal and cultural change can be effected; perhaps we have even been there for a while and have been pitching conservation the wrong way.

Ehrenfeld believes that both our overweaning faith in reason (epitomized by the twin gods of economics and science) and our scoffing at emotion, at nonrational ways of knowing, are killing us and biodiversity. Biologists' pronouncements carry great weight because of the cognitive authority accorded scientists, but when making nonhumanistic arguments for why we should preserve biodiversity, these high priests of humanism are apostates preaching a different gospel: if *they* don't subscribe to humanistic arguments for preserving biodiversity, why should anyone else?

Many biologists making public pronouncements on biodiversity's values try to cope with the contradiction of publicly speaking for the antithesis of what drives them personally. Some wax eloquent over the situation, and at least one, Peter Raven, may have derived a solution that is both pragmatic and morally acceptable:

> "To me, social justice, the fact that the great majority of the people in the world, 77 percent now in developing countries, growing very rapidly, have access to 10 to 15 percent of the world's goods, 80 percent of the world's biodiversity, 6 percent of the world's scientists: if you put that together and roll it around, then it comes back to me that peace, social justice, human order, the protection of biodiversity, the production of, or promotion of, a stable biosphere are all inextricably interwoven. And since I see that all as a unity, I can present that any way that I want for a

particular audience. . . . To me, the whole point is Planet Earth exists as an individual item in space which functions by virtue of its biodiversity and the interactions between the living organisms and the inorganic world here. And if we understand that, then all of the other reasons become valid and can be stressed according to the audience, according to the listener."

Raven refers to Bill McKibben's arguments about the "end of nature."[114] Not merely via direct destruction, but also through global warming and the spread of poisons, humans have sullied every part of the Earth: nothing exists as nature truly apart from us. Therefore Raven thinks "it's spiritually true but almost pragmatically disingenuous to try to divide the world into used and unused, and to try to separate human beings from the rest of creation." Even wilderness, so long rhapsodized by poets and environmentalists, paradoxically persists only as a human creation, managed by us where and how we want it. A callous disregard for the Earth has rendered "corporate ethics" an oxymoron; this, in turn, has created the need for other oxymorons like "wilderness management," "artificial nature," and "sustainable development."

At the same time, Raven thinks that "the only hope for the production of a stable world is indeed spirituality or a realization that there are deeper values. I can't imagine the world becoming stable or biodiversity becoming preserved unless individuals take individual personal responsibility for that." As an ideal, Ehrenfeld's philosophy is fine; Raven is there already, and wants others to join him; "But I don't think I'm going to get there just by saying that people are not engaged with, do not exploit, do not use organisms, or that organisms are not material substances for them that are important to them and their welfare."

Raven has a truly ecological worldview, a holistic perspective often extolled by observers of the science. Donald Worster calls "the ecological ethic of interdependence" a "moral truth" generated by ecological science, part of the marriage between ecology and ethics Worster would like to see.[115] Unfortunately, biologists seldom see things from that perspective. Yet biodiversity embodies and generates moral principles, and Raven adheres to those principles and seeks to spread them. His science has imbued him with a love of the natural world and a deep sense of the interconnection of all things, including all of humanity. If everything really is tied to everything else, Raven thinks, and if he is part of that global whole rather than an "objective observer" standing above it, then he must defend that of which he is a part by whatever means necessary. That defense will comprise all conceivable arguments to preserve a multifaceted symbiosis, as long as these arguments are true and fair. If that means that he applies the salve of economics while awaiting spiritual salvation, perhaps that's life.

COSTA RICA'S NATIONAL INSTITUTE OF BIODIVERSITY (INBio): *BIODIVERSIDAD CENTRAL*

 6

Just north of San José, down an unpaved road, past a paint factory and flanked by coffee fields, the folks who run Costa Rica's Instituto Nacional de Biodiversidad (INBio) are plotting a small revolution. This is not widely known. The taxi driver taking me there for the first time lost some income as he circled the neighborhood asking for the "Instituto Universidad de Algo" (University Institute of Something). People who live on the same road, a fraction of a kilometer away, looked perplexed even when I helpfully interjected INBio's correct name. Yet closer than you could throw a coffee bean, a startling experiment is being conducted where society, nature, and science are being transformed in the name of biodiversity—or, to be more precise, *biodiversidad.*

At the head of the revolutionary army are Dan Janzen, a University of Pennsylvania ecologist and conservation biologist, and Rodrigo

Gámez, a virologist and former director of the Center for Cellular and Molecular Biology at the University of Costa Rica. They are putting a coterie of biologists and others through the paces of re-making society and nature in Costa Rica. *Biodiversity Prospecting*, published in 1993 by the World Resources Institute, is the blueprint for their model (they prefer to call it a "pilot project"),[1] which they hope to export to the rest of the world.

Costa Rica is a reasonable place to develop this pilot project. It gained independence early, and for more than a century, it has played host to foreign biologists, who tend to fall in love with the country and contribute to its ecological collections and reputation. It is a genuine and steadfast democracy and has had no standing army since the late 1940s. Life-expectancy rates and health mea-sures compare to those in Northern nations, and the literacy rate verges on 100 percent.[2]

It is also a small country—about the size of West Virginia—whose dramatic topography gives rise to spectacular habitat diver-sity. In this fine-grained eco-scape dwell perhaps 5 percent of the Earth's species. Biologists and journalists lionize Costa Rica as the Canaan for biodiversity and those who cherish it. Partially because of this portrayal, ecological miracles do sometimes occur there. Sev-eral decades ago, biologists started touting Costa Rica as an ecolo-gist's utopia, and the cash started pouring in. Conservation foun-dations began putting their money where biologists' mouths were; eco-tourists followed suit. A cycle began that shows no signs of stop-ping: money fulfills ecological dreams; eco-paradise found prompts more money. Janzen is speaking literally when, referring to his con-servation work there, he notes that "Costa Rica is a very easy play-ground for this kind of thing."

But there's trouble in paradise: much of Costa Rica is, in fact, an eco-disaster.[3] While official conservation areas cover about a quar-ter of the nation, some of these areas are degraded, and much of what lies between them is a bleak mess from biodiversity's point of view. In the early 1990s, Costa Rica had the highest deforestation rate in the world.[4] International corporations replace huge tracts of rain forest with, among other things, vast monocultures of bananas for U.S. breakfast bowls or single-species plantations of *Gmelina* trees to supply U.S. toilet paper dispensers. Struggling farm families burn and clear other sections of forest to support a few cows and provide subsistence for themselves. At the insistence of USAID and other international aid agencies, Costa Rica's Agricultura de Cambio (Agriculture of Change) program has diversified agriculture; but adding macadamia nuts and bird of paradise flowers to the export roster nevertheless requires widespread land clearing and massive biocide applications. Furthermore, rising standards of living match growth of population, and the nation carries a heavy load of international debt.

Nonetheless, Costa Rica's reputation continues to grow. Tourism doubled between 1989 and 1992, when it pulled in U.S.$421 million to the Costa Rican economy, probably more than coffee or bananas, Costa Rica's previous export champs.[5] Favorable word of mouth and aggressive promotion have prompted international donors to back ideas that might be unthinkable elsewhere, like INBio.

INBio was created to further Costa Rica's ecological reputation and to help battle continued ecological degradation.[6] In 1986, the newly elected administration of President Oscar Arias promulgated a proactive agenda on environmental issues, fomented by the visionaries at the newly created MIRENEM, the Ministry of Natural Resources, Energy, and Mines.[7] Arias sought out his friend Rodrigo Gámez to help the government fashion biodiversity policy, and, in the latter's own words, "as a result of this peculiar political-biological interaction," Gámez created MIRENEM's Biodiversity office and became its director. The office's self-defined mission was to formulate a new and progressive management direction for Costa Rican natural areas. Gámez set about expanding and making the protected areas more secure, while linking their future to the socioeconomic well-being of all Costa Ricans.[8]

In late 1988, Gámez assembled a meeting of Costa Rica's conservation brokers to plot the next step. From all accounts, INBio was an idea whose time had simply come.[9] That is to say, a group of people were of like minds about putting biodiversity to work for society, society to work for biodiversity, scientists to work for both, and both to work for scientists. Janzen informed those assembled of his success in coaxing international donors to fund his dry-forest restoration efforts in Guanacaste National Park; he was convinced that an INBio-like institution could tap into this wealth. The founders opted to form a nongovernmental institute, thus freeing it from government guidelines and fund-raising restrictions. From these and other meetings, INBio (the name was invented by Gámez) was established on 24 October 1989. It was incorporated with the government's blessings, and its charter mandates that it abet Costa Rican society and biodiversity.

Gámez helped convince President Arias of the project's value. According to Janzen, "It wasn't that he [Arias] came at it with a personal knowledge or understanding of biodiversity. What he saw was that, just like good water or good roads or good schools, biodiversity is a sector of society; 'so I'll put it in my platform,' to put it bluntly. And a good politician does that." Arias inaugurated INBio's campus, and he was also there to hand out the diplomas to the first graduating class of parataxonomists (whose pivotal role is outlined below). By 1993, INBio had sixty-six full-time employees at its headquarters, plus forty-one parataxonomists in the field. Its directors are trying to raise tens of millions of dollars, and they have been successful thus far in attracting substantial foreign government, corporate, NGO, and private support; a plaque at INBio lists among its financial contributors the John D. and Catherine T. MacArthur Foundation, the San Francisco Zoological Society, the National Science Foundation, and James Taylor.

Janzen is not quite the Rasputin behind the throne at INBio, as various detractors portray him. Still, his stamp is irradicable. His language and his ideas infiltrate the recesses of INBio. He designs and teaches the parataxonomist courses. He confesses to being INBio's chief cheerleader and fund-raiser—its link to the outside world, upon which INBio depends in its inchoate form for its operating budget. The parataxonomists love and respect him. When you are in the building with him, you know it: he is a presence. He has also kept himself somewhat under wraps in Costa Rica, content at INBio to work out of the public eye. Paul Hanson, a biology professor at the University of Costa Rica, told me that most Costa Ricans, including newspaper reporters covering INBio, have never heard of Dan Janzen.[10] Janzen himself admitted that when playing his "cheerleader or coach" role in Costa Rica, "I do things specifically so that Costa Ricans themselves don't feel it."

Elsewhere in this book, I discuss the philosophical and real-world dangers that David Ehrenfeld and others believe occur when biodiversity is made into just another market commodity.[11] At INBio, their worst fears come true: this is a "pilot project" for the starkest commodification of biodiversity yet seen. The founders of INBio assert that if you do not make biodiversity a commodity, if it is not competitive with other commodities, if it does not outperform other uses of land, it will be lost. To paraphrase Janzen, biodiversity is the product line, Costa Rica Inc. is the corporation, its citizens are the stockholders, and Rodrigo Gámez is the CEO.[12]

Gámez calls biodiversity prospecting "another type of agriculture, a very sophisticated type of agriculture." He supervises cultivation of Costa Rica's vast genetic terrain so that it produces a bumper crop. For centuries, Northern colonialists—explorers, miners, merchants, mercenaries, scientists—have plundered the South's biotic resources and riches at will. The example of vincristine and vinblastine is instructive: from these compounds of Madagascar's rosy periwinkle, Eli Lilly & Company developed drugs that fight lymphatic cancer and childhood leukemia. Lilly has garnered hundreds of millions of dollars from these plants; Madagascar has received nary a cent.[13] Gámez aims to ensure that Costa Rica not be Madagascar, that everything living in or off Costa Rican soil be regarded as a resource that must be paid for if it is to be used.

The philosophy underlying INBio's attitude toward biodiversity is summed up in its oft-used mantra: Save it, know it, use it. The more it is used, the more it will be saved, making the mantra cyclical. Currently, Costa Ricans attempt to save biodiversity in conservation areas that comprise a quarter of their country's land. The integrity of many of these areas is threatened, and the magnitude of these threats may increase if the contents of these areas do not turn a profit.

To know biodiversity, INBio has embarked on a novel and controversial program. It has trained more than forty parataxonomists—

former bartenders, housewives, preachers, poachers, park guards—and seeks eventually to train over one hundred to hunt, collect, sort, and mount invertebrates and plants. The idea is to build on the previous century of "superficial collecting sprees" to build a computer-documented inventory of each of Costa Rica's estimated half million species.[14] Much of the space and personnel at INBio headquarters is devoted to sorting, identifying, and computer-coding these species.

Janzen and others beg expert biologists from around the world to visit INBio to help identify the heaps of specimens carted in by the parataxonomists. As we saw in Chapter 4, he also urges (in no uncertain terms) that these biologists dedicate themselves to biodiversity. Janzen is at the forefront of conservation biologists who would change how science is practiced.[15] It is time academia recognized conservation efforts on biologists' CVs; "it is time to seek a new kind of fine-tuned and responsive contract between tropical academic science, local human problems, and local environmental processes," Janzen says. He pulls no punches in his displeasure at some of his colleagues' behavior: "It is appalling to hear an academic tropical biologist state flatly that he does not know anything about sociology and therefore will only consider biological factors. It is as though a doctor were to state that seat belts are an engineering problem unrelated to medicine." If the "tropical academic scientist studying biodiversity" is to move toward Janzen's wish that she become "the source of life-giving economic and social support for the conservation of tropical wildland biodiversity," then science will have to change.[16] INBio is the experimental locus where new biologists ply their craft.

Not that biology and its practitioners are the only key to tropical biodiversity's future; but they are vital:

> Scientific research, and especially academic scientific research, is only a small—even if critically important—portion of the array of forces, processes and institutions that demand attention and fine-tuning if there is to be sustainable use and development of wildland biodiversity. . . . If the questions and the researchers are isolated from society, the results generally only very slowly find their way into society. . . . I urge the

reader to not confound scientific and academic traditions. The scientific way of thinking (double-checking facts, describing things as the way they are rather than as is politically convenient, posing clear questions, encouraging public discussion of hypotheses and conclusions, conducting controlled experiments, viewing as public the information used to arrive at a decision, placing a philosophy ahead of pecuniary benefit, non-acceptance of contradictions, the separation of a process or idea from the person who happened to think of it, the unwillingness to appeal to higher authority as a justification for a conclusion, etc.) is an extremely important methodology to bring to bear on the problems of the conservation of tropical biodiversity. . . . However, academia and quasi-academia (museums, government research institutes, etc.) have quite reasonably developed their own more specific traditions as to how the tools of science are to be applied and to what. It is these traditions— ranging from what is viewed as an accomplishment to who is a legitimate practitioner of an activity—that require minor to major revision in an effort to put tropical biodiversity to work sustainably.[17]

Despite his dreams of putting science in the service of biodiversity, Janzen still draws a positivist portrait of science and then demarcates it from the rest of the world. Science as depicted here is an untainted set of professional ideologies, procedures, and norms, whose uses are then decided by "academia," another institution. Yet the same scientists who generate "facts" about biodiversity also decide what counts as legitimate professional activities—in other words, the boundary between "science" and "academia" is a myth. Having helped turn nature into biodiversity, Janzen now works to remake science to remake society to remake nature, while attempting, unconvincingly, to maintain science's sanctity.

At INBio, they violate that sanctity for good reason. Janzen is changing the face of systematics and ecology by moving them into society; or, rather, he is demolishing walls that don't really exist, erasing illusions that should not be maintained. Systematics and ecology are obligatory points of passage for those who, in the name of conservation, would blur the boundaries between science and society, nature and culture. INBio is an experiment where transfusions of money are pouring in to make these transformations take place. The boundary bustings occurring there are premeditated; they are part of Janzen's goal, not just a means to an end.

To use biodiversity, INBio is courting outside partners. Thomas Eisner's dream of chemical prospecting is becoming reality.[18] With Eisner, Janzen, and Gámez helping to negotiate the contract, IN-Bio persuaded Merck, the world's largest pharmaceutical company, to pitch in more than $1 million for prospecting start-up costs.[19] In return, INBio is sending Merck thousands of little pouches of pulverized, bar-coded biodiversity. Merck will assay these for biochemical potential and pay INBio an undisclosed (but acknowledged as small) percentage of royalties for any drug developed from Costa Rican species. INBio would then act as biodiversity broker, supplying Merck with as much of the promising species as Costa Rica's ecology could sustain. Although "hits" in chemical prospecting will be few and far between, and development of marketable drugs is extremely costly, even one successful drug could generate as much income as a major export agricultural crop.[20] Part of the resultant money will go to INBio's operating costs; the rest will be channeled to the conservation areas, so that more biodiversity will be saved, to be known, to be used.

Of course, Merck benefits by riding the crest of the biodiversity wave. In a bit of self-promotion in *Merck World*, one of its managers is quoted: "I hope other pharmaceutical companies will follow suit in other parts of the world, even though this is a risky and expensive venture." Risky and expensive? For its $1 million (which comes from a research and development budget of around $1 billion), the pharmaceutical megacorp receives, along with its samples of biodiversity, a wealth of laudatory green press that money usually can't buy. "Merck has never invested a million dollars in a better way than with INBio, if you just look at the press payoff," Alvaro Umaña, MIRENEM minister under Arias, told me. The U.S. National Wildlife Federation awarded Merck its "Corporate Conservation Council Environmental Achievement Award," stating in the banquet program that Merck "stands out as a company that has acted creatively and aggressively to further its economic agenda while simultaneously advancing the causes of biodiversity and sustainable development."[21] This balancing act is hard to sustain: it is the same act Costa Rica Inc. tries to perform as it attempts to profit from biodiversity so that biodiversity may profit.

Alongside its foundation in biodiversity's commercial value, INBio is built on the value of biodiversity as a force for "biocultural restoration."[22] Janzen and Gámez and their colleagues champion the idea that biodiversity used to be—and must again become—a grand intellectual resource for rural tropical people, who otherwise lack intellectual challenges and cultural opportunities. The founders of INBio seek "a local biodiversity literacy. These are Costa Rica's unopened books, written in strange languages. For Costa Rica's socioeconomic survival, and for Costa Rica's mutualism with the rest of the world, these books need to be opened and the languages learned."[23] The parataxonomists are to stand at the vanguard of rural dwellers becoming reacquainted, and thus reenchanted, with biodiversity. Once "born again," it is hoped, they will spread the gospel among their neighbors, triggering a chain reaction of renewed interest in and reverence for biodiversity.

If biodiversity becomes prized this way, the theory goes, rural Costa Ricans will be its staunchest protectors. The sacralization of biodiversity through cultural restoration parallels its desacralization through crass marketing; as Janzen puts it, "If Costa Rica's biodiversity is not firmly introjected into the minds and pocketbooks of Costa Rican society within a decade, the bulk of it will go into constructing cornfields and Miami lawn furniture."[24] At INBio, while literally pulling apart biodiversity in labs, they pull apart biodiversity in different ideological directions. While prospecting for the riches derivable from natural organisms, they prospect for the less tangible but no less important riches minable in the reformulation of the idea of nature that biodiversity represents. If mined deeply, the mother lodes awaiting in both biodiversity the natural object and biodiversity the constructed concept may yield up considerable lucre: by becoming valuable to Costa Rican minds, hearts, and purses, both aspects of biodiversity are to make conservation effective and sustainable.

To get to Elias Rojas Mora's home in Cocorí, you drive north from San José through the pristine Brauillo Carillo National Park,[25] east toward the Caribbean to the rural city of Guapiles. Continue north through banana fields and oil-palm plantations, biocided, stultifying monocultures that are as inauspicious a locale for biodiversity as one is likely to find. As the laterite road deteriorates, a different damaged landscape looms: recently cleared hills and valleys, green grass dotted with charred tree stumps, the occasional "living dead" forest remnant, and everywhere cows placidly chowing down. Lonely cottages sit on hard-won farms. Some of this lies within one of Costa Rica's much-vaunted official conservation areas, although you'd never know it.

Mora lives on one such farm of seventy acres. Compared to his neighbors, he has converted less forest to pasture for his thirty-five cows, and has been especially careful with the riparian area surrounding the stream that meanders through his property. While taking a cup bath at the stream, I was startled by psychedelic green and black frogs, amazed as a troop of howler monkeys whooped it up in the distance.

Elias Mora cleared his original plot of land thirteen years ago. He has since added further acreage and been joined by a wife, Marta. Their two-story wood house is immaculate, the bottom floor devoted to a kitchen and an outdoor dining room, flanked by all manner of forest plants potted in hollowed-out logs, sawed off plastic gas cans, and the remnants of Canada Dry ginger ale bottles. Upstairs are the family living quarters—and Mora's entomology laboratory. For Mora has recently added a full-time occupation to his workload, that of INBio parataxonomist. Most days, he heads out to the neighboring pastures and remnant woodlands with net and vials to stalk insects. He brings forest and field home, sorts through his treasures, bakes them, and pins them to wood and styrofoam boards. Once every month or so, Mora delivers them to INBio, where they will be further sorted, identified, and recorded in computer files.

While most parataxonomists work in or around national parks and reserves, Mora does not. He was originally to have been sponsored by Portico,[26] a forestry company with extensive land holdings that wishes to gain a reputation as a "green" deforester and furniture maker. (One scheme, portrayed in a slick promotional video, has Portico planning to harvest only selected trees and lifting them from the forest by dirigible.) Portico withdrew its sponsorship when

INBio attached certain strings to Mora's employment, and his hunting range was narrowed considerably. This leaves him less land to explore and a bit bored by the repetition. He also has a justified fear of deadly fer-de-lance and bushmaster snakes, which are common in this area, and he fairly dreads night collecting. Three nights before, three nights after, and on the night of the new moon, parataxonomists must set up their car-battery-powered bug lights and collect at 7 P.M., II P.M., and 3 A.M. They then spend the following day mounting their finds. To avoid sleep deprivation (and snakes), one parataxonomist told me he collects only at one of these times each night.

At 7 A.M. we head out and tramp through pastures, streams, and forest scraps for about five hours. Mora has a good eye; he seems to know where to look for what, under which leaves, beneath what bark. He sets up a net in a clearing with several small liquid traps beneath. The work seems interesting, and more than a little odd— at least so it must seem through neighbors' eyes. He exercises discrimination in collecting, pointing out common species, and then, with a quick snatch of his hand or sweep of his net, he nabs something more rare. Still, I'm willing to bet his rate of repetition is staggering.

Mora says that during the parataxonomist training course, Janzen ("El Doctor") emphasized the role of the parataxonomist as rural educator. Parataxonomists are encouraged to go to schools, fairs, and festivals, to set up stands explaining what INBio is and does. One such rural fair brought in 1,300 written comments from folks who stopped by the INBio booth. Mora feels a bit guilty that he doesn't fulfill this role as much as he should. He cites the delicate political situation in the area, which stems from the government's imposition of harsh regulations on free-spirited frontier people. These regulations, designed to make this conservation area live up to its name, have also turned the populace from passively in favor of conservation to dead set against it.

Still, he says, the course taught him a deep appreciation for biodiversity. Before he had never thought twice about, say, the beauty or importance of an insect. When asked why biodiversity was important (or many other questions), he'd respond, "In the course we learned . . ." or "El Doctor said. . . ." In this six-month course,

the parataxonomists are taught everything from the basics of entomology, to rudiments of ecological theory, to how to "fathom and
tolerate foreigners . . . how to work alone at night in the forest without fear, how to lose weight, and how to absorb constructive criticism."[27] They also acquire a new vision, a new way of seeing nature.
Janzen says "for most people, you have to start them out somewhere
else, and then prime the pump. Get them started on something
and then they use the library." During the training course, Janzen
coaches the parataxonomists for their return to nature. He primes
the pump by providing them with the basics of ecological literacy,
which encompass both facts and values stemming from ecological
science. When they are coached to see the world as the ecological
scientist does, Janzen hopes, the parataxonomists will *feel* the full
importance of biodiversity deeply—as the ecological scientist does.
In fact, Elias Rojas Mora seems to have learned his lessons well.

Rosa Gúzman has, too. This eighteen-year-old tracks down her
prey at the Carara Biological Reserve. If it were not for INBio, she
might one day have taken a secretarial course, but she would more
likely have moved to San José and become a maid. Enter Liz Claiborne, the fashion and cosmetics maven, whose foundation sponsored a parataxonomist course for women.[28] Gúzman's aunt, who
knew of her love of nature, spotted a course notice at the local
pulperia (a Costa Rican downscaled 7-Eleven). Against her family's
better judgment, Gúzman applied for and secured a position. She
had never traveled farther than a few miles from her home; she had
to buy all these things she had never had before, like pants.

Before the course, she had never heard of biodiversity. Now her
days and nights are filled with it. Now when she looks at insects, she
studies their differences, looks at their ecology, sees their beauty.
She talks to groups that come to Carara. When she goes to town,
all the kids ask her, "What kind of bug is this? What does it eat?"
When I asked her if she ever got bored, she said, "Bored? No! It's
always changing. There's always different insects. That's what diversity's about!"

Thirty-seven-year-old Gerardo Carballo Carvajal, who has a
grade-school education, once dreamed of becoming a park guard.
For sport, however, he hunted and killed wildlife. Five times he

was caught inside parks; five times he paid a fine. The sixth time landed him in jail. His mentality changed: he never wanted to kill another animal. He made an exception when he became a parataxonomist at the remote Hitoy Cerere Biological Reserve. To catch insects with a net, he says, is similar to his old avocation: "It's like I'm a hunter of insects." And he gets paid for this: what more could he ask? Through his work, he has come to realize that all life forms are interconnected and should be preserved. In local high schools, he found that most of the kids didn't know that caterpillars become butterflies or that tadpoles become frogs; they thought spiders flew. So as an educator and a collector, he gets to pursue what he loves and contributes to Costa Rica's future at the same time.

Nearly all the parataxonomists with whom I spoke were enthusiastic about their new vocation. All were fiercely loyal to Janzen. INBio does seem to be successful in inculcating a spirit, an appreciation, and a body of knowledge into its parataxonomists that they can then spread to others—a love of biodiversity trickling down from the North, the educated, the urban to the parataxonomists, and then through them to the rest of the rural populace, in whose backyards biodiversity must be defended. Janzen and Hallwachs write that "today's parataxonomists are a giant experiment in training, institution-building, and goal-directed science." With the parataxonomists on the front lines, "paraecologists" and "paraecochemists" wait in the wings.[29]

In Costa Rica, biodiversity and INBio, which represents it, sit at the center of a huge resource web with linkages that extend to Cocorí, Carara, Hitoy Cerere—to the humans and nonhumans that dwell in most remote corners of Costa Rica, and beyond. The movers and shakers behind INBio know that to preserve tropical biodiversity "requires creative new structures and collaboration among groups that have traditionally been separate, if not opposed—biologists and businessmen, for example. In Costa Rica, a serious attempt to forge such new socioeconomic and 'socioecological' collaborations is building those bridges."[30] Those bridges join North to South, NGO's to multinational corporate conglomerates, presidents to peasants, biologists to bartenders, pope to poachers, humans to nonhumans, Liz Claiborne to James Taylor, wilderness to Wall Street.

So, for example, the Salomon Brothers investment firm brokered the debt-for-nature swap that endowed INBio. Entomologists from the British Museum named thirteen newly discovered species of Costa Rican parasitic wasps after Salomon corporate executives. Sketches of the wasp *Eruga gutfreundi*—named after then CEO John Gutfreund—graced the pages of *Fortune* magazine. Wall Street went to bat for bugs, and bugs repaid the effort. According to one article, "This is a real example of mutualism. For Salomon Brothers money is not a scarce resource but immortality is. For a taxonomist cash is scarce but he faces immortality with every publication."[31] These insects caught in a collector's net are used to catch wealthy investors in an ever-widening resource net.

In 1983, Pope John Paul II visited Costa Rica. In 1990, Janzen and Hallwachs participated in the Vatican Study Week "Tropical Forests and the Conservation of Species," touting INBio's and Costa Rica's groundbreaking efforts on biodiversity's behalf. In 1991, INBio and Costa Rica won the St. Francis Award for Nature.[32] According to Gámez, who went to the Vatican to accept the award, the pope was pleased that Costa Rica had been so honored: "He said, 'Look after that beautiful biodiversity of Costa Rica.'. . . It was interesting to see how important Costa Rica was to him."

Costa Rica's international reputation has become linked to the conservation of biodiversity. Reputation enables conservation, conservation reinforces reputation. INBio is positioned as a generator for the cycle. It stands as a clearinghouse for a multitude of forces that, in turn, garner even more forces that must continue to meet at the locus of INBio. INBio's strategists seek to package its mechanics and philosophy and then export their system where moths become chemical pulp becomes cancer-fighting drugs become funds for conservation, funneled back through INBio to become more parataxonomists and more land preserved, both of which safeguard the original source of wealth and "the very things that give meaning to our lives."[33] Figure 6.1 shows but a small part of this process. It leaves out, for example, the diverse international forces that have

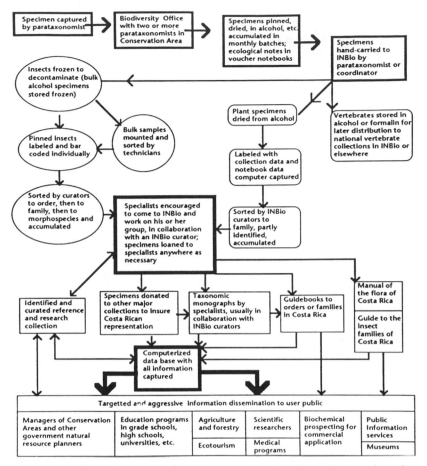

FIG. 6.1 Biodiversity moves from nature to society. From Janzen (1992); used with permission.

been brought together so that the parataxonomist can collect her specimen.

Throughout his writings, Janzen lays bare his machinations, revealing the diverse linkages in his Latourian resource net. Despite his occasionally disingenuous tone,[34] he can be frank, because he believes he works for public and ecological good, and he must be honest, because he wants this very explicit system to be exported far and wide. His revolution seems to be taking hold in Costa Rica. And it is being exported to Mexico and Taiwan, the Philippines,

Indonesia, and Kenya.[35] It has also captured imaginations closer to home, if presidential candidate Bill Clinton's 1992 Earth Day speech is any indication: "We should explore establishing the international equivalent of The Nature Conservancy. . . . A fund contributed to by developed nations and pharmaceutical companies to purchase easements in the rainforests for medical research. These easements and the profits from new drugs could make not developing the forests more profitable than tearing them down."[36]

To turn society and nature inside out—and to have the world laud you for it—is to invite criticism, and INBio has received plenty. Much of it comes from the upper echelons of the University of Costa Rica and the National Museums of Costa Rica, institutions that were previously in charge of collecting and understanding Costa Rica's biodiversity. Turf battles are to be expected, and they have been joined. Some of my interviewees questioned INBio's collecting methodologies (Paul Hanson), the role of Janzen in Costa Rican affairs (Maria Morales), and the role of the parataxonomist as the "reincarnation of the noble savage" (Bill Eberhart). In response to INBio's usurpation of the spotlight and resources, the director of the Biology School at UCR published a letter redirecting attention to the contributions that UCR has made and is likely to make to Costa Rica's biodiversity.[37] More cogent criticism can be pared down to this kernel: INBio is a private institution that is selling off Costa Rica's heritage for a very low price, in an agreement signed behind closed doors, with little benefit to local communities or indigenous knowledge holders.[38]

During interviews, I put these criticisms to Gámez and to Ana Sittenfeld, INBio's director of biodiversity prospecting. Both responded that, while INBio is a private institution, it was created by a governmental task force, which decided it should be that way. Furthermore, it is a private institution that has a mandate to work in the public interest, which it does. Gámez notes that many private entities—fishing companies, coffee growers, eco-tourism outfits, timber cutters—commercialize Costa Rican biodiversity: "Why can a timber company extract a biodiversity product we call timber and destroy the rest of the forest?" If the agreement was signed behind closed doors, both Gámez and Sittenfeld note, that is the way most business deals are concluded, and it had to be done that way to protect Merck and INBio if either wishes to negotiate similar agreements in the future.

As for lack of rural benefit, one major philosophy behind IN-Bio is to restore ecological literacy to local communities; and if that isn't enough, according to Sittenfeld, parataxonomists spend nearly their entire salaries in rural areas. Finally, they have decided not to use indigenous knowledge (although two parataxonomists are from indigenous groups in the southeast of the country), since "we don't have the right to take our information from our indigenous people. . . . We need to respect them, and we are respecting them."

Further cogent critiques of biodiversity prospecting come from anthropologist Arturo Escobar. Efforts such as INBIO's, according to Escobar, "amount to placing the wolf in charge of the sheep; those responsible for the destruction of nature—the World Bank, the international development apparatus, multinational corporations— are entrusted with the task of saving it!"[39] This paradox has the discomfiting ring of truth to it. Although the alliances outlined by Escobar increasingly destroy biodiversity, it remains true that efforts to reverse that destruction will have to exact cooperation from these very same institutions. For that reason alone, INBio is an experiment worth watching, with vigilance.

I find two other criticisms, which I have not heard or seen elsewhere, quite compelling. Gámez has said of INBio, "This is true sustainable development. This gives us parity with the developed nations, instead of carrying bananas on our backs for them."[40] As a caveat, the whole notion of what constitutes "sustainable development" is as thorny a thicket to untangle as the issue of what biodiversity actually represents. Still, I cannot but wonder whether anything about INBio is truly sustainable. It has required massive infusions of Northern cash to get under way. Even should it one day become self-sufficient, it will be because of financial arrangements with technocapital giants such as Merck. INBio will always be at the mercy of more powerful Northern forces and inextricably tied to fickle Northern economies, just as Costa Rica's other forms of export agriculture are. Furthermore, the entire inventory of the country's biological wealth crucially depends on Northern expertise; Costa Rica does not have nearly enough scientists to identify the specimens snagged by the parataxonomists.

Jorge Jimenez, director of the Biodiversity Inventory, responds that "every year the world is more mixed," and that some systematists must be regarded as "world heritage persons," belonging not just to the Smithsonian Institution, say. Gámez notes, too, that the world is interdependent, and that dependency "I think, is going to increase among nations rather than decrease." He points out that the developed world depends on the developing world to preserve biodiversity. And this is true. But I cannot help think that the resource net that attempts to put INBio at the locus of power in reality retains Northern mega-institutions at their traditional focus of power. INBio lies at a distant node, dependent rather than focal. It fosters a kind of development that may be more ecologically friendly, less physically onerous, and more socially benign than more traditional export agriculture, but is not really any more independent or sustainable. Even if this is a move away from carrying bananas on their backs, does it make Costa Rica any less of a banana republic?

Underlying all of this is the question of biodiversity's value. At INBio, they are pinning their hopes on pinning bugs to boards and pinning down those willing to pay for the privilege of using biodiversity. They are trapping, snaring, netting, etherizing, drowning, pressing, baking, pulverizing, grinding, centrifuging, and otherwise molesting nature in order that some of it may remain unmolested. They are selling extracts of its chemicals and snippets of its genes to preserve the totality of its aesthetic richness. Many at INBio—the directors as well as the parataxonomists—expressed feelings about biodiversity similar to those I have chronicled for North American biologists: they *love* biodiversity. It is their life's blood. But to sustain this love, they need to sell off the objects of their affection, and fast. Can such stark contradictions continue to run in parallel? Is our world really such that the only way to conserve wild riches is to rend and auction them to the highest international bidder? This troubles me, as it has troubled others, like Ehrenfeld, who protest the commodification of diversity. But perhaps we do live in a brave new world, which INBio and the biologists who run it are poised to capitalize on—and to perpetuate.

At this institute dedicated to biodiversity, so much of interest is being transformed in its name: nature, for one thing, is being transformed. By trapping, mashing, and converting fragments of nature to chemical pulp, then bar-coding them, injecting them into Ziploc bags, and sending them to international corporations, INBio seeks to preserve large chunks of nature inviolate and ensure that ecological and evolutionary processes continue unabated. The hope is that the conspecifics of the entities so transformed in INBio's collection and production process will be able to live unmolested lives—unless their value becomes known to us and they become widely harvested, or more land is turned over to plant newly lucrative export monocultures.

The way we think of the natural world is being transformed in Costa Rica, with forces pulling in opposing directions. Currently, most Costa Ricans think of nature either as something vaguely benign or as vaguely threatening but largely not of pressing interest. Through a campaign of biocultural restoration, the forces behind INBio desire that Costa Ricans conceive of nature instead as a source of vital intellectual stimulation, a cornucopia of aesthetic and cultural riches waiting to be explored, and consequently revered. Simultaneously, nature would be regarded as a larder of economic resources, a stockpile of economic bounty. Nature so conceived becomes something to be violated, something to be torn apart and pried open in search of marketable chemicals and genes. If INBio has its way, nature will keep its distance from humans, inasmuch as it is viewed as a resource that is ultimately to be subjugated; simultaneously, however, nature will be much more a part of us, a resource ultimately to be venerated.

Costa Rican society is being transformed by these two antithetical ideas of nature. On the one hand, INBio generates new professions and new sources of economic wealth depending on extreme exploitation, the ultimate separation of humans from nature; on the other, INBio hopes to reintroduce campesinos to the vast source of intellectual wealth their ancestors enjoyed, bringing them back to nature by injecting biodiversity into their hearts and minds. Either way, biodiversity is a resource for societal renewal.

Conservation is being transformed in Costa Rica and elsewhere. INBio is a program of conservation action and a system of conservation thought open for export. One part of INBio's conservation philosophy avers that citizens must come to value biodiversity deeply on an intellectual and aesthetic plane, forgetting economics. It says that unless biodiversity is injected into the hearts and minds of the people, it has no chance. Simultaneously, INBio's other cardinal conservation tenet takes it that the above is sentimental, that society doesn't work like that, and unless biodiversity can be transformed into cash, it has no chance. This is a dichotomous conservation philosophy generated in the Third World for export elsewhere to the Third World, informed by the antipodes of First World idealism and capitalism.

And science is being transformed at INBio as biologists abandon their random research in order to be at the controls of a new conservation system, to change a society's and a world's ideas of nature, and therefore to come to the rescue of biodiversity. Biologists are put in the service of society and of the natural world, while simultaneously controlling economic and natural resources that at very least are said to improve the quality of everyone's lives, and possibly are essential to all our continued existences. Biologists broker a new contract between biodiversity and society whose details are realized at INBio and whose ripples lap at the most distant shores, including the rarefied world of academic biology, which may never be the same again.

If ever there was "science in action,"[41] it is at INBio, where scientists serve peasants, and peasants become scientists. Here a U.S. biologist uses Costa Rican biodiversity to change Costa Rican society so that its members will guard biodiversity. INBio is the physical embodiment of the state of mind and the set of values that have driven biologists to make a name for biodiversity. It is a place where specimens of moths arrive from the rain forest to get ground into chemicals and spliced into genes to be sold to international bidders;

where eco-illiterate peasants enter and emerge again as born-again conservationists; where scientists wander in, afraid of the light of day, and come out born-again politicians; where a nation is investing in a bold experiment in redefining culture, nature, and science, and in redrawing geographical, normative, professional, behavioral, and ideological boundaries to make more space for the values that inhere in the term *biodiversity*, and therefore for the natural world to which it refers.

EDWARD O. WILSON

 7

> My truths, then, three in number, are the following: first, humanity is ultimately the product of biological evolution; second, the diversity of life is the cradle and greatest natural heritage of the human species; and third, philosophy and religion make no sense without taking account of these first two images.
>
> E. O. WILSON, "The Return to Natural Philosophy" (1991–92 Dudleian Lecture to the Harvard Divinity School)

Biodiversity is a revolutionary term: its makers and promoters aim to foment radical changes on several fronts. They wage battle in the contested realm of how we view, and thus value, and thus treat the Earth. Biologists seek revolutions not only in our environmental ethic but also in who should be the spokespersons for that ethic, the shapers of our ideas about the natural world, and the policies that stem from those ideas.

I have argued that a cadre of elite ecologists and conservation biologists stand at the "cutting edge" of their fields. I use this phrase differently than we normally might in a scientific context. These biologists are at the cutting edge of the battle to redefine the role of the scientist in society via their quest to conserve the planet's biological wealth. One of them, Edward O. Wilson, is at the cutting edge of the cutting edge of the dual mission to preserve biodiversity and to relocate the biologist's place in the world.[1]

When I examine the corpus of Wilson's work, it strikes me that by traditional standards, he will be remembered as one of the greatest twentieth-century scientists: his life and work await appropriate hagiography. Among other honors, the Harvard professor is a member of the National Academy of Sciences and has won two Pulitzer prizes, the National Medal of Science, and the Swedish Academy's Crafoord Prize, the world's most distinguished environmental biology award. He is a (if not *the*) leading expert on ants, co-derived the theory of island biogeography (one of the cardinal principles of conservation biology), and nearly single-handedly created the controversial discipline of sociobiology.[2]

Yet his current endeavor, the culmination of his career, is his most daring and most comprehensive. Selecting strands from the loom of his life's work—swatches of conservation biology, ecology, evolutionary biology, entomology, systematics, the genetic bases of human behavior—Wilson is stitching together nothing less than a new "natural" religion, with biodiversity as the icon of worship. He seeks to lead a corps of renaissance biologist acolytes in the mission to spread the new eco-gospel. In breadth, scope, and ambition, Wilson's endeavors are unparalleled in the history of biologists working for conservation. His activism has unprecedented implications, which deserve closer examination.

E. O. Wilson has built his illustrious career on the backs of ants. He is a "world authority" on their taxonomy, behavior, ecology, and evolution. His gorgeous preponderant scientific-compendium-cum-coffee-table-book *The Ants*, written in collaboration with Bert Hölldobler, won the 1991 Pulitzer Prize for nonfiction.[3] In an autobiographical essay, Wilson notes that "in a sense, the ants gave me everything, and to them I will always return, like a shaman reconsecrating the tribal totem." He can wax eloquent or humorous on his focal organisms: "The question I'm asked most often about ants is: 'What do I do about the ones in my kitchen?' And my answer is always the same: 'Watch where you step.' Be careful of little lives. Feed them crumbs of coffeecake. They also like bits of tuna and whipped cream. Get a magnifying glass. Watch them closely. And you will be as close as any person may ever come to seeing social life as it might evolve on another planet."[4]

Wilson staunchly supports the ecosystem services argument for biodiversity conservation. "It needs to be repeatedly stressed that invertebrates as a whole are even more important to the maintenance of ecosystems than are vertebrates," he emphasizes. The disappearance of invertebrates would be followed by the extinction of most of the rest of the multicellular world; he notes that "if invertebrates were to disappear, I doubt that the human species could last more than a few months."[5]

Despite the best efforts of Wilson and his colleagues, these ubiquitous, pivotal denizens of Earth are virtually invisible: scientists know nearly nothing about them, and they rarely impinge on the conscious thoughts of the rest of us. In 1985, Wilson proposed an "International Decade for the Study of Life on Earth" to fill the knowledge chasm. He has been tireless in his calls to revitalize the moribund discipline of systematics, to which he has contributed so much, and upon which so much of his research depends, for "if systematics is an indispensable handmaiden of other branches of research, it is also a fountainhead of discoveries and new ideas."[6]

His calls for increased training of systematists and massive surveys of biodiversity are key elements of his conservation-oriented writing. We must realize the full importance of invertebrates, so that their sustenance becomes incorporated into our ethical code: "A hundred years ago few people thought of saving any kind of animal or plant. The circle of concern has expanded steadily since, and it is just now beginning to encompass the invertebrates. For reasons that have to do with every facet of human welfare, we should welcome this new development."[7]

Wilson has been instrumental in instigating this development. He insists that not only does the teeming infinitude of invertebrates silently fulfill our ecological needs; they could help meet our genetic needs as well.

In 1975, Wilson detonated a bombshell called *Sociobiology: The New Synthesis*.[8] Building upwards from population genetics, he sought to develop a new paradigm to explain animal social behavior: virtually everything animals do expresses their genetic makeup, which has been molded by the exigencies of evolution through natural selection. The drive to pass on genetic information to future generations ultimately explains aggression, altruism, mating systems, and elaborate, even bizarre social organization.

Wilson's ambitious synthesis probably would have aroused some controversy within the cloistered worlds of evolutionary biology, animal behavior, and population genetics, and gone largely unnoticed elsewhere, had he not appended Chapter 27. Here Wilson extends the same rules of genetic determinism to speculate on the evolution of the social behavior of that oddest of species, *Homo sapiens*. Wilson elaborated on this chapter in *On Human Nature* (1978), which won him his first Pulitzer Prize. Subsequently, in *Genes, Mind, and Culture* (1981), written in collaboration with Charles Lumsden, Wilson attempted to put his hypothesis of how genes get expressed as mind and behavior on a firmer footing. Lumsden and Wilson suggest that innate "epigenetic rules" constrain individual behavioral and cultural choices. Through a multiplier effect, very small changes in a population's gene frequencies for these epigenetic rules may result in very big changes in group behavior, which are expressed as changes in culture.[9]

In *Genes, Mind, and Culture*, Wilson shows his belief that the profound cultural changes he seeks are possible. In fact, he may believe that he can steer society in the direction he wants: by appealing to what he believes is innate in people, to the genetically coded learning rules that constrain and govern social and moral choices, he feels he can help effect major and widespread change. This belief facilitates his daring gambit. Wilson acts as a macro-scale multiplier effect, wishing to amplify genetic predilections into cultural revolutions. Simultaneously, he advocates intensified research into these biological rules of cultural transmission. Otherwise, "a society that chooses to ignore the implications of the innate epigenetic rules will still navigate by them and at each moment of decision yield to their dictates by default. Economic policy, moral tenets, the practices of child rearing, and virtually every other social activity will continue to be guided by inner feelings whose origins are not examined." This, according to Wilson, is no way to run twenty-first-century society. We must investigate the biology of cultural choice, for "although they cannot escape the inborn rules of epigenesis, and indeed would attempt to do so at the risk of losing the very essence of humanness, societies can employ knowledge of the rules to guide individual behavior and cultural evolution to the ends upon which they agree."

Once again, Wilson would be poised to broker such an agreement: he understands what evolutionary imperatives dictate. He knows what the world needs for ecological sustenance. He knows how our minds work. He can even, perhaps, see into the future, as he seeks "The Explanation of History": when the relationships between genes, minds, and culture are sufficiently understood, predictions of cultural change become statistical calculations, and "the envelope of possible futures can thereby be estimated and analyzed."[10]

In response to these works, Wilson's praises were sung with renewed vigor in some quarters. Elsewhere, vituperative criticism was hurled in very public forums. The unseemly controversy landed Wilson and his opponents on the cover of *Time*. At a 1978 AAAS meeting, a demonstrative protester dumped a pitcher of ice water on Wilson's head.[11]

Two principle categories of objection to Wilson's work pertain to the biodiversity story. First, opponents have criticized sociobiology's lack of scientific rigor. Some such claims invoke the arcana of population genetics and need not concern us here.[12] But others have declared that Wilson's scientific explanations, particularly when applied to humans, amount to little more than just-so stories. One could take any current trait of human beings and seek to "explain" it by sociobiology. Yet no independent way exists to *test* most of these explanations.

Say you want to explain the phenomenon of concealed ovulation in human females. Why do neither women nor men know for sure when women are ovulating? To apply sociobiology to this problem is to take an imaginary trip back to the time of protohumans to explain why it would be adaptive to possess this trait: maybe women who concealed ovulation secured the allegiance and resources of men who might otherwise stray from woman to woman; maybe men who chose mates who tended to conceal ovulation spent less energy fending off competitors. Lacking a time machine, these hypotheses cannot be tested: they are just so many just-so stories. Furthermore, the "adaptation" chosen—say, concealed ovulation—may merely be a side effect of some other adaptation: perhaps our sense of smell has been impoverished to the point where we no longer can detect olfactory cues. Or perhaps concealed ovulation is an ancillary product of a general trend toward greater plasticity in human social and sexual behavior. Something like concealed ovulation may have only accidental reality as an ontological concept.[13]

Political beliefs, often of a Marxist stripe, fuel the other class of criticisms. Such vocal opponents as his Harvard colleagues Stephen Jay Gould and Richard Lewontin have attacked Wilson for both his science and his politics. Specifically, Wilson's foes object strenuously to his biological determinism. When Wilson or other sociobiologists stress strong genetic predilections for such phenomena as sex-role differences or rape, it may be deduced that such phenomena have resulted from extremely powerful selective forces, which we are nearly helpless to contravene. If our behavior is in our genes, then the will of a human to change or improve the conditions of her life may be for nought: you can't fight the evolutionary history that made you who you are. This is anathema to Marxists, who place supreme faith in the power of human will to effect personal and historical change. Furthermore, critics have suggested that such genetic determinism facilitates institutional discrimination, as sociobiology can be called upon to explain why some groups really are inferior or intractable in some way or other. The historical precedent of the ravages of social Darwinism gave these criticisms some weight.[14]

Wilson pressed on despite the attacks. Having invested so much work in developing, presenting, and defending the sociobiological synthesis, he was not about to let it rest before putting it to work in "the arena of my greatest passion of all . . . the stewardship of the living environment."[15]

As noted in Chapter 5, Wilson believes that as a result of our evolutionary origins as savanna-dwelling primates, completely dependent on the processes and organisms of the natural environment, humans retain "the innate tendency to focus on life and lifelike processes," which he calls "biophilia." Although he waited until 1984 to explain his biophilia hypothesis more fully, the ideas behind it were with him from the launching of the sociobiological synthesis. In a 1976 interview in *House and Garden*, Wilson explained: "If we can gain a sense of emotional reward from contemplating our environment and living closer to it, perhaps our descendants might evolve further in that direction and discover deeper emotional rewards. Providing an environment in which we can develop fully our capacity for spiritual enrichment is the most important duty we can perform for future generations—an enrichment strongly centered upon our affiliation with natural, diverse living environments."[16]

Here we encounter the rudiments of Wilson's ambitious plan to tie our genetic endowment to an ethic, a religion of biodiversity conservation. We gain "an emotional reward from contemplating our environment and living closer to it" because biophilia resides in our genes, awaiting stimulation and expression. This contemplation provides "spiritual enrichment," a deep sense of fulfillment when in the thrall of something grander than ourselves. Such spiritual fulfillment can only occur in biodiverse natural environments, and thus we have a responsibility to bequeath these environments to future generations so that they, too, can fulfill the spiritual needs bequeathed to them by the evolutionary process.

As explained by Wilson in 1978, in the first use of the term I have found, our *biophilic* need is as deeply rooted in our evolutionary past, and therefore in our genetic endowment, as any of "our most intense emotions: enthusiasm and a sharpening of the senses from exploration; exaltation from discovery; triumph in battle and competitive sports; the restful satisfaction from an altruistic act well and truly placed; the stirring of ethnic and national pride; the strength from family ties; and the secure biophilic pleasure from the nearness of animals and growing plants."[17]

Sociobiology explains that all of these phenomena result from the unique set of environmental circumstances that faced our ancestors as they struggled for survival on the African plains where humans descended from the trees and thus descended from other apes. Millions of years of natural selection for the traits that conferred survival in hand-to-hand struggle against an unforgiving environment cannot have been wiped out in the blink of an eye that has

THE COLUMN

Capital ideas from people who publish with Harvard.

BIOPHILIA

Human genius has the power to thread the needles of technology and politics: it is within our reach to avert nuclear war, feed a stabilized population, and generate a permanent supply of energy and materials. What then? We can strive toward personal fulfillment and the realization of individual potential. But what is fulfillment? Potential to what ends?

The answers to these surprisingly difficult questions must eventually be drawn from a deeper understanding of human nature, based in good part on knowledge of the brain's machinery and the processes that guided its evolution for millions of years. Our deepest needs stem from ancient and still poorly understood biological adaptations. Among them is biophilia:

the rich, natural pleasure that comes from being surrounded by living organisms, not just other human beings but a diversity of plants and animals that live in gardens and woodlots, in zoos, around the home, and in the wilderness.

Other creatures not only satisfy innate emotional needs, but also present unending intellectual challenge. More complexity exists in a single butterfly than in all the machines on earth. and even more in its ecosystem. The one truly irreparable damage we can inflict on ourselves is eliminating a large fraction of the earth's species, through careless destruction of the natural environment.

Our biophilic descendants will regard species extermination as the greatest sin of the twentieth century.

Edward O. Wilson

Author of *The Insect Societies, Sociobiology* and *On Human Nature*

HARVARD UNIVERSITY PRESS

FIG. 7.1 An early explication of E. O. Wilson's (1979) "biophilia" concept. From *New York Times Book Review* 14 (January 1979), 43; used with permission of Harvard University Press.

marked the onset of modern civilization. We convince ourselves that we no longer intimately depend on the natural environment; according to Wilson, our genes intimate something very different indeed.

In a 1979 promotional ad for Harvard University Press in the *New York Times Book Review*, Wilson spells out most of the elements of his program, elements that revolve around his unique understanding and expertise (see fig. 7.1). To reach true fulfillment—

fulfillment of a spiritual kind—we must first understand human nature: this means we must understand the sociobiological precepts that explain, ultimately, the function and functioning of the human brain. Biophilia is one of the manifestations of our evolutionary history. By indulging biophilia, we achieve a deep fulfillment that natural selection has shaped the brain to crave and register. This resembles the philosophy of biocultural restoration upon which Janzen and Gámez have built INBio, but with a genetic explanation for why their plan of action is destined to work.

Biophilia has several components, as the third paragraph of the ad suggests. Nature's beauty salves our genetically based emotional needs, just as nature's information content answers our genetically based intellectual needs. Wilson developed this idea with the passage of time, postulating another genetic need for an unending frontier, a wilderness that unexplored biodiversity provides for us. In *Biophilia*, he writes: "To the extent that each person can feel like a naturalist, the old excitement of the untrammeled world will be regained. I offer this as a formula of reenchantment to invigorate poetry and myth: mysterious and little known organisms live within walking distance of where you sit. Splendor awaits in minute proportions." We crave the challenge of the unexplored, as Wilson evocatively claims in the poetic first chapter of *The Diversity of Life*: "In our hearts we hope we will never discover everything. We pray there will always be a world like this one at whose edge I sat in darkness. The rainforest in its richness is one of the last repositories of that timeless dream."[18]

The last paragraph of the Harvard Press ad trumpets the prophetic moral of Wilson's story. In diminishing biodiversity, we deny those who will come after us the fullest opportunity of and joy in fulfilling their genetic needs for nature's beauty, information, and uncharted frontier. Wilson brings this point home in an oft-cited line from *Biophilia*: "The one process now going on that will take millions of years to correct is the loss of genetic and species diversity by the destruction of natural habitats. This is the folly our descendents are least likely to forgive us."[19]

As we saw in Chapter 3, Wilson and his peers wield fear of the unknown consequences of the human assault on biodiversity to argue that we curb this destruction. We shall see that Wilson couples these fears with his own expertise, which we must heed to assuage those fears:

> The manifold ways by which human beings are tied to the remainder of life are very poorly understood, crying for new scientific inquiry and a boldness of aesthetic interpretation. The portmanteau expressions "biophilia" and "biophilia hypothesis" will serve well if they do no more than call attention to psychological phenomena that rose from deep human history, that stemmed from interaction with the natural environment, and that are now quite likely resident in the genes themselves. The search is rendered more urgent by the rapid disappearance of the living part of that environment, creating a need not only for a better understanding of human nature but for a more powerful and intellectually convincing environmental ethic based upon it.[20]

Wilson sees the necessity for strong measures if we are to avoid our terrible folly. It is not enough to extol the economic or ecosystem values of biodiversity, although he certainly does this at every chance he gets.[21] Wilson, like many of the biologists I have portrayed here, wishes to change the way we think about nature: "If humanity is to have a satisfying creation myth consistent with scientific knowledge—a myth that itself seems to be an essential part of the human spirit—the narrative will draw to its conclusion in the origin of the diversity of life."[22] Somehow our genes demand this creation myth, and Wilson tries to translate his science into a myth that carries strong conservation maxims. In this attempt to formulate a creation myth that is simultaneously a conservation ethic—and more than that, a conservation religion—Wilson calls into play his sociobiological understanding of human needs and aspirations.

For him, "the goal is to join emotion with the rational analysis of emotion in order to create a deeper and more enduring conservation ethic." Or, to put it another way, "the more the mind is fathomed in its own right, as an organ of survival, the greater will be the reverence for life for purely rational reasons." He believes that sociobiological analysis provides the pieces of the puzzle to formulate this ethic. In fact, there is little sociobiology cannot eventually do: "The important point is that modern biology can already account for many of the unique properties of our species. Research on that subject is accelerating, quickly enough to lend plausibility to the proposition that more complex forms of social behavior, including religious belief and moral reasoning, will eventually be understood to their foundations."[23]

The conservation ethic Wilson seeks to create must be consistent with his materialist worldview and must be informed by his scientific precepts. An ethic can draw its fundaments from two sources. Early in his life, Wilson turned away from his Southern Baptist roots to become a born-again biologist: he would banish the ethical systems to which adherents of traditional religions subscribe. These postulate free-floating ethical principles, strictures that exist in the universe, placed there by some divine fiat, awaiting our discovery and obeisance. He is no fan of metaphysics, declaring, "I don't like the ineffable." To Wilson, species rights and the intrinsic value of biodiversity fall into this category of transcendental—and thus false—ethical precepts.[24]

Wilson subscribes to an alternate view of ethical foundations. The necessary precepts for ethical systems, and the need for these systems in the first place, are rooted in human evolutionary history: "Morality, or more strictly our belief in morality, is merely an adaptation put in place to further our reproductive ends. . . . In an important sense, ethics as we understand it is an illusion fobbed off on us by our genes to get us to cooperate."

Abstruse metaphysics yield to concrete materialism in Wilson's worldview. He seeks a scientific (and hence, in his view, objective and ultimate) basis for conservation ethics, one that cannot be refuted by the political, ethical, or cultural tenets prevailing at a given time or place. This ethic will endure because it is "true": it is grounded in the real world that Wilson's scientific investigations have revealed to him. For Wilson, the building blocks for a conservation ethic are not to be discovered in the cosmos; they are to be discovered in our genes. Our environmental ethic will be "based on the hereditary needs of our own species."[25]

Not just biophilia lurks within us. Faith, too, originates in our genes: "But did blind natural selection also lead to the human mind, including moral behavior and spirituality? That is the grandmother of questions in both biology and the humanities. Common sense would seem at first to dictate the answer to be no. But I and many other scientists, and especially evolutionary biologists, believe that the answer may be yes." Even so, Wilson describes himself as "intensely" religious or spiritual, although not in the sense of believing any particular creation story:

> "Not in the sense of accepting a patriarchal supernatural being, or matriarchal. Not in the sense of going through that exquisitely pleasurable experience of surrendering to the tribe known as religious conversion, or being born again, or giving yourself to Jesus, or whatever. Not in any of those senses. But in the sense of recognizing that at the core of it all is the set of deep, almost mystical motivations to rise above ordinary human experience. To find meaning in life that transcends individual mortality. And it is the burden and also the extraordinary opportunity of the humanist to try and travel that route in a way that honestly embraces what we can know with certainty about the human condition."

Wilson believes that "it is an undeniable fact that faith is in our bones, that religious belief is a part of human nature and seemingly vital to social existence. Take away one faith, and another rushes in to fill the void." "Men, it appears," according to Wilson, "would rather believe than know."[26]

Wilson the iconoclast aspires first to sweep away the false faith systems to which most humans mistakenly adhere, and then to fill the resultant void. He believes "that traditional religious belief and scientific knowledge depict the universe in radically different ways, that at bedrock they are incompatible and mutually exclusive."[27] Rather than finding our required spiritual fulfillment by worshipping false gods, we should seek "that deep and uniquely human spiritual strength that comes from witnessing the earth as it was before the coming of man." To destroy biodiversity, therefore, is to "court spiritual disaster."[28]

The love of nature, our need for intellectual challenge and unending wilderness, our yearning for religion: all these stem from our genes, manifestations of the past forces of natural selection. Biodiversity fulfills us on all these levels. And in fulfilling our deep genetic urges, it and humanity, which depends on it, find salvation. Wilson believes "that the more you understand organisms, each species in turn, its natural history, behavior, the more involved people become with them. . . . Familiarity produces if not love, at least affection. And affection produces a desire to salvage, to hold on, not to be reckless." The more we understand biodiversity—the more we sate our genetically based biophilic longing—the more we appreciate biodiversity, the more we become spiritually attached to biodiversity and thus the more we conserve biodiversity. The more we conserve biodiversity, the more opportunities we and others will have to fulfill our genetic longings. For this positive feedback between genes and conservation to work at maximum efficiency, Wilson and colleagues must work to get as many people as possible into the loop. Once in the loop, biodiversity tugs at our genes, transforming us by reawakening deep dormant urges.

Recall the boundary work conservation biologists perform to validate both the environmental values they hold and their privileged position to speak for those values. Reasoning that scientists' (and particularly systematists') data and views derive from a process that is objective and thus above the fray of partisan interests and cultural bias, Terry Erwin argues that we should heed scientists' conservation prescriptions (see figure 4.1).[29] Wilson also proposes a transcultural argument for conservation, albeit a more radical one. His arguments transcend culture both in respect to the specific interests they serve and in legitimizing the transcendent right of scientists to judge. If biophilia lies in our genes, and if all peoples share the needs for an endless frontier and a close affiliation with nature, then saving biodiversity to satisfy primeval urges is a universal value transcending the interests of any individual or group. It is ironic that in attempting to circumvent transcendent religious ethics, Wilson espouses his own transcendent conservation rationale. Biodiversity conservation may not be metaphysically ordained; instead, it is evolutionarily ordained.

Evolution also ordains that we should listen to the policy prescriptions of Wilson and his fellow conservation biologists. Erwin says scientists have authority to make transcultural policy prescriptions because those prescriptions spring from a process shorn of values and subjectivity. Wilson's position, if valid, is even stronger: biophilia, an affiliation for living things, has been molded by eons of natural selection. The more attuned our ancestors were to the biota around them, the more adept they were at procuring food and avoiding harm. According to Wilson, this same predilection for tuning in to the living world around us inheres in all of us today. Yet like many selected traits, it is expressed more strongly in some of us than in others. And if some of us are more genetically predisposed to love the natural world, none could be more so than ecological biologists, who devote their lives to studying life.

For example, Reed Noss told me: "I think Ed Wilson is correct, that people have an inherent biophilia, but some people more than others. And I think probably the vast majority of biologists have an inherent biophilia that's stronger than [that of] much of the rest of the population." So, then, does this suggest that the conservation biologist's calling is in some sense evolutionarily ordained, analogous in a sense to the divine calling traditionally claimed by priesthoods? Noss laughed and replied, "I know it sounds elitist probably. . . . I don't see what other reason why people would go into a relatively low-paying field unless they had some kind of calling. I think there is some kind of calling, if you want to call it that."

I did not ask Wilson this question as directly. But when I asked him what it was about biodiversity that made him so committed to its conservation, he responded: "I'm congenitally attracted to biodiversity. I think there's a naturalist spirit that draws a fraction of the people that become professional biologists into biology in the first place. It causes you to gravitate towards studies of biodiversity. . . . The important thing is delight in the abundance of nature, the desire to participate in it, share it. It's a naturalist's instinct." For Wilson, *congenital* and *instinct* are synonymous with *genetically based*. That he has become a conservation prophet—an apostle preaching the biodiversity gospel—is part of the evolutionary process: in that sense, it is fate.

Furthermore, this trait that drives Wilson, Noss, and their peers is newly adaptive today. These genes do not just whisper to their bearers that they should study life—they also urge them to fight for it. Noss believes this natural inclination led him into eco-sabotage at an early age, and causes him and his colleagues to continue to work for conservation, and to forge and promote a conservation ethic: "I would think biologists would naturally play a large role in articulating such an ethic, or refining such an ethic. You know, it makes sense. It doesn't come from our professional training. Again, I think it comes from what we talked about earlier, the predisposition, biophilia or whatever, that we have that also . . . brought us to seek that kind of training." In struggling for the preservation of the biotic world, those who bear the alleged biophilic genes are fighting for the survival of their own offspring, as well as those of their human and nonhuman kin.

Wilson circumvents those who point out the fraud of invoking the authority of nature to validate human preferences.[30] In Wilson's system of belief, nature is no longer an external referent. Rather, it is part of us and we are part of it. A strong affiliation for living things, a persistent drive to explore uncharted wilderness, a universal need for faith in something grander than our ephemeral lives: all these stem from the nature within us, from the magisterial evolutionary process that shaped the genetic endowment shared by all members of the species *Homo sapiens*. We are nature self-reflexive, nature that can understand, even if inchoately, both the natural evolutionary processes that created us and the natural ecological processes that sustain us today. Our inextricability from nature is its own rationale for preservation: by preserving biodiversity, we save ourselves. Our genes, the result of nonhuman and human evolution, murmur to us to study and save biodiversity. Conservation of nature becomes self-defense. The process by which biologists and those who heed them come to love and defend nature is itself a natural process.

Wilson and his co-thinkers believe that if we would only listen to what our genes whisper to us and what conservation biologists— who are naturally ordained to preach this gospel—proclaim to us, we would be ultimately fulfilled as humans, while simultaneously reforming the way we treat the Earth. Here is a twist on apocalyptic religions: rather than waiting for the end of the world, we are urged to avoid the end by worshiping that which sustains us ecologically, psychically, emotionally, and spiritually. By following Wilson's evolutionarily based conservation ethic, we save biodiversity, save our souls, and save humankind.

In one dizzying swoop, Wilson tries to explain the universal cultural phenomena of faith in a higher force, love of the natural world, and common cultural precepts of ethical behavior. He seeks to harness his understanding of our evolutionary destiny to form a new religion that would translate our genetic code into ethical change, and thus into practical action for biodiversity's sake. Were we to convert, Wilson and those of his fellow biologists who have promoted biodiversity, to whom we have granted cognitive authority to speak on its behalf, who claim to know most about biodiversity and have the most finely honed appreciation of it, could attain the level of high priests. Just as those most religious in a conventional sense claim to be closer to their god, so could biodiversity biologists claim to be closer to the source of spiritual renewal and worship. They seek to become missionaries, and more than that, prophets, preaching the gospel of biodiversity to the genetically unfulfilled, and saving the Earth in the process.

As we have seen, Wilson was instrumental in coining the term *biodiversity* and was a keynote presenter at the National Forum that launched it. He edited the conference volume that synthesized the new intellectual and political compacts and worldviews necessary to promote a new spirit of conservation, that "documents a new alliance between scientific, governmental, and commercial forces— one that can be expected to reshape the international conservation movement for decades to come." Furthermore, with Paul Ehrlich's help, Wilson has been trying to institute biodiversity studies programs to promote the biodiversity synthesis among a core of apostles who will fan out to preach to the unconverted to spread the new conservation ethic and the values that inform it.[31] *BioDiversity* is a how-to book: it is a successful example of how alliances can be forged to save biodiversity, as well as a model of how senior biologists can successfully extend the purview of science.

Wilson's canon for biodiversity studies programs would probably include his multipoint plan to solve the biodiversity crisis. Having created the term *biodiversity* and promoted it and the values subsumed under its aegis, having warned of the ecological, emotional, and spiritual crises that biodiversity's diminishment presents to humanity, it is only reasonable that Wilson should present us with a concrete formula for what we can do to avert the catastrophes that biodiversity's reduction portends.

In his contribution to *Conservation for the Twenty-First Century* (1989), Wilson offers a seven-point response to his own question: "What is to be done?" The main elements of his plan are: (1) undertake a complete worldwide biotic survey; (2) promote ex situ conservation (gene banks, zoo and botanical-garden breeding, and the like); (3) link conservation with economic development; (4) have international development and lending agencies exert pressure on Third World nations to promote conservation planning; (5) promote restoration ecology; (6) foster engagement by social scientists: economists must come up with more realistic models to account for environmental externalities, psychologists must study the need of the human mind for nature (read: biophilia), and "sociology is so ideological and unaware of any environment except the 'social' as to be almost useless in addressing the root problems of overpopulation and environmental degradation in the third world"; and, (7) develop aesthetic and moral reasoning, since "when all the accounting is done, conservation will boil down to a decision of ethics based on empirical knowledge."[32]

Three years later, in *The Diversity of Life*, Wilson presents a slightly retooled version of this agenda, one upon which those enlisted in the enterprise of biodiversity studies "might agree." Points 2, 6 and 7 above are played down, and points 3 and 4 are expanded; we are urged to "create biological wealth" in the tropics by arrangements such as the one between Merck and Costa Rica outlined in Chapter 6, and to "promote sustainable development," even though "the raging monster upon the land is population growth. In its presence, sustainability is but a fragile theoretical construct. To say, as many do, that the difficulties of nations are not due to people but to poor ideology or land-use management is sophistic."[33]

Just as Wilson has defined and publicized the "biodiversity crisis," so, too, he promotes a specific agenda for ending the crisis. Notice that many of his plans revolve around his unique areas of exper-

tise. For example, the global biological survey involves massive infusions of capital into training systematists to make us more aware
of the prevalence of, importance of, and threats to "the little things
that run the world," especially invertebrates. His recommendations,
which seem to go beyond the realm of the biological, in fact cover
areas that sociobiology can help explain, derive, and change, such
as the understanding of our biophilic urges and the need for an
empirically based conservation ethic. Wilson argues that the solutions must naturally be effected largely with biologists' guidance,
because the problems are biological in the first place: "The problems of human beings in the tropics are primarily biological in
origin: overpopulation, habitat destruction, soil deterioration, malnutrition, disease, and even, for hundreds of millions, the uncertainty of food and shelter from one day to the next. These problems
can be solved in part by making biodiversity a source of economic
wealth."[34]

But are these problems really biological in nature? Let's look for
a moment at another scourge of humanity. Cancer is often considered a biological problem. After all, isn't cancer just about failures
of biological control systems within a body, metastasis of cells, in-
hibition of normal bodily functions? Not necessarily. Through different lenses, we can focus on cancer as a social or a political problem. To prevent cancer, or to treat it, we must examine what makes
people abuse their bodies, import carcinogens into their systems.
We can examine the political and economic decisions that facilitate
the spewing of corporate toxins—carcinogenic chemicals, tobacco
advertisements—into our environment.

Similarly, in some narrow sense, biodiversity loss may be a biological problem, as Wilson posits. But "overpopulation, habitat destruction, soil deterioration, malnutrition, disease," and so on—
the forces that drive the engines of biodiversity destruction—have
social, political, and economic roots. Thus the only way to curb
these forces—the only chance to stop the engine of biodiversity destruction permanently—is by social, political, and economic solutions.

So in his analysis of the problem and agenda for its solution, Wil-

son avoids the hard questions and the hard answers. He urges the
World Bank and other development agencies to put the conservation
screws on Southern nations. He reifies overpopulation as a biological
problem—too many people having too much sex—that is destroying
the world. What of the other side of the coin, the gluttonous overconsumption by Northern nations that devours worldwide resources,
leaving the have-nots with even less? What of crushing Third World
debt loads that cause the governments of these nations furiously to
exploit their biotic resources as their only hope of paying up? Such
arguments, Wilson says, are "sophistic," specious diversionary tactics thrown up by those who refuse to acknowledge biologists' facts.

I mentioned to Wilson the growing critique from what might for
lack of a better name be called the "sustainable development community," which says the problems of tropical humanity are not biological but sociopolitical. These voices suggest that a grossly inequal
pattern of resource distribution drives biodiversity loss—that overconsumption, not overpopulation, is destroying biodiversity. How
did he respond to these people?

> "Well again, I don't want to play with words. When you go down to the
> tropics and find Brazilian peasants on the edge of the rain forest [who]
> barely can get enough to eat from one day to the next, it's hard to speak
> of their problems as being anything other than biological. What those
> people need right now, sure, are sociopolitical changes. That's a given. I
> never meant to imply that to achieve the powerful, essentially biologi-
> cal changes needed to improve the lot of the Third World, you have to
> have a workable sociopolitical structure and leadership. But by 'biologi-
> cal,' I meant overpopulation, which is undeniable. The disease, which
> is biological, which is devastating. . . . And it's biological in the sense
> that a lot of this can be alleviated by biological solutions: antibiotics,
> proper medical care, and improved agricultural land use management:
> all biological. . . . They're going to have to live on the land, especially
> with their enormous populations. So that the way to settle them on the
> land and keep them from destroying the land is by popular attention to
> their biological nature, their biological needs—getting their population
> growth slowed down by one means or another, giving them the type of
> soil restoration and superior land management and new crops that will

allow them to make a decent living on the land they've already cleared, and holding them off from the remaining wild lands."

Wilson's proposed solutions may seem to him to fall solely within the domain of the biological; but they, too, are strongly political. He has long railed against adversaries—whether opposed to sociobiology or to conservation—who he claims are motivated by politics and ideology.[35] We have avoided seeing the true nature of our problems in the twentieth century, since "to a substantial degree, we got diverted and distracted by ideology. And especially communist ideology, socialist ideology, Marxist-Leninist ideology. . . . And it so distracted us that we didn't become aware of the real problems of the world till the 1990s."

Wilson shows us the "real problems," as well as the real solutions to these problems. By presenting the problems and their solutions as biological, Wilson attempts to put all of science's ideological clout in the service of a passionate defense of biodiversity. His opponents may be tainted by politics and ideology, but his analyses and prescriptions spring from the objective, value-neutral, pure realm of science. Yet Wilson is every bit as political—in three different senses of the word—as the opponents that plague him.

First, he is political in a conventional sense. He testifies before Congress. He has gone to the White House "to address a dozen heads of America's leading corporations." He attempts to influence the official policies of governments here and abroad. He notes: "If you're a scientist, you've got the ear of the media, the ear of Congressmen."[36] One imagines that given the chance, he would put his analyses (such as the one above on Brazil, or the ones offered in his papers) into action, or at least offer strong counsel to those that could.

Next, he is political in the ideological sense. He claims that ideology drives his opponents: he points out the Marxist motivations of sociobiology's enemies. And he is correct. But Wilson's sociobiology efforts carry with them an ideological commitment to rid the social sciences of spurious logic and sweep out the metaphysical commitments that militate against a proper understanding of spirit and ethics. His solutions to the biodiversity crisis are imbued with politics and ideology as well. His commitment to international capitalism as saviour is every bit as feverish as his Harvard opponents' commitment to Marxism. By choosing to ignore the sociopolitical problems that destroy biodiversity, he tacitly reinforces the current international political and economic order. This inexhaustible, and up to this point incontrovertible, drive of the international capitalist juggernaut is destroying biodiversity in the first place. Creating new wealth in the tropics from biodiversity evidently and ironically destroys this biodiversity. Choosing to create new wealth from biodiversity rather than to redistribute the old wealth that has come from the transnational exploitation of biodiversity is a political, not biological, choice. It may not sound this way to the average reader; it may appear nonpolitical because the average Northern reader likely thinks of capitalism as inevitable, as God-given, as *natural*, and therefore as correct.

Wilson is also consummately political in the Latourian sense. He has woven an elaborate resource web, at whose center he sits, at the nexus of invertebrate biology, sociobiology, biophilia, biodiversity studies, conservation policy, and rational analysis of the human mind, including our need for faith. In analyzing the system ecologist H. T. Odum's calls for biologists to control the dials of social engineering, Peter Taylor offers the pertinent caveat that we should beware the sweeping social prescriptions of anyone who holds the power to effect their own recommended prescriptions.[37]

Such is the case of Wilson, who controls so many of the resources we would need to solve the problem he has defined, who launched the term that has been so successful in promoting his worldview, and who is organizing the training of future biologist-proselytizers to effect his solutions. Wilson advocates a new conservation ethic "uncoupled from other systems of belief."[38] Instead, he is attempting to hitch this ethic to his philosophy, his expertise. He has presented the problem, his analysis of the problem, his solutions to the problem, and his right to speak for the problem as value-neutral, irrefutable, inevitable results, not only of the scientific process, but of the evolutionary process as well.

The critiques leveled at sociobiology pertain here as well. Wilson would instill into us a new conservation ethic based on the playing out of biophilia on a massive scale. But too much about biophilia is coincidental in a way that gives one uncomfortable pause. Could biophilia be just an elaborate just-so story about the plausible but essentially unprovable alleged needs of savanna-dwelling proto-humans? We are said to have a deep fear of snakes, an affinity for scenic vistas, a fascination with other forms of life. Our plains ancestors may well have shared these traits; they may be genetically constrained adaptations that have been passed down to us today. It is plausible but not provable: alternate just-so stories also fit the evidence. Wilson has taken his own love of the natural world, which saw him through a difficult youth and became a life passion, and has made it, not a personal quirky preference, but a reified, universal, inevitable, genetically determined trait. If we believe him, evolution dictates how we should formulate the environmental ethic, and also who should do the formulating—namely, the biologists, who have this adaptive trait so strongly expressed.

The political critiques of sociobiology apply here as well. The opposition to the "population problem" approach I offer here is hardly new. Wilson must be aware of it, and has made the choice to avert it. Instead, he has chosen a different political program. When biologists define social problems as biological, and concomitantly avoid truly difficult radical, long-lasting solutions that might tackle those problems, warning bells must ring. Why, for example, does Wilson choose to avoid combating "overconsumption" or "inequal resource distribution" and instead rail against "the population problem"? He can more easily label the latter a biological problem, one within the parameters of biologists' expertise, and one in which the powers of the North—and Wilson advocates and predicts explicit linkages between biologists and the prevailing powers that be—can interfere, which they can control without making corresponding, fundamental, but more excruciating changes closer to home. Circumscribing the problems as biological, he arrogates the power to define and solve them, to apply his unique, but unfortunately not universal, perspective.

Wilson promotes a social, political, ethical, and spiritual program whose tenets spring directly from his scientific theories. At the program's core lies the derivation of a new, genetically determined conservation ethic that will save the biodiversity that Wilson loves. Wilson wishes to infiltrate hearts and minds of everyone everywhere with the love of biodiversity by tripping the stirring impulses that crave both contact with the natural world and faith in a grander force. As part of this program, despite his claims to the contrary (see Chapter 4), Wilson is attempting to redefine what it will mean to be a scientist in the twenty-first century. He is vastly expanding the boundaries of science, a program of disciplinary imperialism that began in 1975 with the appearance of *Sociobiology*.

To acknowledge this is not to fall into Julian Simon's camp and to label all prognostications about biodiversity loss as erroneous and hysterical, or to say that Wilson and cronies seek to profit by inventing the biodiversity crisis and proposing solutions. I do not even question Wilson's sincerity in defining the problems as he has and proposing the solutions he has. He may be responding in the only way he knows how to what I believe is for him—and for all of us—a true, devastating crisis. The campaign he has orchestrated is fully and completely dedicated to the protection of wild species and wild places.

So of course we should listen to Wilson. But we should listen with sophisticated ears. We need to understand the values, the ideological commitments, and the assumptions that underlie his analyses of the problem and the solutions he proposes. We must be wary of resource webs that give so much control to Wilson's own unique expertise, that afford him so much power to shape minds and policies. At the same time, however, we must remain riveted to the possibility that part of what he is saying may be true.

CONCLUSION

 8

By their ideas, words, and actions, the biologists who promote biodiversity's values are putting an end to the frankly metaphysical notion of an objective, value-neutral search for knowledge. Each of biodiversity's values, with the possible exception of economic value,[1] comes *from* their work, from doing what they do, from the time they spend in contact with and learning about biodiversity. These values shape their worldview and continuously affect their work. Values drive, and derive from, this unbroken, spinning cycle.

Biologists' relationships to their subjects of study run so deep and are so complex that for some, the dualisms disappear, and they identify with what they study; no objective separation is conceivable. The pioneer geneticist Barbara McClintock encountered such novelties as jumping genes in part because her scientific method led her to identify with her corn plants: "I even was able to see the internal parts of the chromosomes—actually everything was there. It surprised me because I actually felt as if I was right down there and these were my friends. . . . As you look at these things, they become part of you. And you forget yourself."[2]

Some conservation biologists become part of biodiversity, not despite the scientific process in which they're engaged, but because of it. They identify with—they become part of—the organisms, species, landscapes, and processes they have labeled *biodiversity*. What they learn, what they feel, they feel *must* be transmitted to us. An overwhelming sensation of love and a foreboding sense of crisis lead them to redefine what it means to be a scientist; they do so to save the source of their work, the fount of their professional, emotional, and, perhaps, genetic sustenance.

· · · · · · · · · · ·

Contra Terry Erwin and others, science is not transcultural: it's completely, distinctly cultural. Precisely because of these cultural characteristics, or at least the characteristics of this subculture, biologists fight for biodiversity conservation. They need not, should not, present themselves as beyond culture (see Chapter 4). Rather, conservation biologists can make it plain how their culture informs the veracity of their pronouncements and mandates that they speak out loudly and clearly as part of their postmodern, preapocalyptic science.

Ultimately, a number of things will be required to save biodiversity—and, if many of these biologists are right, to save humanity. First, each person on Earth must have the basic necessities of life and the modest amenities required for comfort and dignity. Of course, all "goods," almost all our "needs," are ultimately derived from biotic resources. A small minority of Earth's residents are gluttons, voraciously devouring all resources in our paths. And more than that, we exploit what we can seize from far away, out of view, where we need not be bothered by the havoc its extraction and refinement into goods wreaks, both on local people and on the nonhumans that share their environments.

And, despite the recent trend in conservation and development circles to downplay this, the Earth's ability to nurture humans is finite. Many of the biologists I interviewed cited overpopulation as the major threat to biodiversity, although often they did not balance the equation by including overconsumption.[3] But even were we to awaken tomorrow to equitable distribution of wealth throughout the world, could that wealth support the crushing numbers of men and women alive today? What about twice that number? Biodiversity stands no chance while the world's population inexorably increases, and while consumption imbalance remains as it is today, where capital's net flow is still overwhelmingly from *the South to the North*, where we wage wars to defend resources to fuel our unslakable desires.[4]

Here's a danger to biologists' pronouncements: contra Wilson, these are not merely biological problems. They demand social solutions, widespread dialogue, and fundamental overhaul of our notions of social justice. We need eloquent articulations of the profuse and insidious ways in which the haves benefit at the expense of the have-nots, whether the latter be human beings or nonhuman entities and processes. And we need new laws and new lifeways that will debilitate the forces destroying biodiversity and human lives.

I do not see that scientists have the sole right to analyze or solve these kinds of problems, except to the extent that they participate as concerned citizens with a distinctive way of understanding the world and a distinctive interest in the outcome. So when we listen to biologists define problems and offer solutions, we must be wary, because they often tell us only part of the story, the part they understand. If implemented, their solutions or policy pronouncements will only get us part of the way to a world where biodiversity and the humanity that depends on it and is inextricable from it will continue long and noble existences. When biologists emphasize the need for new nature reserves with better borders, for example, they evade the sociopolitical factors that require us to fence off and protect land in the first place; they do nothing to mitigate the forces of destruction. Other suggestions for saving biodiversity, such as the proposed infusion of resources into insect taxonomy, seem an even more obvious avoidance of the deeper problems.

Some biologists portrayed here believe, and announce, that their assessments stem from natural imperatives, that nature has revealed its secrets to them—that their values have nothing to do with it. This continues a historical trend in which scientists have legitimated their own hegemony by promoting their status as arbiters of truth and black-boxing science's social origins, the values that underlie and drive it. This is as counterproductive as it is dishonest, even if the scientists engaged in such rhetoric do not deliberately dissemble but merely heed their instinct or tradition.

Other biologists talk about their values, but often separate them from their science: "This is no longer me talking as a scientist; I am speaking as a private citizen." They thus renounce a potentially potent weapon: values may, in fact, have some of the same natural validation as facts, and the life scientist may have license to speak about both.

None of us should forget that the bounds of science are shifting and contingent, and that scientists and the public that observes and supports science constantly renegotiate the boundaries. Some scientists are aware of this and manipulate the boundaries accordingly. Others should be more aware so that they can manipulate those same boundaries. And observers of science should especially be aware that scientists may be engaged in this process.

Ecologist Daniel Simberloff has been quoted as saying: "Some of my colleagues will argue that biodiversity is an instrument to human welfare, even if they don't really believe it. . . . I won't. I just won't make that argument. My real temptation is to say that if a person doesn't know the value of a species, even if there's no practical or economic concern, then I can't say anything that will convince the person. It would really be something like a religious conversion. It's essentially a moral or ethical kind of thing."[5] I agree that we have a moral obligation not to beard our heartfelt reasons for valuing biodiversity with insincere reasons we think society will appreciate. The biologists I have portrayed here may, however, play a very real role in effecting this "conversion" to which Simberloff alludes, but from which he excludes himself. A handful of biologists are on the cutting edge in expanding the boundaries of science to incorporate this nontraditional role for scientists.

If some healthy, renewable chunk of biodiversity is to endure in the world, it must become part of world consciousness, part of the fundamental ethical core of each woman and man. Each of us must come to love it as deeply as do the biologists I portray here. When our basic needs are met, why shouldn't we selfishly devour more if we have not been transformed by biodiversity? And when government will fails and sacrosanct park boundaries fall, those surrounding biodiversity's remaining refuges will respect its sanctity only if they know and *love* what's inside.

We devalue the opinions of biologists when we grasp how little they really know about the very subject on which they so publicly pronounce. Nonetheless, conservation biologists may have an expanded role to play in providing us with a vocabulary for understanding nature, a taxonomy of reasons for valuing biodiversity, and an emotional message to the same effect. They could and do help derive an ethic for nurturing, loving, and stewarding the biological organisms and processes with which we share the Earth. The creation and dissemination of the term *biodiversity* lies at the heart of this effort to put our ideas of nature into new conceptual frames.

Still, biologists might wish to become more circumspect. Words are potent weapons that vanquish the recalcitrant, silky lures that woo the reluctant. The discipline of conservation biology and the current strategy for preserving Earth's myriad biological entities and interactions are built on the wobbly foundation of biodiversity. The term *biodiversity*'s phenomenal success as avatar of a scientific discipline and a recharged conservation movement, which that discipline seeks to buttress, may be its own undoing unless biologists circumscribe what *biodiversity* represents, unless they lay bare the values that inform their pronouncements, and unless they build a stronger case for the authority that allows them to make those pronouncements in the first place.

Biologists must find a way to communicate biodiversity's complexity lucidly to the lay public. If they fail to convey clearly what biodiversity is, why it is important to them, and why they should be permitted to speak for it, not only may biologists lose status as trusted spokespersons for conservation; they also jeopardize the enormous amount of conservation momentum that has gathered behind biodiversity. If it's nature we want, if we have good reasons to care about relatively untrammeled places full of beautiful living things to admire and investigate and exult in, then let's say so, and let's say it loudly and with passion, but let's do so clearly. This is a defining moment in conservation history.

Those who listen to biologists' pronouncements need to be wary. We heed scientists, we tell ourselves, because they are objective: allegedly, they offer us facts that result from the distance they keep from their objects of study, and from a process that is cold in its calculations. But the conservation biologists I have portrayed here are so deep in their subject matter: do we trust their perspective? Should we pledge allegiance to the biologists going through such Herculean efforts to remake Costa Rican biodiversity and Costa Rican society? What do we make of E. O. Wilson, who has all the answers, who has woven a rich, multilayered resource web, at the center of which he sits, implacable, spinning alluring wisdom? Do we snip the web's strands, or do we fall into it as one of our last, best hopes? Where do we use our powers of analysis to guide us in skepticism? And when and where do we allow ourselves to fall into biodiversity's thrall?

Despite my environmentalist leanings, I get a little nervous when someone calls biodiversity conservation a "moral imperative," as Reed Noss did several times during our interview. I believe many of Noss's co-thinkers share this view of their own crusade. Yet so much so evil has been justified over the ages in terms of moral imperative. When a group becomes the mouthpiece, the obligatory point of passage for moral imperatives it has defined, we are forced into a skeptical position; we must closely examine who stands to gain from the promulgation of the imperative. In an open letter to his colleagues in *Conservation Biology*, Noss advises: "Granting that a biodiversity crisis exists and that conservation biologists are charged with helping society find a solution, we have a duty to make our science relevant to policy. . . . We need to stop arguing over esoteric details, stop declining to comment when we do not have all the data, and pull together to offer strong guidance on how to save the Earth."[6]

We must remember that Noss and his peers defined the crisis and coined the term that represents it. They charged themselves with helping society find a solution; they called themselves to duty. We must ask what constitutes an "esoteric detail." We should consider what happens when biologists intercede in public life without all the data in hand. What will result from such "strong guidance"? In a rebuttal to Noss, Ian Desmukh warns that "eminent scientists confound their professional integrity with omniscient statements on loss of biodiversity. As scientists we should not pronounce as though we are eco-ayatollahs."[7] Boundary work abounds as these two men, many of their colleagues, and interested observers such as myself angle to define the proper territory for scientists, with biodiversity the pivot around which our actions revolve.

But once again: what if Noss, Wilson, and their co-thinkers are right? The terrifying possibility remains that the faith so many invest in the pronouncements of these "eco-ayatollahs" is well earned: the moral imperative may be entrenched in nature. As we destroy our resource base, humans may well be so many lemmings approaching the abyss of self-destruction. Furthermore, it strikes me as incontrovertible that if everyone loved biodiversity as much as do the biologists profiled here, the violence with which we are destroying it would be abated—and that would be a fine thing. David Ehrenfeld's defense of conservation biology may then obtain: "If 'the true test' of conservation biology is the preservation of biodiversity, there will be no shortage of theoreticians and empiricists who feel morally obliged to suggest and try one conservation method after another as long as the resolution of the issue remains in doubt."[8]

One new conservation method is to expand the biologist's role way out into society, which many of the biologists I have profiled here are attempting to do. Their work sets an example for their more demure peers who remain cloistered in the laboratory or rain forest, and for the up-and-coming scholars who are so often urged to hide their values rather than assert them and allow them to motivate their work. And the biologists I have profiled here would establish their ethic as a touchstone for those members of the general public who seek spiritual salvation, the survival of the human race, or merely the company of creatures and processes they neither understand nor control.

To help us wend our way through the maze of claims made on behalf of biodiversity, we need to understand who is making which claims. What is the basis for the claim? Who is this person to speak for it? If we heed the warnings of biodiversity's votaries, how we answer these questions is not part of the minutiae of academic give-and-take: life may hang in the balance. We must discern whom to disarm when dangerous and whom to support when advancing a cause worthy and righteous. Fledgling academic disciplines such as environmental history and science studies can help us determine where we stand with respect to biodiversity and the declarations made on its behalf. Simultaneously, these disciplines—along with conservation biology—may exemplify new directions for scholars who would have their research help interested citizens make informed decisions about their places in the world and about the kind of world in which they wish to live.

Conservation biologists have woven the idea of biodiversity from diverse strands with the explicit purpose of changing how people far and near see the natural world, value the natural world, and therefore treat the natural world. Maps get altered to preserve biodiversity to the extent that people's hearts and minds get altered to want it preserved. Of course, for it to be preserved, some of it must live on to show people, to make them feel why it is important. A positive feedback loop cycles here between the idea of biodiversity and the natural bounty to which the idea (in part) refers.

The extent to which the idea has prevailed or will prevail rides on a number of factors: the ardor with which it is promoted; the specific strategies used in its promotion; the audiences to which it is targeted; the amount of natural reality that inheres in the idea, or is perceived as adhering in the idea; the degree to which the threat alluded to is real or is perceived as real; the degree to which the love biologists feel for biodiversity really is potentially universal and not merely part of the predisposition that drove them to study biology in the first place; the degree of success biologists have in getting people into contact with biodiversity so that it can potentially transform; the extent to which people have a comfortable margin to their lives so that this idea has a chance; and the honesty with which biologists promote the idea, for credibilities have a way of collapsing when rationalizations are revealed.

In this work, I have discussed various positions on the ultimate sources for facts and values. We can be realists, believing that we may discover facts about the natural world or values that inhere in the natural world. On the other hand, we may be constructivists, propounding that facts and values are elaborate social constructions that acquire reality when we all agree they are real. Or, as I do, we can take some intermediate position, believing that both facts and values are woven by human desires from heterogeneous strands drawn from rich fabrics afforded us by both nature and culture.

Constructivist analysts of science, deep-ecological environmentalists, and some of the biologists portrayed in this book may, in fact, concur that the demarcation of facts from values is as impossible as it is unwise. For science studiers, all facts are value-laden; facts have as much to do with what really happens in nature as values do, and neither has very much to do with natural dictates. Deep ecologists believe that the divisions drawn between facts and values are artificial, the result of historical processes stemming from the rise of natural philosophy in the seventeenth century, and they assert that this division is largely responsible for our current environmental crisis. And some conservation biologists embrace a set of values in their search for facts, use their facts to promote a set of values, derive both from (or impose both on) nature, and generally blur the bounds so that no concrete distinctions persist between facts and values.

We are often inclined to believe both facts and values to the extent that both derive from a natural world that stands opposed to our human social world. Nature provides us with external validation, transcending the relativism of human desires, human values. That values could come from nature is, however, the antithesis of the science studies perspective.[9] But what if, for the sake of argument, values (and even facts) are 100 percent culturally derived? What if all values really are relative, are nothing more than individual caprice? In that case—really, in any case—it comes down to this: *How on Earth do we want to live our lives?* What kind of planet do we want to live on now and bequeath to future generations? What do we feel we need for survival and for happiness?

Even with complete cultural solipsism, it behooves us to heed the voices of those who have found delight, for whatever multiplicity of reasons, in biodiversity. If a spring of stimulation and delight is flowing, as E. O. Wilson suggests, "only a bicycle ride away,"[10] then perhaps we should drink from it, bathe in it, and decide whether we wish to keep it cascading as it has for over a billion years. If the universe is devoid of inherent moral tenets, we all acquire the responsibility for designing the rules by which we wish to live. We can continue to measure our lives by the yardstick of economic expediency, or we may decide that the maelstrom of killing in which we are engaged does not befit the human prospect, and so derive personal and cultural ethics to stop the carnage and to steward rather than squander.

I came to love nature while growing up in a suburban purgatory where a few, dwindling scraps of green space remained. What was there, I felt then and I feel now, transformed me. My training as a biologist strengthened this love. The more deeply I dug into the

anatomy that moves us, the physiology that sustains us, the biochemistry that drives the reactions, the genetics that dictates how
so much is going to play itself out, the evolution that shaped everything from the most minute biochemical reaction to the panoply of
life forms that exist and that we are destroying, the more spiritual
I became about life. It has little to do with humanistic values. It is
about complexity and beauty and interconnectedness, a peace I feel
when in nature, an astonishment I feel when confronted with its
labyrinthine intricacies, a deep love that may or may not represent
a universal biophilia. A complex, unfathomable connection drives
people to study biology—where rigid reductionism and professional
priggishness may squash this love, except in those biologists who
nurture it, whose connections to life replenish them. They love biodiversity, not because you can make Rainforest Crunch from macadamia nuts, but because biodiversity is there: exuberant, lush, apart
from us but a part of us.

Even those who proclaim the inherent value or intrinsic rights
of all organisms nonetheless use the term *biodiversity* as if it had
humanistic-scientific reality; I do, too. The complexity of the biodiversity concept does not only mirror the natural world it supposedly represents; it is that plus the complexity of human interactions
with the natural world, the inextricable skein of our values and its
value, of our inability to separate our concept of a thing from the
thing itself. Don't know what biodiversity is? You can't. Perhaps
biodiversity is an appropriate term. The confusion it conveys reveals our pathetic weakness in thinking we can define, know, and
control a nature that will always dance just beyond our grasp. Social
studies of science analysts may be right that our facts about nature
are social constructions. But they cannot say that what we have
labeled *biodiversity* has value and this infects biologists and other
environmentalists and makes us grope for language, for rationales
to society that will make people think of it differently, care for it,
nurture it. We search for ways to preserve it, which means preserving intact rain forests as well as preserving our value systems, our
awe and wonder: we want to pass all on to future generations, and
biodiversity as a term encompasses all of this. So let us keep this all
in mind when promoting biodiversity to society.

Scientists have a big role to play here, as some who have labeled themselves conservation biologists have discerned. They, inevitably, perhaps ironically, are experts on love. They can share this expertise at many levels. They can continue to try and persuade the powers that be of our need for connection to our evolutionary relatives, our biotic matrix, our emotional and physical source. They can educate for biocultural restoration, introduce us to ways of seeing and feeling, fire the biophilic neurons, prepare the ground for transformation. They can try to make us feel what they feel so deeply that it becomes intuition; they can help us reach a place where we do not attempt to observe or transcend the natural world, but where we identify with, become a part of that world. When inculcated with new ideas about our magnificent, threatened planet, when asked whether we should save this or that species, that or the other ecosystem, we would respond, as Michael Soulé would: Yes! If today's crop of biologists fail, we lose a little more hope for beautiful organisms, unique species, complex communities, creative evolution, dynamic ecosystems, for the resplendent, mind-boggling fount of joy and wonder we have come to call *biodiversity*.

APPENDIX: INTERVIEWS

United States

Biologist	Date in 1992	Place of Interview
Peter Brussard	24 Nov.	U. of Nevada, Reno
David Ehrenfeld	11 Apr.	Rutgers U., New Brunswick, N.J.
Paul Ehrlich	10 Nov.	Stanford U., Palo Alto, Calif.
Thomas Eisner	23 Mar.	Cornell U., Ithaca, N.Y.
Terry Erwin	27 Apr.	Smithsonian Institution, Washington, D.C.
Donald Falk	19 May	Center for Plant Conservation, St. Louis, Mo.
Jerry Franklin	17 Nov.	U. of Washington, Seattle
Vickie Funk	31 Mar.	Smithsonian Institution, Washington, D.C.
Hugh Iltis	15 May	U. of Wisconsin, Madison
Daniel Janzen	2 Nov.	U. of Pennsylvania, Philadelphia
K. C. Kim	7 Apr.	Pennsylvania State U., State College, Pa.
Thomas Lovejoy	1, 27 Apr.	Smithsonian Institution, Washington, D.C.
Jane Lubchenco	12 Nov.	Oregon State U., Corvalis
S. J. McNaughton	6 Mar.	Syracuse U., Syracuse, N Y
Reed Noss	13 Nov.	his home, Corvalis, Ore.
Gordon Orians	16 Nov.	U. of Washington, Seattle
David Pimentel	28 Feb.	Cornell U., Ithaca, N.Y.
Peter Raven	20 May	Missouri Botanical Gardens, St. Louis
G. Carleton Ray	28 Apr.	U. of Virginia, Charlottesville
Walter G. Rosen	30 Mar.	his home, Washington, D.C.
Michael Soulé	11 Dec.	U. of California, Santa Cruz
Edward O. Wilson	15 June	Harvard U., Cambridge, Mass.
David Woodruff	4 Dec.	U. of California, San Diego

Costa Rica

Person	Title	Place of Interview
Gerardo Carballo Carvajal	parataxonomist	INBio
Isidro Chacon*	curator	National Museums of Costa Rica
Maria Marta Chavarría	coordinator of parataxonomists	INbio
Bill Eberhart	professor of biology	UCR**
Rodrigo Gámez	director of INBio	INBio
Rosa Gúzman	parataxonomist	Carara Biological Reserve
Paul Hanson	professor of biology	UCR
Jorge Jimenez	director of biodiversity inventory	INBio
Elias Rojas Mora	parataxonomist	his home, Cocorí
Maria Morales	professor of biology	UCR
Petrona Rios	parataxonomist	INBio
Gladys Rodrigues Ramirez	parataxonomist	INBio
Juan Carlos Saborio	parataxonomist	Carara Biological Reserve
Elvira Sanchez	director, public relations	INBio
Ana Sittenfeld	director, chemical prospecting	INBio
Alvaro Umaña Quesada	former minister***	Centro de Estudios Ambientales
Manual Zumbado	curator, Diptera, INBio	INBio

Notes: All interviews in Costa Rica were conducted in May–June 1993.
   * Helped train parataxonomists.
   ** University of Costa Rica.
   *** Ministry of Natural Resources, Energy, and Mines (MIRENEM).

NOTES

Preface

1. Janzen 1989.
2. See Lincoln and Guba 1985 on qualitative research methods.

Chapter 1. Tensions at the Crossroads of Science, Nature, and Conservation

1. I discuss these tensions all too briefly in Chapter 6 and elsewhere. Please see also, e.g., Escobar 1994; Shiva et al. 1991; Reid, Barber, and Miller 1992; Hecht and Cockburn 1990; and various articles in *Cultural Survival Quarterly* and *The Ecologist*.
2. Soulé 1985.
3. Leibhardt 1988, 23; Worster, ed. 1988b, 303.
4. French, ed. 1985, 58.
5. Martin 1993.

Chapter 2. The Making of Biodiversity

1. To do justice to the history and complexity of this phenomenon is beyond the scope of this project. Tjøssem 1993 discusses the discord surrounding activism by members of the Ecological Society of America. Allison 1995 looks at the way conservation activism by biologists is channeled into public representations of cherished icons. See also Hays 1987; Shabecoff 1993; Stephen Fox 1981; McCormick 1989; Nelkin 1977.
2. Stephen Fox 1981.
3. Oelschlaeger 1991; Worster 1985; Norton 1991; Mitman 1996; Botkin 1990. Two collections that introduce the reader to deep ecology are Devall and Sessions 1985 and Tobias 1985
4. Ehrenfeld 1981; Norton 1991.
5. For two fine works on Leopold's life and ideas, see Meine 1988 and Flader 1974.
6. Among them Norton 1991; Oelschlaeger 1991; Worster 1985; Nash 1973, 1989; Callicott 1987, 1989.
7. The portrayal of Muir in Stephen Fox 1981 and Norton 1991 closely resembles my depiction of Leopold here, as a proto-ecologist bent on combining the prosaic and the ecstatic to inculcate the values of nature upon his readers.
8. Leopold 1970, 251.
9. For an insightful deconstruction of Leopold's mastery in concocting his convincing book, see John Tallmadge, "Anatomy of a Classic," in Callicott 1987.
10. Leopold 1970, 258, 264, 246–47.
11. Ibid., 227.
12. Ibid., 190.
13. Flader 1974, 50, 167.

14. Leopold 1970, 146; Stephen Fox 1981 and Meine 1988 identify this unnamed philosopher as P. D. Ouspensky. For Ouspensky's thinking on the *noumenon*, see his *Tertium Organum: A Key to the Enigmas of the World* (1920; Random House, Vintage Books, 1982), ch. 16.

15. Leopold 1970, 189.

16. Ibid., 162.

17. Ibid., 116–17.

18. Ibid., 278. The metaphors of the "theater" and the "achievement" would both eventually be subsumed under the biodiversity concept.

19. Flader 1974, 2, makes a similar point.

20. Leopold 1970, 200, 253.

21. Gould 1989.

22. Oelschlaeger 1991; Nash 1987, 1989; Norton 1991.

23. Leopold 1970, 258, 260–61.   24. Ibid., 178.

25. Ibid., 117.   26. Nash 1987, 82.

27. Notwithstanding Muir's wish that he not be labeled a scientist, Norton 1991 calls him a protoecologist.

28. Leopold 1970, 230–31.

29. Meine 1988, 27, believes Leopold was, in fact, referring to himself.

30. Leopold 1970, 238.   31. Ibid., 214, 202, 220.

32. Ibid., 246.   33. Ibid., 261.

34. Ibid., 263.   35. Ibid., 262.

36. Ibid., 290. Callicott 1987, 194, makes a similar point.

37. Leopold 1970, 295.

38. Worster 1985.

39. Elton 1958, 142. Meine 1988 notes that Leopold and Elton were friends and respected each other.

40. Elton 1958, 15.   41. Ibid., 143.

42. Ibid., 143–44.   43. Ibid., 144.

44. Ibid.   45. Ibid., 144, 145.

46. Ibid., 145. Emphasis and capitalization in original.

47. Ibid., 154, 153, 155.

48. Ibid., 155, 151.

49. Ibid., 158, 159.

50. Carson 1987 [1962]. Her earlier books were *Under the Sea-Wind* (New York: Simon & Schuster, 1941); *The Sea Around Us* (New York: Oxford University Press, 1951), an even bigger best-seller than *Silent Spring* and winner of the National Book Award; and *The Edge of the Sea* (Boston: Houghton Mifflin, 1955). See Stephen Fox 1981, 292ff., and Nash 1989, 78ff., for brief discussions of Carson's earlier works.

51. Carson 1987, 189, 23, 67.

52. Ibid., 275. Ellipsis in the original.

53. Ibid., 297.

54. Oddly, although Carson's scientific and environmentalist beliefs closely mirror Leopold's, she does not cite him in *Silent Spring*.

55. Carson 1987 [1962], 1, 2.

56. Ibid., 79. One maddening thing about *Silent Spring* is Carson's haphazard citations: who, we may wonder, was the scientist who proposed such a revolutionary idea?

57. Ibid., 117.

58. Ibid., 72, e.g.

59. Ibid., 73, 86, 71.

60. Ibid., 99, 127.

61. Ibid., 189.

62. Ibid., 6, 10.

63. Today's ecologists paint a different picture, rendering suspect Carson's ecology-based arguments. See Chapter 5.

64. For an overview of the history and provisions of the Endangered Species Act, see Kathryn A. Kohm, "The Act's History and Framework," in Kohm 1991.

65. Ehrenfeld 1972, xi; 1981.

66. Ehrenfeld 1972, xi, 3, 49.

67. Ibid., xii.

68. Ibid., 11, 55. Ehrenfeld in fact alludes to the phenomenon that E. O. Wilson was later to label "biophilia": "Most humans find natural surroundings pleasant, which is hardly surprising, since this is the setting in which our species evolved and to which we are genetically attuned" (ibid., 7).

69. "In the years since Leopold's death there has been scant progress toward his goal of a meaningful and binding land ethic," Ehrenfeld notes (ibid., 79).

70. Ibid., 11, 316.

71. Ibid., 20

72. Ibid., 328.

73. Ehrenfeld 1981, 193, 194. Ehrenfeld 1981 also provides further bibliographic references on the debunking of the diversity-stability hypothesis.

74. Ehrenfeld 1972, 4.

75. Ehrenfeld 1981, 5.

76. Ibid., 189, 177.

77. Ibid., 206–7.

78. Ibid., 211.

79. Myers 1979; Ehrlich and Ehrlich 1981.

80. Although I have not interviewed Myers formally, I spoke with him briefly after a talk he gave at Cornell in 1990 and at some length during two extended visits to campus during his tenure as an A. D. White Distinguished Visiting Professor.

81. Myers 1983, xiii.

82. Ehrlich and Ehrlich, 58–59.

83. Soulé 1985, 1986.

84. "Biodiversity was not even a word when I was doing my graduate work. Biological diversity was two words: you just assumed it. . . . And all of a sudden it became an objective for conservation, and a lot of the objective for science," the marine biologist G. Carleton Ray told me.

85. It is difficult to trace exact events and degrees of participation in the National Forum on BioDiversity. The participants' memories are fuzzy, and the record in the National Academy's archives is sealed until twenty-five years after the event. Check back in 2011.

86. Tangley 1986, 708.

87. The teleconference featured E. O. Wilson, Paul Ehrlich, Thomas Lovejoy, Peter Raven, Michael Robinson (director of the National Zoo), and Joan Martin

Brown (senior liaison in Washington, D.C., for the UN Environment Program).

88. Wetzler 1988.

89. National Academy of Sciences 1991.

90. *Biodiversity* made its appearance in *Biological Abstracts* in 1989, based on Cairns 1988.

91. Sawhill 1994, 7. Sawhill is president of the Nature Conservancy.

Chapter 3. Why and Whence the Term *Biodiversity*?

1. Oelschlaeger 1991, 281.

2. Nash 1973, xii.

3. Guha 1989. See also Callicott 1994–95. I use the term *Northerners* throughout for citizens of the (over)developed industrialized nations, which usually lie in the northern latitudes, as opposed to the less-industrialized, developing, Southern, or Third World, nations.

4. Cronon 1983; Denevan 1992; Hecht and Cockburn 1990; Yoon 1993; McKibben 1989. For a particularly delightful piece on the dubiousness of wilderness, see Pollan 1991, esp. ch. 10, "The Idea of a Garden." For a rebuttal of this "myth," see Soulé 1995.

5. Cronon 1995; Hirt 1994; Pollan 1991.

6. Wilson 1993, 39.   7. Noss 1991b, 51.

8. Murray and Takacs 1993.   9. Devall 1988, 57.

10. Laura Murray pointed out something that I—who share some of Raven's ideological goals and cultural identity—had missed. Notice who "you" are here. You are not black, female, urban, elderly, or young. You are "everybody else," the possessor of privileged knowledge, which you share with others in a kind of trickle-down beneficence. Raven and others share a culturally and politically rooted power base that allows them to offer seemingly universal proclamations in the name of biodiversity. Does this mean we should ignore these proclamations? Not necessarily, but we should be wary. See Chapter 4.

11. Vandana Shiva, "Introduction," in Shiva et al. 1991; Nabhan 1995.

12. Reid, Barber, and Miller 1992, 23.

13. E.g., Shiva 1991; Thrupp 1990.

14. Reid, Barber, and Miller 1992, 5.

15. Norton 1987, 117; Iltis 1972a. I explore the concept of biophilia in subsequent chapters.

16. Lovejoy 1985; Ehrenfeld 1988.

17. Ehrenfeld 1988, 1991.

18. Latour 1983, 157–58.

19. Daniel A. Alexandrov of the Leningrad Branch of Institute of History of Science and Technology of the (then) USSR Academy of Sciences told me this story.

20. Ehrlich and Ehrlich 1992, 219.

21. Erwin 1991b, 3.

22. Kirkland, Rhoads, and Kim 1990, 156.

23. Reid, Barber, and Miller 1992, 2, 3.

24. Raven 1994, 11.

25. Gould 1989; Petulla 1980, 101; Tierney 1985, 57.

26. May 1988, 1448.

27. Simon 1986; Simon and Wildavsky 1993. See also Stevens 1991; Mann and Plummer 1995, 64–65; Easterbrook 1995.

28. Michael Mares in Stevens 1991; Lugo 1988; Solbrig 1992.

29. Thanks to an anonymous reviewer who made this point.

30. Rojas 1992, 170.

31. Woodruff 1989; Shen 1987.

32. Nowak 1992; Dowling et al. 1992; Dold 1995.

33. Dowling et al. 1992.

34. Woodruff 1989; Daily and Ehrlich 1992.

35. Rolston 1985, 720.

36. Pound, personal communication, Monteverde Cloud Forest Reserve, 1990; Norton 1987.

37. Sober 1986, 175–76; Ehrenfeld 1986; Orians 1990.

38. Norton 1987, 115; 1986.

39. Myers 1991a, 20.

40. Ehrlich and Ehrlich 1991, xi; Lovejoy 1986.

41. Rolston 1985, 723; Sober 1986, 177.

42. Of course, the appearance of ecological health may be illusionary; who knows what effects will cascade over time from the loss of the American chestnut?

43. Pimentel et al. 1992; Ehrlich 1988; Murphy 1991; Wilson 1987.

44. Stevens 1992; Peters and Lovejoy 1992.

45. Daily and Ehrlich 1992; Brussard interview.

46. Hutto, Reel, and Londres 1987.

47. For example, as I write this the peregrine falcon is about to be "delisted" (Eure 1995).

48. Angel 1991; Agardy 1994.

49. Franklin 1993, 203.

50. Mann and Plummer 1995.

51. See, e.g., McMichael 1982; Myers 1991a, 19–20; 1979, 1982.

52. Tudge 1987, 75; Orians 1990, 13.

53. May 1990, 130.

54. Daily and Ehrlich 1992.

55. Norton 1987, 259; Gardner M. Brown 1990, 207.

56. Challinor 1985, 4; Brussard in Gibbons 1992; Hutto, Reel, and Londres 1987.

57. See, e.g., Mills, Soulé, and Doak 1993. They point out that, like *biodiversity*, the term *keystone species* may mean different things to different people.

58. Kellert 1986, 59, 62; Dobson and May 1986, 345.

59. Western 1989a, xi; see also Woodruff 1989; Einarsson 1993.

60. Koshland 1991.

61. Noss 1990; Graber 1976, 97; tamarin, Challinor 1985, 4; Mittermeier 1988, 149.

62. Murphy 1991; Diamond 1986. Some studies suggest that this might not actually be the case; for an overview, see Stevens 1993.

63. For a history of the ESA, see Kohm 1991. Mann and Plummer 1995, 149 ff., provides a good anecdotal history.

64. Mann and Plummer 1992, 52, 48.

65. Ehrlich and Ehrlich 1981. After the Republican takeover of Congress in late 1994, the ESA became a favorite target. Kanamine 1994; Cushman 1995a, 1995b; Noss and Murphy 1995.

66. Reid 1992.

67. Scott et al. 1987, 783; Pitelka 1981, 634; Malcolm L. Hunter 1991, 268; Ehrlich 1985.

68. Scott et al. 1987; Humphrey 1985; Reid and Miller 1989; Brussard 1994.

69. Blockstein 1989, 64; Salwasser 1991, 247.

70. Iltis 1970, 2.

71. Salwasser 1991, 281; Rojas 1992; Malcolm L. Hunter 1991; Pickett, Parker, and Fiedler 1992; Christine M. Schonewald-Cox, "Preface," in Schonewald-Cox et al. 1983.

72. Frankel and Soulé 1981; Willers 1992; Western 1989c; Botkin 1990.

73. Rolston 1985, 722; di Castri and Younes 1989, 14, iii; Erwin 1991a. Erwin 1991b makes the point that process, instead of species preservation, must be the main goal.

74. Quoted in Rohlf 1991, 274. Rohlf 1991 also cites deficiencies of the ESA and provides background on how it is designed to work.

75. Reid and Miller 1989, 88 (this is also the view in Reid, Barber, and Miller 1992); Malcolm L. Hunter 1991, 268; Scott et al. 1991, 284.

76. Noss and Harris 1986, 302; Noss 1983; Murphy 1991, 193, 187.

77. Scheuer 1991, 7.

78. Some biologists use *habitat* as a generic term, equivalent to *land*, or perhaps *ecosystem*. For others, habitat only exists in relation to the species using it: "We need to save as much habitat [i.e., land] as possible in Michigan for birds" is a less specific statement than "We need to save the Kirtland warbler's habitat." Further compounding the problem, some biologists use *ecosystem* as a generic term for land, like the first definition of *habitat* above.

79. Norton 1986b, 269.

80. Hunt 1989, 1; Malcolm L. Hunter 1991, 278.

81. Malcolm L. Hunter 1991, 277–78; Noss 1991a, 228; Rolston 1991, 89.

82. Soulé 1991, 749.

83. Wilson 1988a, vi.

84. Noss 1991a, 241; Aplet, Laven, and Fiedler 1992, 299.

85. Raymond Williams 1976, 187; 1980. I will return to this in Chapter 4.

86. Evernden 1992, 26.

87. "Maintaining dynamic patterns of biodiversity requires attention to underlying functions such as hydrological and climatological processes, nutrient cycles, disturbance regimes, dispersal of seeds and spores, and adaptation," Erwin goes on to say. "The ESA and other environmental legislation, even if fully enforced, are inadequate to maintain the myriad patterns and processes of biodiversity—especially at higher levels of organization. Clearly we need to think bigger." His definition of *biodiversity* here is clearly aimed at what strate-

gies we need to take if we are going to preserve as much of nature in its integrity as possible.

88. "La biodiversidad, o riqueza natural si se prefiere, es un vocablo nuevo que describe algo muy viejo. . . . Lo incorporaremos a nuestro diccionario junto con: democracia, ecología, libre comercio, y otros complicados productos de la civilización moderna" (Ortega 1992).

89. Soulé 1995, 140.

90. U.S. House 1989, 226; Ford 1989, 103–4

91. Leonard 1989, 50.

92. Hays 1987, 253.

93. May 1988; Erwin 1991b; Wilson 1988b, 1992a; C. B. Williams 1964.

94. Erwin 1982, 1983, 1988; Wilson, 1993, 35.

95. Reid, Barber, and Miller 1992, 1.

96. Wilson 1987; May 1990; Lovejoy 1992a; Wilson 1985a; Dourojeanni 1990, 88.

97. Latour 1987, 72. See Chapter 4 for more on Latour and science studies.

98. Wilson 1989, 4; 1992a, 280; final figure cited in Mann and Plummer 1992. See Lugo 1988 for a (skeptical) overview on various estimates of species loss.

99. Myers 1989, 1991b; Erwin 1991b, 2.

100. Erwin 1991c.

101. Myers 1991a, 20–21.

102. Iltis 1988, 99.

103. Norton 1987; Daily and Ehrlich 1992; Angel 1991; Lovejoy and Orians interviews.

104. See, e.g., Lovejoy 1988b; Hutto, Reel, and Londres 1987.

105. Lubchenco et al. 1991, 374, 373, 390.

106. Lovejoy 1988b, 724, 725.

107. Leopold 1970, 196; Ehrlich and Ehrlich 1981; Raven, interview.

108. Daily and Ehrlich 1992, 26; Wilson 1992a, 182, 351.

109. Lovejoy 1988a, 424.

110. Noss 1991, 239, 241.

111. Simon quoted in Stevens 1991, C1, Lugo and Marcs quoted in Mann 1991, 736; Solbrig 1992. Easterbrook 1995 brings these views to a broader audience.

112. Ehrlich 1991.   113. Devall 1988, 92–93.

114. Merchant 1980, 293.   115. Dunlap 1993, 5.

116. Craige 1992, 112, 113; Slack and Whitt 1992, 572.

117. Hutto, Reel, and Londres 1987, 3; McNaughton 1989, 112.

118. Erwin 1991b, 3. Emphasis in original.

119. Hutto, Reel, and Londres 1987; Norton 1987.

120. Worster 1985, 338, 332.

121. Noss 1991, 230.

122. Warwick Fox 1990, 106. I return to this in Chapter 5.

123. Nash 1973, 260.

Chapter 4. Examining Biodiversity

1. This chapter draws on Takacs 1993, 1994a, and 1994b.
2. Lovejoy 1988b, 726.   3. Cronon 1983, 13.
4. Schrepfer 1983, 238.   5. Worster 1988b, 303.
6. Evernden 1992, xi; Worster 1985, 1988b; Lukács 1986, 234; Raymond Williams 1980.
7. Glacken 1967.
8. Graber 1976; Nash 1973, 73; Oelschlaeger 1991, 7.
9. Evernden 1992, xi.
10. Raymond Williams 1976, 184; Winner 1986, 121.
11. Hayles 1995; Bird 1987; Raymond Williams 1976.
12. Evernden 1992, 28. Price 1993, a delightful natural history of the plastic pink flamingo, shows how moderns disconnected from nature self-consciously make meanings of nature.
13. Lease 1995, 4.
14. Warwick Fox 1990, 31.   15. Worster 1990b; Dunlap 1988, x.
16. Dunlap 1993, 4.   17. Merchant 1987, 272.
18. Nicholson 1959, 1.
19. Worster 1990a; see also Worster 1995.
20. Botkin 1990, 5, 189.
21. See the papers in Bijker, Hughes, and Pinch 1987.
22. Peter J. Taylor 1988.
23. Evernden 1992, 15–16.
24. Jasanoff et al. 1995 provides a fine introduction to what science studies is all about.
25. Worster 1988a, viii.
26. I owe a debt of gratitude here to Larry Carbone and Kavita Philip. We have shared the joys and frustrations of attempting to apply science studies to real-world problems to which we wished to contribute analyses and solutions. See Carbone 1995 and Philip 1996.
27. Yearley 1995.
28. Yearley 1991; also Yearley 1995.
29. Martin 1988.
30. See, e.g., Jasanoff 1990; Peter J. Taylor 1992.
31. Yearley 1989a, 354; Yearley 1991, 19.
32. Gieryn 1983, 1995.
33. Soulé 1985, 730. For more on biologists and intrinsic value, see Chapter 5.
34. Gieryn 1995, 393–94.
35. Rudwick 1985; Lincoln and Guba 1985.
36. Latour 1987.
37. Collins 1985.
38. Soulé 1995, 146, 138. See also Gary Lease's introduction in the same volume.
39. Harding 1992.
40. Sismondo 1993, 548; Hayles 1995. See also Sismondo 1996.
41. Peter J. Taylor 1992; Peter J. Taylor and García-Barrios 1994.

42. Marx 1988; essay originally written in 1970.

43. Dunlap 1991, 197, 216; 1988, 176.

44. Bawa and Wilkes 1992, 474; Noss 1989, 202–3.

45. Eisner et al. 1981; Benz 1990, 82; Iltis 1972b, 205.

46. Iltis 1970, 11. Emphasis in original.

47. Naess 1990, 169; Devall 1988, 91.

48. Sherman 1991; Raven 1990, 773–74; Ehrlich interview.

49. See also Wilson 1994, ch. 18.

50. Brussard 1995, 1; Brussard 1991, 9.

51. Ehrlich 1988, 26; Janzen 1988a, 136–37; Cade 1988, 287; Lester P. Brown 1988, 448–49.

52. Soulé 1988, 467–68.

53. Soulé 1986, 12, 6, and 9.

54. Ugalde 1989; Angier and Eisner 1988; Salwasser 1991, 263.

55. Brussard, Murphy, and Noss 1992, 157.

56. Noss 1991–92, 56, 58, and 60.

57. Janzen 1986a, 302.

58. Janzen 1986b, 305–6. Emphasis in original.

59. Ehrenfeld 1992.

60. Ehrlich and Ehrlich 1986, xii xiii.

61. W. Franklin Harris, quoted in Gibbons 1992.

62. For an adapted text of their Crafoord Prize acceptance speech, see Ehrlich and Wilson 1991. See also Pennisi 1991.

63. Latour 1983, 157–58. 64. Soulé 1986, 11.

65. Gutiérrez 1992. 66. Eisner 1991b.

67. Botkin 1990, 5.

68. "Interview with the 'Father of Biodiversity,'" 25, 29.

69. Quoted in Pacchioli 1991, 10.

70. Mann and Plummer 1995, 206.

71. Challinor 1988, 496; Leopold 1970, 246.

72. Warwick Fox 1990, 247, 268, and 256.

73. Ehrlich 1985, 159.

74. Dunlap 1988, 102; Ehrenfeld 1972.

75. See also Chapter 5, where I discuss biodiversity's transformative value.

76. Orians 1990, 35.

77. Worster 1985, 1990b.

78. International Council for Bird Preservation 1992.

79. Naess 1986, 512, 513.

80. Soulé 1988, 486, 469.

81. Lewontin, Rose, and Kamin 1984, 31.

82. Western 1989b; Desmukh 1989.

83. Yearley 1989b, 435.

84. Windle 1992, 364, 366; Orr 1992, 486.

85. Murphy 1990, 203–4.

86. Noss 1991–92, 58.

87. On systematists and conservation, see Wilson 1991a.

88. Merton 1973; a paper originally published in 1942.

89. See also Soulé 1995, 154. Moreover, Soulé broadens the concept of peer review to cross cultures (ibid., 151ff.). That is to say, if many cultures share views about how the natural world really is, and agree with views expressed by scientists, this makes it more likely that scientists have objective purchase on reality.

90. Simon 1986; Lugo 1988.

91. Noss 1991–92, 58.

92. Ibid. Deep ecologists capitalize names of organisms ("Elk") as a gesture of respect.

93. See also Brussard 1995, in which he warns that conservation biologists should only advocate when they are certain: "For example, we know more than enough to assert that humankind's excesses of reproduction and consumption and our habit of maximizing short-term economic gains at the expense of natural capital, if not curtailed promptly, will result in world-wide ecological armageddon. . . . Explaining this clearly to anyone who will listen is a duty incumbent on every SCB member."

94. Erwin 1991a.

95. Note a bit of boundary work by Funk here—they don't realize they're cheating. She has to say this to protect the scientific sanctity of her colleagues and the institution they represent.

96. Soulé 1995.

97. Cronon 1993.

98. Cronon 1993 makes a similar point.

99. Worster 1988b, 293.

Chapter 5. Values

1. Evernden 1992, 29.

2. Winner 1986.

3. See the chapter "The Conservation Dilemma" in Ehrenfeld 1981; and see esp. chs. 6 and 7 of Warwick Fox 1990, where Fox presents a (sometimes confusing) typology of approaches to eco-philosophy, leading to his theory of "transpersonal ecology" as a novel, and best, approach. Norton 1987 details these arguments at great length, leading to advocacy of "transformative value" as a novel, and best, approach. See also Soulé 1995; Kellert 1993, 1995; Sagoff 1985; Rolston 1985; Tierney 1985.

4. I am avoiding two dichotomies used inconsistently in environmental ethics literature. The first is *instrumental/noninstrumental values*. These roughly map out on selfish/nonselfish values. That which is instrumental is that which has use for something else; noninstrumental values are values not usable by others. However such "use" need not apply to humans. For example, a given species of insect may have no discernible use to humans, but it may nonetheless have instrumental value for birds who eat it. Next is *anthropocentric/nonanthropocentric values*. I avoid these because they have two conflated meanings. Some use *anthropocentric* to mean that the locus of value is always the human valuer; others use it to mean that the human is always who benefits from a given value.

5. Lovejoy 1986, 17; 1991a, 18.

6. Lovejoy 1992a; National Science Foundation 1977; Wilson 1992a.

7. Leopold 1970, 258.

8. Wilson 1987 is a good introduction to this line of argument. See also Ehrlich and Ehrlich 1991; Ray 1988; Ehrlich 1990; Raven 1990; Myers 1991a; Pimentel et al. 1992.

9. Soulé 1985, 730; Wilson 1992a, 308; Ehrlich and Wilson 1991; Ehrlich 1985; Ray 1988.

10. Ehrlich 1985.   11. Erwin 1991b, 1.

12. Wilson 1992a, 182.   13. Ehrlich and Ehrlich 1981, xiv.

14. Ehrlich 1988, 25.   15. Gould 1990, 30.

16. Norton 1986a.

17. Warwick Fox 1990, 157; Kirkland et al. 1990, 156; Ehrlich 1983, 30.

18. Solbrig 1992; Gouyon 1990.

19. Simberloff 1988, 502; Mills et al. 1993, 219.

20. Leopold 1970, 262.

21. Ehrenfeld 1993.

22. Dunlap 1988; Sagoff 1985. These writers also provide further references on the biological arguments for and against the diversity/stability hypothesis.

23. Contra: Lawton and Brown 1993. Pro: Tilman and Downing 1994.

24. Pimm 1991.

25. See, e.g., Lovejoy and Woodwell 1992; Reid and Miller 1989; Stevens 1992; Cade 1988.

26. Wilson 1992, 340; Soulé 1989, 300, 303. See Mitman 1996 for a fascinating view of past efforts to create artificial natures.

27. Wise 1993.

28. Including those that believe that artificially manipulating diversity completely misses the point that "natural" biodiversity generates unique, irreplaceable benefits. See Angermeier 1994.

29. Janzen 1986b, 316.

30. Wilson 1992, 244, 281, 282, 283.

31. Lovejoy 1992b, 1; 1992a, A27; 1995.

32. Although I did not interview Myers formally, he has communicated this to me extensively on several occasions during his visits to Cornell

33. Myers 1983, viii.

34. Myers 1979, 6; Iltis 1959, 3.

35. Aspirin, Lovejoy 1986; Raven 1981; vincristine, Wilson 1992a, 283; Raven 1987, 15; squalamine, Altman 1993; 40 percent, Wilson 1992a, 286–87, and Myers 1983 provide other examples, and McNaughton 1989, 116, provides further suggestions for where to look. For an overview, see also Baladrin et al. 1985.

36. Eisner 1981, 297. See also Eisner 1990, 1991a.

37. Angier 1995; Zuniga 1995.   38. Gardner M. Brown, Jr., 1990.

39. Ehrlich and Ehrlich 1981, 77.   40. Wilson 1988b.

41. Few: Myers 1984; many: Prescott-Allen and Prescott-Allen 1990.

42. Wilson 1992a, 288ff., quotation, 291. See also Myers 1983; Plotkin 1988.

43. Clifford D. May 1993.

44. Iltis et al. 1979; Santana, Guzman, and Jardel 1989; McNeely 1988; Iltis 1988, 101. For an overview of crop-germ-plasm preservation, see J. Trevor Williams 1988.

45. Pimentel et al. 1992; Plotkin 1988.

46. U.S. Fish & Wildlife Service, poster, 1992; Wilson 1992; Brennan 1993.

47. Ehrlich and Ehrlich 1992; McNeely 1988; Norton 1988.

48. But see, esp., McNeely 1988; Brown 1990; Wilson 1988a; Bormann and Kellert 1991.

49. Janzen, Hallwachs, Gámez, Sittenfeld, and Jimenez 1993, 156. Later in this chapter I discuss more philosophical objections to economic valuations of biodiversity.

50. Reid and Miller 1989, 4; Reid, Barber, and Miller 1992. See also McNeely 1988, 6.

51. Myers 1992; Raven 1990, 770.

52. Western 1989b, 24-25; Ugalde 1989; Myers 1991b; Reid, Barber, and Miller 1992; Western et al. 1989.

53. Baker 1990.

54. Janzen and Hallwachs 1990, 8.

55. It is questionable whether scientists were really pushing sustainability in tandem with biodiversity, at least at first; many hopped on when sustainable development was hitched to biodiversity's rapidly moving bandwagon. Redford and Sanderson 1992, 36.

56. Ehrlich and Ehrlich 1992, 225.

57. For proponents of this argument, see Kenneth J. Taylor 1988; Hecht and Cockburn 1989; Lohmann 1991. See also Chapter 3 above on the parallels between cultural and biological diversity.

58. Soulé 1995, 159.

59. Kellert 1986, 67; Salinas de Gortari 1992; on Costa Rican pride, see, e.g., Hovore 1991; on pride in developing nations, Sagoff 1974; Wilson 1984; Wilson 1992a; on pride and diplomacy, see Myers 1979; Atkinson 1989.

60. Janzen 1988b, 243.

61. Janzen 1990, xi-xiv.

62. For more on cultural restoration, see Janzen 1986a, 1986b, 1988b, 1992; Janzen and Hallwachs 1990.

63. Wilson 1991, 11.

64. Wilson 1979, 43; Wilson 1984, 1.

65. Iltis, Loucks, and Andrews 1970, 4.

66. Wilson 1984, 1992a. The statistic is from Boyd 1990-91. See also Ulrich 1993; Heerwagen and Orians 1993; Lawrence 1993. I return to Wilson and biophilia in Chapter 7.

67. Heerwagen and Orians 1986; Orians 1986, 17.

68. Kellert 1993, 66. Soulé 1993, his summarizing contribution to *The Biophilia Hypothesis* (Kellert and Wilson 1993), takes a skeptical view and concludes that if biophilia does exist, it is shallow and unlikely to form the basis for a deep conservation ethic.

69. Norton 1987. The book was written before the neologism *biodiversity* was coined. When Norton talks about this theory now, he substitutes *biodiversity* for *natural variety*.

70. This took place during an informal conversation at the American Society for Environmental History Meetings, Pittsburgh, 5 Mar. 1993.

71. A similar idea is proposed in Norton 1990.

72. Graber 1976; Dunlap 1988.

73. Stephen Fox 1981; Mittermeier 1988.

74. Naess 1986, 505.

75. Noss 1991–92, 56.

76. Graber 1976; Guha 1989; Haraway 1989.

77. Stephen Fox 1981, 317.

78. Stephen Fox 1981 (quoting Muir), 28, 335.

79. Wilson 1985b, 465. See also Wilson's autobiography, *Naturalist* (1994).

80. Iltis 1974, 289, 291

81. Sherman 1991, 319.

82. In a letter to the editor of the *New York Times* Travel Section (28 July 1991), Jean Colvin, an eco-tour leader, says: "The tourist must be a partner in the goal of conservation, not just a passive observer of nature. For example, he or she should understand the value of biodiversity. . . . A person who has experienced the rain forest, seen the diversity of species, learned about traditional medicinal plants and, one hopes, also understands local cultural values and economic needs becomes an ambassador for conservation ideals and goals. . . . We have seen this conversion in hundreds of people."

83. Philosophical explications of intrinsic-value theory can be found throughout the pages of the journal *Environmental Ethics* and in O'Neill 1993; Norton 1987; Fox 1990; Rolston 1991; Nash 1989; and Callicott 1986.

84. Ehrenfeld 1993, ix.

85. Ehrlich and Ehrlich 1981, 58. Note that having a right to exist is slightly different than having intrinsic value. Some who will assert that things other than humans have intrinsic value will also assert that "rights" are a uniquely human construct.

86. Naess 1986, 504.

87. Soulé 1985, 730. Emphasis in original.

88. Ehrenfeld 1988, 214.

89. Orians 1990, 11.

90. Ehrlich 1988, 26, 22; Soulé 1993, 454; Ehrenfeld 1986, 43.

91. Oelschlaeger 1991; Stephen Fox 1981; White 1985; Graber 1976; Engel 1983; Schrepfer 1983.

92. Engel 1983, 21.

93. Myers 1985, 56; Ehrenfeld 1981, 207–8; Ehrlich and Ehrlich 1992, 220.

94. Myers 1991a, 22; Kirkland et al. 1990, 156; McNaughton 1989, 110; Kaufman and Mallory 1986, vii.

95. Muir quoted in Fox 1981, 12.

96. Iltis 1974, 308; 1988.

97. Naess 1973, 1986; Tobias 1985; Devall and Sessions 1985; Foreman and Haywood 1987.

98. Ehrlich 1986, 17.

99. Hargrove 1989, 227.

100. Orians 1986, 18.

101. Pimentel 1982, 44; Carr 1982, 59. See also, e.g., Ehrlich and Wilson 1991.

102. Ehrlich and Ehrlich 1981, 50ff.

103. Norton 1986a; Worster 1990b; Botkin 1990, 127.
104. Orians 1990, 34. Ehrenfeld 1981 makes the same point (180).
105. Diamond 1986, 503.   106. Janzen 1991, 159–60.
107. Norton 1991, 141.   108. Daily and Ehrlich 1992.
109. Winner 1986, 135.   110. Soulé 1982, 61.
111. Hardin 1968.
112. Kellert 1993 makes a similar point (62).
113. Ehrenfeld 1981, 210.
114. McKibben 1989.
115. Worster 1985, 338.

Chapter 6. Costa Rica's National Institute of Biodiversity

1. Gámez et al. 1993, 58.
2. Umaña 1990; Gámez and Ugalde 1988.
3. Power 1989; Christian 1992; J. Robert Hunter 1994.
4. Shabecoff 1990.
5. Brennan 1993.
6. A brief history of INBio can be found in Gámez et al. 1993.
7. When Arias won the 1987 Nobel Peace Prize for his leadership in Central American peace negotiations, it certainly did nothing to hurt Costa Rica's high reputation in the eyes of the world.
8. Information from interviews with Gámez and Alvaro Umaña (MIRENEM minister under Arias); Gámez and Ugalde 1988.
9. In interviews, Gámez, Umaña, Janzen, and Isidro Chacon independently suggested that the idea of INBio emerged by consensus during a series of meetings.
10. In the United States, however, Janzen is increasingly the subject of hagiographic profiles. See, e.g., Langrath 1994.
11. See, e.g., Ehrenfeld 1981, 1986.
12. Janzen 1991; Janzen and Hallwachs 1990.
13. "¿Cuánto vale la biodiversidad?" (1993).
14. Janzen et al. 1993b, 232; Yoon 1995.
15. Janzen 1991; Janzen and Hallwachs 1992; Janzen 1986b. In Chapter 4, I show how biologists advocating for biodiversity are simultaneously attempting to change the definition of what it means to be a biologist.
16. Janzen 1992, 28, 31, 51.   17. Janzen 1992, 28–29.
18. Eisner 1990, 1991a&b.   19. Joyce 1991.
20. Sittenfeld and Gámez 1993. The rate for "hits" is estimated at somewhere between 1 in 5,000 and 1 in 25,000. In the United States, it costs an average of $231 million to develop one successful drug.
21. Dr. Georg Albers-Schönberg, executive director, natural products chemistry, quoted in "A Modern-Day Noah's Ark" (1991), 7; Vargas Mena 1992; National Wildlife Federation 1993.
22. Janzen 1988b.
23. "What is INBio, the National Biodiversity Institute of Costa Rica?" (1993).
24. Janzen 1991, 162.

25. As is often the case in conservation, areas chosen for protection are those that have no other uses and have escaped despoilment due to inaccessibility. Brauillo Carillo is "pristine" because of its treacherous topography, which makes it unfit for crops, cattle, or commercial forestry.

26. The word *portico* means "porch," but also may be construed as "For Tico," the nickname Costa Ricans give themselves.

27. Janzen et al. 1993b, 226.

28. Officially, the Liz Claiborne and Art Ortenberg Foundation. Three men were trained alongside the eighteen women in the course. Janzen et al. 1993b, 231.

29. Janzen and Hallwachs 1992, 11.

30. Gámez et al. 1993, 53.   31. Valerio Gutierrez 1992, 106.

32. McKean 1991.   33. Janzen 1991, 160.

34. See, e.g., Janzen and Hallwachs 1992, which states: "Unexpectedly, the parataxonomist program has caught the attention of a tropical international community confronted with wildland biodiversity management and use problems similar to those of Costa Rica" (2). This is, however, the point of the parataxonomist program; such an outcome could not have been unexpected.

35. Gámez et al. 1993; Eberhart 1993; "Societies Sound Alarm on Biodiversity" (1992); "RI, Costa Rica Agree in Biodiversity Cooperation" (1992).

36. Quoted in Eisner and Chapela 1993.

37. Letter from Ramiro Barrantes recounted in *Seminario Universidad*, 21 May 1993, 9.

38. "¿Cuánto vale la biodiversidad?" (1993); Vargas Mena 1992; Nilsson 1992; González and Chuprine 1992; Bermúdez 1992; Rojas 1992.

39. Escobar 1994, 2.

40. Wille 1993, 8.

41. Latour 1987.

Chapter 7. Edward O. Wilson

1. The epigraph to this chapter is from Wilson 1992b, 12.

2. See Wilson 1994 for the full scope of his scientific accomplishments.

3. Lovejoy 1991b; Hölldobler and Wilson 1990.

4. Wilson 1985b, 465; 1991b, 4.

5. Wilson 1987, 345.

6. Wilson 1985a, 29, 27.

7. On systematics training, see Wilson 1988b, 1989, 1992a; Wilson 1987, 346.

8. Wilson 1975.

9. Wilson 1978; Lumsden and Wilson 1981.

10. Lumsden and Wilson 1981, 358, 360, 361.

11. See Wilson 1985b and 1994, 348ff., for brief personal accounts of the sociobiology debate, and Segerstrale 1986 for a more detailed analysis of why the debate occurred, and with such vehemence.

12. See references in Segerstrale 1986.

13. One of the more widely cited critiques of doing science by just-so anecdotes is Gould and Lewontin 1979. For an overview of sociobiological speculation on human concealed ovulation, see Strassmann 1981 or Small 1995.

14. Such critiques are clearly and eloquently spelled out in Lewontin, Rose, and Kamin 1984 and Gould 1981.

15. Wilson 1992b, 15.

16. Wilson 1984, 1; 1976.

17. Wilson 1978, 199.

18. Wilson 1984, 139; Wilson 1992a, 7.

19. Wilson 1984, 121. Note that Wilson also used the phrase "the folly our descendents are least likely to forgive" when testifying before a congressional panel on the Endangered Species Act (Wilson 1981, 289). In Wilson 1994 (355), he assigns its first use to Wilson 1980, the "article [that] marked my debut as an environmental activist."

20. Wilson 1993, 40.

21. Much of Wilson 1992a is dedicated to these arguments.

22. Wilson 1993, 39.

23. Wilson 1984, 119; 1991c, 10; 1992b, 13.

24. See Wilson 1994 for engaging accounts of his religious background and subsequent rejection thereof.

25. Ruse and Wilson 1985, 51; Wilson 1993, 38.

26. Wilson 1992b, 13, 12; 1978, 171.

27. Wilson 1992b, 12.

28. Ibid., 15; Betsy Carpenter with Bob Holmes, "E.O. Wilson: Living with Nature," *U.S. News and World Report*, 30 Nov. 1992, 65.

29. Erwin 1991a.

30. See Chapter 4; Evernden 1992; Bird 1987; Haraway 1989; Williams 1976; Winner 1986.

31. Wilson 1988a, vi. See also Chapter 4 and Ehrlich and Wilson 1991.

32. Wilson 1989, 4, 7.

33. Wilson 1992a, 312, 319, 328.

34. Wilson 1988b, 3.

35. See, e.g., Wilson 1985b, 480ff.

36. Quoted in Royte 1990, 39.

37. Latour 1983, 157–58; Taylor 1988.

38. Wilson 1992a, 351.

Chapter 8. Conclusion

1. On the other hand, as we saw in Chapter 6, what Janzen, Gámez, Eisner et al. do in Costa Rica, they do as scientists. Their promotion and fostering of biodiversity's economic values is science in action.

2. McClintock quoted in Keller 1985, 165. Ellipsis in original.

3. Overpopulation was cited by Lovejoy, Ray, Pimentel, McNaughton, Eisner, Funk, Kim, Ehrenfeld, Erwin, Raven, Brussard, and Soulé in interviews, and in Ehrlich 1988, Iltis 1988, and Wilson 1992a. Those citing both overpopulation and overconsumption were Raven, Ray, Lovejoy, McNaughton, Kim, and Brussard. Eisner and Pimentel cited overconsumption, but identified overpopulation as the root problem.

4. Norman Myers notes that in the late 1980s, Northern nations contributed about $90 billion/year to Southern nations in the form of aid and World Bank loans. Yet these same Southern nations were paying $150 billion/year to service their debts. As Myers puts it, this is "akin to a blood transfusion from the sick to the healthy" (Myers 1993, 234).

5. Simberloff quoted in Yoon 1991.
6. Noss 1989.
7. Desmukh 1989.
8. Ehrenfeld 1992.
9. Merchant 1992 makes this point as well.
10. Wilson 1985b, 483.

LITERATURE CITED

Agardy, M. Tundi. 1994. "Advances in Marine Conservation: The Role of Marine Protected Areas." *Trends in Ecology and Evolution* 9, 7: 267–70.

Allison, Steven. 1995. "Transplanting a Rain Forest to the Exhibit Hall: Natural History Research and Public Representation at the Smithsonian Institution, 1960–1975." Ph.D. diss., Cornell University.

Altman, Lawrence K. 1993. "Sharks Yield Possible Weapon Against Infection." *New York Times*, 15 Feb., A8.

Angel, Martin V. 1991. "Biodiversity in the Oceans." *Ocean Challenge* 2: 28–36.

Angermeier, Paul L. 1994. "Does Biodiversity Include Artificial Diversity?" *Conservation Biology* 8, 2: 600–602.

Angier, Natalie. 1995. "Rx for Endangered Species Law: Empty Medicine Bottles." *New York Times*, 7 Mar., C4.

Angier, Natalie, and Thomas Eisner. 1988. "Use the Media for Your Message." *The Scientist*, 2 May, 18–19.

Aplet, Gregory H., Richard D. Laven, and Peggy L. Fiedler. 1992. "The Relevance of Conservation Biology to Natural Resource Management." *Conservation Biology* 6, 2: 298–300.

Atkinson, Ian. 1989. "Introduced Animals and Extinctions." In Western and Pearl 1989.

Bailes, Kendall E., ed. 1985. *Environmental History: Critical Issues in Comparative Perspective*. Lanham, Md.: University Press of America.

Baker, James A., III. 1990. "Diplomacy for the Environment." Address to Winter Meeting of the National Governors' Association, 26 Feb.

Baladrin, Manuel F., James A. Klocke, Eve Syrkin Wurtele, and William Hugh Bollinger. 1985. "Natural Plant Chemicals: Sources of Industrial and Medicinal Materials." *Science* 228: 1154–60.

Bawa, Kamaljit, and Garrison Wilkes. 1991. "Who Shall Speak for Biodiversity?" *Conservation Biology* 6, 3: 473–74.

Benz, Bruce F. 1990. "Hugh H. Iltis." *Maydica* 35: 81–84.

Bermúdez, Kattia. 1992. "Pésimo negocio ecológico." *Panorama Internacional*, 10 Feb.

Bijker, Wiebe E., Thomas P. Hughes, and Trevor J. Pinch, eds. 1987. *The Social Construction of Technological Systems: New Directions in the Sociology and History of Technology*. Cambridge, Mass.: MIT Press.

Bird, Elizabeth Ann R. 1987. "The Social Construction of Nature: Theoretical Approaches to the History of Environmental Problems." *Environmental Review* 2, 4: 255–64.

Blockstein, David. 1989. "Toward a Federal Plan for Biological Diversity." *Issues in Science and Technology* 5, 4: 63–67.

Bormann, F. Herbert, and Stephen R. Kellert, eds. 1991. *Ecology, Economics, Ethics: The Broken Circle*. New Haven, Conn.: Yale University Press.

Botkin, Daniel B. 1990. *Discordant Harmonies: A New Ecology for the Twenty-First Century.* New York: Oxford University Press.

Boyd, Linda, ed. 1990–91. *Directory of the American Association of Zoological Parks and Aquaria.* Wheeling, W.Va.: Ogle Bay Park.

Brennan, Peter. 1993. "EXPOTUR's Opening Session Features Lawsuit, Debates." *Tico Times,* 11 June, 1, 5.

Brown, Gardner M., Jr. 1990. "Valuation of Genetic Resources." In *The Preservation of Biological Resources,* ed. Gordon H. Orians, Gardner M. Brown Jr., William E. Kunin, and Joseph E. Swierzbinski. Seattle: University of Washington Press.

Brown, Lester P. 1988. "And today we're going to talk about biodiversity . . . that's right, biodiversity." In Wilson 1988a.

Brussard, Peter F. 1991. "The Role of Ecology in Biological Conservation." *Ecological Applications* 1, 1: 6–12.

———. 1994. "Why Do We Want to Conserve Biodiversity Anyway?" *Society for Conservation Biology Newsletter* 1, 4: 1.

———. 1995. "Thoughts from an Outgoing President." *Society for Conservation Biology Newsletter* 2, 2: 1.

Brussard, Peter F., Dennis D. Murphy, and Reed F. Noss. 1992. "Strategy and Tactics for Conserving Biological Diversity in the United States." *Conservation Biology* 6, 2: 157–59.

Cade, Tom J. 1988. "Science and Technology to Reestablish Species Lost in Nature." In Wilson 1988a.

Cairns, John, Jr. 1988. "Can the Global Loss of Species Be Stopped?" *Speculations in Science and Technology* 11, 3: 196–204.

Callicott, J. Baird. 1986. "On the Intrinsic Value of Nonhuman Species." In Norton 1986b.

———, ed. 1987. *Companion to "A Sand County Almanac": Interpretive and Critical Essays.* Madison: University of Wisconsin Press.

———. 1989. *In Defense of the Land Ethic: Essays in Environmental Philosophy.* Albany: State University of New York Press.

———. 1994–95. "A Critique of and an Alternative to the Wilderness Idea." *Wild Earth* 4, 4 (Winter): 54–59.

Carbone, Larry. 1995. "Death by Decapitation: Veterinarians, Scientists and Animal Euthanasia." Conference, International Society for the History, Philosophy, and Social Studies of Biology, Leuven, Belgium, 23 July.

Carr, Archie, III. 1982. Comments. In *Proceedings of the U.S. Strategy Conference on Biological Diversity, November 16–18, 1981.* Department of State Publication 9262. Washington, D.C.: Department of State.

Carson, Rachel. 1941. *Under the Sea-Wind.* New York: Simon & Schuster.

———. 1951. *The Sea Around Us.* New York: Oxford University Press.

———. 1955. *The Edge of the Sea.* Boston: Houghton Mifflin.

———. 1987 [1962]. *Silent Spring.* Boston: Houghton Mifflin.

Challinor, David. 1985. "Introductory Address: What Everyone Should Know about Animal Extinctions." In Hoage 1985.

———. 1988. "Epilogue." In Wilson 1988a.

Christian, Shirley. 1992. "There's a Bonanza in Nature for Costa Rica, Its Forests, Too, Are Besieged." *New York Times*, 29 May, A4.

Collins, Harry M. 1985. *Changing Order: Replication and Induction in Scientific Practice*. London: Sage.

Craige, Betty Jean. 1992. *Laying the Ladder Down*. Amherst: University of Massachusetts Press.

Cronon, William. 1983. *Changes in the Land: Indians, Colonists, and the Ecology of New England*. New York: Hill & Wang.

———. 1993. "The Uses of Environmental History." *Environmental History Review* 17, 3: 1–22.

———. 1995. "The Trouble with Wilderness, or, Getting Back to the Wrong Nature." In *Uncommon Ground: Toward Reinventing Nature*, ed. William J. Cronon. New York: Norton.

"¿Cuánto vale la biodiversidad?" 1993. *Panorama International*, 17 May, 37–39.

Cushman, John H., Jr. 1995a. "Babbitt Seeks to Ease Rules in Bid to Rescue Imperiled Species Law." *New York Times*, 7 Mar., C4.

———. 1995b. "Conservatives Tug at Endangered Species Act." *New York Times*, 28 May, 26.

Daily, Gretchen C., and Paul Ehrlich. 1992. "Extinction and the Biodiversity Crisis." MS.

Denevan, William M. 1992. "The Pristine Myth: The Landscape of the Americas in 1492." *Annals of the Association of American Geographers* 82, 3: 369–85."

Desmukh, Ian. 1989. "On the Limited Role of Biologists in Biological Conservation." *Conservation Biology* 3, 3: 321.

Devall, Bill. 1988. *Simple in Means, Rich in Ends: Practicing Deep Ecology*. Salt Lake City: Peregrine Smith Books.

Devall, Bill, and George Sessions. 1985. *Deep Ecology: Living As If Nature Mattered*. Salt Lake City: Peregrine Smith.

Diamond, Jared. 1986. "The Design of a Nature Reserve System for Indonesian New Guinea." In Soulé 1986.

di Castri, F., and T. Younes. 1989. "Ecosystem Function of Biological Diversity." *Biology International: The International Union of Biological Sciences News Magazine*, special issue #22.

Dobson, Andrew P., and Robert M. May. 1986. "Disease and Conservation." In Soulé 1986.

Dold, Catherine. 1995. "Florida Panthers Get Some Outside Genes." *New York Times*, 20 June, C4.

Dourojeanni, Marc J. 1990. "Entomology and Biodiversity Conservation in Latin America." *American Entomologist* 36: 88–93.

Dowling, Thomas E., W. L. Minckley, Michael E. Douglas, Paul C. Marsh, and Bruce D. Demarais. 1992. "Response to Wayne, Nowak and Phillips and Henry: Use of Molecular Characters in Conservation." *Conservation Biology* 6, 4: 600–603.

Dunlap, Thomas R. 1988. *Saving America's Wildlife: Ecology and the American Mind, 1850–1990*. Princeton, N.J.: Princeton University Press.

————. 1991. "Organization and Wildlife Preservation: The Case of the Whooping Crane in North America." *Social Studies of Science* 21: 197–221.

————. 1993. "Environmental History at the Crossroads." MS. American Society for Environmental History Conference, 6 Mar.

Easterbrook, Gregg. 1995. *A Moment on the Earth: The Coming Age of Environmental Optimism*. New York: Viking Press.

Eberhart, Rob. 1993. "Costa Rica's INBio: An Example for Conservation." *Mesoamerica*, May, 6–7.

Ehrenfeld, David W. 1970. *Biological Conservation*. New York: Holt, Rinehart & Winston.

————. 1972. *Conserving Life on Earth*. New York: Oxford University Press.

————. 1981. *The Arrogance of Humanism*. New York: Oxford University Press.

————. 1986. "Thirty Million Cheers for Diversity." *New Scientist* 110: 38–43.

————. 1988. "Hard Times for Diversity." In Wilson 1988a.

————. 1991. "The Management of Diversity: A Conservation Paradox." In Bormann and Kellert 1991.

————. 1992. Letter. *Science* 255: 1625–26.

————. 1993. *Beginning Again: People and Nature in the New Millenium*. New York: Oxford University Press.

Ehrlich, Anne H., and Paul R. Ehrlich. 1991. "Needed: An Endangered Humanity Act?" In Kohm 1991.

Ehrlich, Paul R. 1983. "Genetics and the Extinction of Butterfly Populations." In Schonewald-Cox et al. 1983.

————. 1985. "Extinctions and Ecosystem Functions: Implications for Humankind." In Hoage 1985.

————. 1986. *The Machinery of Nature*. New York: Simon & Schuster.

————. 1988. "The Loss of Diversity: Causes and Consequences." In Wilson 1988a.

————. 1990. "Diversity and Humanity: Science and Public Policy." *Center for Conservation Biology UPDATE* 4, 2: 2–3.

————. 1991. Letter. *Science* 254: 175.

Ehrlich, Paul R., and Anne H. Ehrlich. 1981. *Extinction: The Causes and Consequences of the Disappearance of Species*. New York: Ballantine Books.

————. 1986. *Healing the Planet: Strategies for Resolving the Environmental Crisis*. Reading, Mass: Addison-Wesley.

————. 1992. "The Value of Biodiversity." *Ambio* 21, 3: 219–26.

Ehrlich, Paul R., and Edward O. Wilson. 1991. "Biodiversity Studies: Science and Policy." *Science* 253: 758–62.

Einarsson, Neils. 1993. "All Animals Are Equal but Some Are Cetaceans: Conservation and Culture Conflict." *Environmentalism: The View from Anthropology*, ed. Kay Milton. London: Routledge.

Eisner, Thomas. 1981. Statement. Endangered Species Act Oversight. Hearings Before the Subcommittee on Environmental Pollution of the Committee on Environment and Public Works. U.S. Senate, 97th Cong., 1st sess., 10 Dec., 295–97.

————. 1990. "Prospecting for Nature's Chemical Riches." *Issues in Science and Technology* 6, 2: 31–34.

———. 1991a. "Chemical Prospecting: A Call for Action." In Bormann and Kellert 1991.

———. 1991b. "Chemical Prospecting and the Conservation of Biodiversity: Financial Considerations." Conference, "Sustainable Development and Biodiversity: Conflicts and Complimentarities," Cornell University, 20 Sept.

Eisner, Thomas, Hans Eisner, Jerrold Meinwald, Carl Sagan, Charles Walcott, Ernst Mayr, Edward O. Wilson, Peter H. Raven, Anne Ehrlich, Paul R. Ehrlich, Archie Carr, Eugene P. Odum, and Carl Gans. 1981. "Conservation of Tropical Forests." *Science* 213: 1314.

Eisner, Thomas, and Ignacio Chapela. 1993. "Conservation: Should Drug Companies Share in the Costs?" *Science* 259: 294-95.

Elton, Charles S. 1958. *The Ecology of Invasions by Animals and Plants*. London: Methuen.

Engel, J. Ronald. 1983. *Sacred Sands: The Struggle for Community in the Indiana Dunes*. Middletown, Conn.: Wesleyan University Press.

Erwin, Terry L. 1982. "Tropical Forests: Their Richness in Coleoptera and Other Arthropod Species." *Coleoptera Bulletin* 36, 1: 74-75.

———. 1983. "Tropical Forest Canopies, the Last Biotic Frontier." *Bulletin of the Entomological Society of America* 29, 1: 14-19.

———. 1988. "The Tropical Forest Canopy: The Heart of Biotic Diversity." In Wilson 1988a.

———. 1991a. "An Evolutionary Basis for Conservation Strategies." *Science* 253: 750-52.

———. 1991b. "A Plan for Developing Consistent Biotic Inventories in Temperate and Tropical Habitats." Part 1 of *Establishing a Tropical Species Co-occurrence Database*, ed. Terry L. Erwin. Memorias del Museo de Historia Natural, parts 1-3. Lima: Universidad Nacional Mayor de San Marcos, 16 Dec.

———. 1991c. "How Many Species Are There? Revisited." *Conservation Biology* 5, 3: 330-33.

Escobar, Arturo. 1994. "Notes on Science and Biodiversity." Panel on "Theorizing Invention, Reimagining Technoscience," 93d American Association of Anthropologists Annual Meeting, Atlanta, Nov.-Dec.

Eure, Rob. 1995. "Falcons May Fly off Endangered List." *Oregonian*, 1 July, A1, A12.

Evernden, Neil. 1992. *The Social Creation of Nature*. Baltimore: Johns Hopkins University Press.

Fergus, Chuck. 1991. "The Florida Panther Verges on Extinction." *Science* 251: 1178-80.

Flader, Susan. 1974. *Thinking Like a Mountain: Aldo Leopold and the Evolution of an Ecological Attitude Toward Deer, Wolves, and Forests*. Columbia: University of Missouri Press.

Ford, David. 1989. Statement. National Biological Diversity Conservation and Environmental Research Act. Hearing Before the Subcommittee on Natural Resources, Agricultural Research, and Environment of the Committee on Science, Space, and Technology. U.S. Senate, 101st Cong., 1st sess., 17 May.

Foreman, Dave. 1994–95. "Wilderness Areas Are Vital: A Response to Callicott." *Wild Earth* 4, 4 (Winter): 64–68.

Foreman, Dave, and Bill Haywood, eds. 1987. *Ecodefense: A Field Guide to Monkeywrenching.* Tucson, Ariz.: Ned Ludd Books.

Fox, Stephen. 1981. *John Muir and His Legacy: The American Conservation Movement.* Boston: Little, Brown.

Fox, Warwick. 1990. *Toward a Transpersonal Ecology: Developing New Foundations for Environmentalism.* Boston: Shambhala.

Frankel, O. H., and Michael E. Soulé. 1981. *Conservation and Evolution.* Cambridge: Cambridge University Press.

Franklin, Jerry F. 1993. "Preserving Biodiversity: Species, Ecosystems, or Landscapes?" *Ecological Applications* 3, 2: 202–5.

French, Roderick S. 1985. "Comment: Environmental Values and History." In Bailes 1985.

Gámez, Rodrigo, Alfio Piva, Ana Sittenfeld, Eugenia Leon, Jorge Jimenez, and Gerardo Mirabelli. 1993. "Costa Rica's Conservation Program and National Biodiversity Institute." In Reid et al. 1993.

Gámez, Rodrigo, and Alvaro Ugalde. 1988. "Costa Rica's National Park System and the Preservation of Biological Diversity: Linking Conservation with Socio-economic Development." In *Tropical Rain Forests: Diversity and Conservation,* ed. Frank Almeda and Catherine M. Pringle. San Francisco: California Academy of Sciences.

Gibbons, Ann. 1992. "Conservation Biology in the Fast Lane." *Science* 255: 20–22.

Gieryn, Thomas F. 1983. "Boundary Work and the Demarcation of Science from Non-Science: Strains and Interests in Professional Ideologies of Scientists." *American Sociological Review* 48: 781–95.

———. 1995. "Boundaries of Science." In Jasanoff et al. 1995.

Glacken, Clarence. 1967. *Traces on the Rhodian Shore.* Berkeley: University of California Press.

González, Herman, and Alekoey Chuprine. 1992. "¡Alerta! El legado de la ciencia indigena, bancos geneticos y el comercio farmacologico internacional." *Boletín Talamanca* 5, 5 (Oct.).

Gould, Stephen Jay. 1981. *The Mismeasure of Man.* New York: Norton.

———. 1989. *Wonderful Life: The Burgess Shale and the Nature of History.* New York: Norton.

———. 1990. "The Golden Rule: A Proper Scale for Our Environmental Crisis." *Natural History,* Sept., 24–30.

Gould, Stephen Jay, and Richard C. Lewontin. 1979. "The Spandrels of San Marco and the Panglossian Paradigm: A Critique of the Adaptationist Programme." *Proceedings of the Royal Society of London,* B 205: 581–98.

Gouyon, Pierre-Henri. 1990. Interview. *Libération,* 12 Feb., 32.

Graber, Linda H. 1976. *Wilderness as Sacred Space.* Washington, D.C.: Association of American Geographers.

Guha, Ramachandra. 1989. "Radical American Environmentalism and Wilderness Preservation: A Third World Critique." *Environmental Ethics* 11: 71–83.

Gutiérrez, Rocky. 1992. "The Spotted Owl: A Unique Application of the Endangered Species Act." Conference, "Endangered Species/Endangered Future: Preserving Biodiversity for the 21st Century," Cornell University, 3 Apr.

Haraway, Donna. 1989. *Primate Visions: Gender, Race, and Nature in the World of Modern Science*. New York: Routledge.

Hardin, Garrett. "The Tragedy of the Commons." *Science* 162: 1243–48.

Harding, Sandra. 1992. "After the Neutrality Ideal: Science, Politics, and 'Strong Objectivity.'" *Social Research* 59, 3: 567–87.

Hargrove, Eugene C. 1989. "An Overview of Conservation and Human Values: Are Conservation Goals Merely Human Attitudes?" In Western and Pearl 1989.

Hayles, N. Katherine. 1995. "Searching for Common Ground." In Soulé and Lease 1995.

Hays, Samuel P. 1987. *Beauty, Health, and Permanence: Environmental Politics in the United States, 1955–1985*. Cambridge: Cambridge University Press.

Hecht, Susanna, and Alexander Cockburn. 1990. *The Fate of the Forest: Developers, Destroyers, and Defenders of the Amazon*. New York: Harper Perennial.

Heerwagen, Judith H., and Gordon H. Orians. 1986. "Adaptations to Windowlessness: A Study of the Use of Visual Decor in Windowed and Windowless Offices." *Environment and Behavior* 18, 5: 623–29.

———. 1992. "Humans, Habitats, and Aesthetics." In Kellert and Wilson 1993.

Hirt, Paul W. 1994. *A Conspiracy of Optimism: Management of the National Forests since World War Two*. Lincoln: University of Nebraska Press.

———. 1995. "The Professional *Is* Political: Environmental History as Public Discourse." MS.

Hoage, R. J., ed. 1985. *Animal Extinctions: What Everyone Should Know*. Washington, D.C.: Smithsonian Institution Press.

Hölldobler, Bert, and E. O. Wilson. 1990. *The Ants*. Cambridge, Mass.: Harvard University Press.

Hovore, Frank T. 1991. "INBio: By Biologists, for Biologists." *American Entomologist* 37, 3: 157–58.

Humphrey, Stephen R. 1985. "How Species Become Vulnerable to Extinction." In Hoage 1985.

Hunt, Constance E. 1989. "Creating an Endangered Ecosystems Act." *Endangered Species Update* 6, 3 and 4: 1–5.

Hunter, J. Robert. 1994. "Is Costa Rica Truly Conservation-Minded?" *Conservation Biology* 8, 2: 592–95.

Hunter, Malcolm L., Jr. 1991. "Coping with Ignorance: The Coarse-Filter Strategy for Monitoring Biodiversity." In Kohm 1991.

Hutto, Richard L., Susan Reel, and Peter B. Londres. 1987. "A Critical Evaluation of the Species Approach to Biological Conservation." *Endangered Species Update* 4, 12: 1–4.

Iltis, Hugh H. 1959. "We Need Many More Scientific Areas." *Wisconsin Conservation Bulletin* 24, 9: 3–8.

———. 1970. "Biological Diversity, and the Social Responsibility of the Systematic Biologist." MS.

———. 1972a. "Conservation, Contraception, and Catholicism, a Twentieth-Century Trinity." *Biologist* 54, 1: 35–47.

———. 1972b. "Shepherds Leading Sheep to Slaughter: The Extinction of Species and the Destruction of Ecosystems." *American Biology Teacher* 34, 4: 201–6.

———. 1974. "Flowers and Human Ecology." In *New Movements in the Study and Teaching of Biology*, ed. Cyril Selmes. London: Temple Smith.

———. 1988. "Serendipity in the Exploration of Biodiversity: What Good Are Weedy Tomatoes?" In Wilson 1988a.

Iltis, Hugh H., J. F. Doebley, R. Gúzman, M. Pazy, and B. Pazy. 1979. "*Zea diploperennis* (Gramineae): A New Teosinte from Mexico." *Science* 203: 186–88.

Iltis, Hugh H., Orie L. Loucks, and Peter Andrews. 1970. "Criteria for an Optimum Human Nature." *Bulletin of the Atomic Scientists*, Jan., 2–6.

International Council for Bird Preservation. 1992. *Putting Biodiversity on the Map: Priority Areas for Global Conservation*. Foreword by Edward O. Wilson. Cambridge: ICBP.

"An Interview with the 'Father of Biodiversity,' E. O. Wilson." 1994. *Nature Conservancy*, July–Aug., 25–29.

Janzen, Daniel H. 1986a. "The Eternal External Threat." In Soulé 1986.

———. 1986b. "The Future of Tropical Ecology." *Annual Review of Ecology and Systematics* 17: 305–24.

———. 1988a. "Tropical Dry Forests: The Most Endangered Major Tropical Ecosystem." In Wilson 1988a.

———. 1988b. "Tropical Ecological and Biocultural Restoration." *Science* 239: 243–44.

———. 1989. "From Caterpillars to Moths to the Costa Rican National Biodiversity Institute: A Progression in Complexity." Boyce Thompson Distinguished Lecture Series, Cornell University, 11 Oct.

———. 1990. "Sustainable Society Through Applied Ecology: The Reinvention of the Village." In *Race to Save the Tropics*, ed. R. Goodland. Washington, D.C.: Island Press.

———. 1991. "The National Biodiversity Institute of Costa Rica: How to Save Tropical Biodiversity." *American Entomologist* 37, 3: 159–71.

———. 1992. "A South-North Perspective on Science in the Management, Use, and Economic Development of Biodiversity." In *Conservation of Biodiversity for Sustainable Development*, ed. O. T. Sandlund, K. Hindar, and A. H. D. Brown. Oslo: Scandinavian University Press.

Janzen, Daniel H., and Winifred Hallwachs. 1990. "Ethical Aspects of the Impact of Humans on Biodiversity." *Pontificae academiae scientiarum scripta varia*, Vatican Study Week on "Tropical Forests and the Conservation of Species," 14–18 May.

———. 1992. "Costa Rica's National Biodiversity Inventory: The Role of the Parataxonomists and the Experiences of the First Two Parataxonomist Training Courses, 1989 and 1990." Final Narrative Report to U.S. AID.

Janzen, Daniel H., Winnie Hallwachs, Rodrigo Gámez, Ana Sittenfeld, and Jorge Jimenez. 1993a. "Research Management Policies: Permits for Collecting and Research in the Tropics." In Reid et al. 1993.

Janzen, Daniel H., Winnie Hallwachs, Jorge Jimenez, and Rodrigo Gámez. 1993b. "The Role of the Parataxonomists, Inventory Managers, and Taxonomists in Costa Rica's National Biodiversity Institute." In Reid et al. 1993.

Jasanoff, Sheila. 1990. *The Fifth Branch: Science Advisors as Policymakers.* Cambridge, Mass.: Harvard University Press.

Jasanoff, Sheila, Gerald T. Markle, James Peterson, and Trevor Pinch, eds. 1995. *Handbook on Science, Technology, and Society.* Thousand Oaks, Calif.: Sage.

Joyce, Christopher. 1991. "Prospectors for Tropical Medicines." *New Scientist* 132 (19 Oct.): 36–40.

Kanamine, Linda. 1994. "Support for Controversial Program Slips." *USA Today,* 2 Dec., 1.

Kaufman, Les, and Kenneth Mallory, eds. 1986. *The Last Extinction.* Cambridge, Mass.: MIT Press.

Keller, Evelyn Fox. 1985. *Reflections on Gender and Science.* New Haven, Conn.: Yale University Press.

Kellert, Stephen R. 1986. "Social and Perceptual Factors in the Preservation of Animal Species." In Norton 1986b.

———. 1993. "The Biological Basis for Human Values of Nature." In Kellert and Wilson 1993.

———. 1995. "Concepts of Nature East and West." In Soulé and Lease 1995.

Kellert, Stephen R., and E. O. Wilson, eds. 1993. *The Biophilia Hypothesis.* Washington, D.C.: Island Press.

Kirkland, Gordon L., Jr., Ann F. Rhoads, and Ke Chung Kim. 1990. "Perspectives on Biodiversity in Pennsylvania and Its Maintenance." *Journal of the Pennsylvania Academy of Science* 64, 3: 155–59.

Kohm, Kathryn A., ed. 1991. *Balancing on the Brink of Extinction: The Endangered Species Act and Lessons for the Future.* Washington, D.C.: Island Press.

Koshland, Daniel E. 1991. "Preserving Biodiversity." *Science* 253: 717.

Langrath, Robert. 1994. "The World According to Dan Janzen." *Popular Science,* Dec., 79–82, 112–14.

Latour, Bruno. 1983. "Give Me a Laboratory and I Will Raise the World." In *Science Observed: Perspectives in the Social Study of Science,* ed. Karin Knorr-Cetina and Michael Mulkay. London: Sage.

———. 1987. *Science in Action.* Cambridge, Mass.: Harvard University Press.

Lawrence, Elizabeth Atwood. 1993. "The Sacred Bee, the Filthy Pig, and the Bat out of Hell: Animal Symbolism as Cognitive Biophilia." In Kellert and Wilson 1993.

Lawton, J. H., and V. K. Brown. 1993. "Redundancy in Ecosystems." In *Biodiversity and Ecosystem Function,* ed. E. D. Schulze and H. A. Mooney. New York: Springer-Verlag.

Lease, Gary. 1995. "Introduction: Nature under Fire." In Soulé and Lease 1995.

Leibhardt, Barbara. 1988. "Interpretation and Causal Analysis: Theories in Environmental History." *Environmental Review* 12: 23–36.

Leonard, George M. 1989. Statement. National Biological Diversity Conservation and Environmental Research Act. Hearing Before the Subcommittee on Natural Resources, Agricultural Research, and Environment of the Committee on Science, Space, and Technology. U.S. Senate, 101st Cong., 1st sess., 17 May.

Leopold, Aldo. 1970 [1949]. *A Sand County Almanac.* San Francisco: Sierra Club.

Lewontin, R. C., Steven Rose, and Leon J. Kamin. 1984. *Not in Our Genes: Biology, Ideology, and Human Nature.* New York: Pantheon Books.

Lincoln, Yvonne S., and Egon G. Guba. 1985. *Naturalistic Inquiry.* Beverly Hills, Calif.: Sage Publications.

Lohmann, Larry. 1991. "Who Defends Biological Diversity: Conservation Strategies and the case of Thailand." *The Ecologist* 21, 1: 5–13.

Lovejoy, Thomas E. 1985. "Conservation: The Province of All Nations." *BioScience* 35, 5: 269.

———. 1986. "Species Leave the Ark." In Norton 1986b.

———. 1988a. "Diverse Considerations." In Wilson 1988a.

———. 1988b. "Will Unexpectedly the Top Blow Off?" *BioScience* 38, 10: 722–26.

———. 1991a. Testimony. National Biological Diversity Conservation and Environmental Research Act. Hearing Before the Subcommittee on Environmental Protection of the Committee on Environment and Public Works. U.S. Senate, 102d Cong., 1st sess., 26 July.

———. 1991b. "Edward O. Wilson: An Introduction." *Wings* 16, 2: 3.

———. 1992a. "Earth's Living Library: Check It Out." *Washington Post*, 19 Mar., A27.

———. 1992b. "Biological Diversity: Options for UNCED." MS.

———. 1995. "Bugs, Plants, and Progress." *New York Times*, 28 May, E19.

Lovejoy, Thomas E., and George Woodwell. 1992. *Global Warming and Biological Diversity.* New Haven, Conn.: Yale University Press.

Lubchenco, Jane, Annette M. Olson, Linda B. Brubaker, Stephen R. Carpenter, Marjorie M. Holland, Stephen P. Hubbell, Simon A. Levin, James A. MacMahon, Pamela A. Matson, Terry M. Melillo, Harold A. Mooney, Charles H. Peterson, H. Ronald Pulliam, Leslie A. Real, Philip T. Regal, and Paul G. Risser. 1991. "The Sustainable Biosphere Initiative: An Ecological Research Agenda. A Report from the Ecological Society of America." *Ecology* 72, 2: 371–412.

Lugo, Ariel. 1988. "Estimating Reductions in the Diversity of Tropical Forest Species." In Wilson 1988a.

Lukács, Georg. 1986. *History and Class Consciousness: Studies in Marxist Dialectics.* Cambridge, Mass.: MIT Press.

Lumsden, Charles, and Edward O. Wilson. 1981. *Genes, Mind, and Culture: The Coevolutionary Process.* Cambridge, Mass.: Harvard University Press.

Lyotard, Jean-François. 1984. *The Postmodern Condition: A Report on Knowledge.* Minneapolis: University of Minnesota Press.

Mann, Charles C. 1991. "Extinction: Are Ecologists Crying Wolf?" *Science* 253: 736–38.

Mann, Charles C., and Mark L. Plummer. 1992. "The Butterfly Problem." *Atlantic Monthly*, Jan., 47–70.

———. 1995. *Noah's Choice: The Future of Endangered Species.* New York: Knopf.

Martin, Brian. 1988. "Coherency of Viewpoints among Fluoridation Partisans." *Metascience* 6, 1: 2–19.

———. 1993. "The Critique of Science Becomes Academic." *Science, Technology, and Human Values* 18, 2: 247–59.

Marx, Leo. 1988. "American Institutions and Ecological Ideals." In *The Pilot and the Passenger: Essays on Literature, Technology, and Culture in the United States.* New York: Oxford University Press. An essay originally written in 1970.

May, Clifford D. 1993. "The Buffalo Returns: This Time as Dinner." *New York Times Magazine,* 16 Sept., 30–34.

May, Robert M. 1988. "How Many Species Are There on Earth?" *Science* 241: 1441–48.

———. 1990. "Taxonomy as Destiny." *Nature* 347: 129–30.

McCormick, John. 1989. *Reclaiming Paradise: The Global Environmental Movement.* Bloomington: Indiana University Press.

McKean, Karen. 1991. "C.R. collects St. Francis Award for Nature." *Tico Times,* 8 Nov., 21.

McKibben, Bill. 1989. *The End of Nature.* New York: Random House.

McMichael, D. F. 1982. "What Species, What Risk?" In *Species at Risk: Research in Australia,* ed. R. H. Graves and W. D. L. Ride. Berlin: Springer-Verlag.

McNaughton, S. J. 1989. "Ecosystems and Conservation in the Twenty-First Century." In Western and Pearl 1989.

McNeely, Jeffrey A. 1988. *Economics and Biological Diversity: Developing and Using Economic Incentives to Conserve Biological Resources.* Gland, Switzerland: International Union for Conservation of Nature and Natural Resources.

Meine, Curt. 1988. *Aldo Leopold: His Life and Work.* Madison: University of Wisconsin Press.

Merchant, Carolyn. 1980. *The Death of Nature: Women, Ecology, and the Scientific Revolution.* New York: Harper & Row.

———. 1987. "The Theoretical Structure of Ecological Revolutions." *Environmental Review* 2, 4: 265–74.

———. 1992. *Radical Ecology: The Search for a Livable World.* New York: Routledge.

Merton, Robert K. 1973. "The Normative Structure of Science." In *The Sociology of Science: Theoretical and Empirical Investigations.* Chicago: University of Chicago Press. A paper originally published in 1942.

Mills, L. Scott, Michael E. Soulé, and Daniel F. Doak. 1993. "The Keystone-Species Concept in Ecology and Conservation." *BioScience* 43, 4: 219–24.

Mitman, Gregg. 1996. "When Nature *Is* the Zoo: Vision and Power in the Art and Science of Natural History." *Osiris,* 2d ser., 11. Forthcoming.

Mittermeier, Russell A. 1988. "Primate Diversity and the Tropical Forest: Case Studies from Brazil and Madagascar and the Importance of the Megadiversity Countries." In Wilson 1988a.

"A Modern-Day Noah's Ark." 1991. *Merck World,* Nov.

Murphy, Dennis D. 1990. "Conservation Biology and the Scientific Method." *Conservation Biology* 4, 2: 203–4.

———. 1991. "Invertebrate Conservation." In Kohm 1991.

Murray, Laura, and David Takacs. 1993. "Cultural Diversity and Biodiversity: Mutually Authorizing Metaphors." Conference, "Science as Metaphor," Cornell University, 12 Mar.

Myers, Norman. 1979. *The Sinking Ark: A New Look at the Problem of Disappearing Species*. Oxford: Pergamon Press.

———. 1982. Comments. In *Proceedings of the U.S. Strategy Conference on Biological Diversity, November 16–18, 1981*. Department of State Publication 9262. Washington, D.C.: Department of State.

———. 1983. *A Wealth of Wild Species: Storehouse for Human Welfare*. Boulder, Colo.: Westview Press.

———, ed. 1984. *Gaia: An Atlas of Planetary Management*. Garden City, N.Y.: Anchor Books.

———. 1985. "A Look at the Present Extinction Spasm and What It Means for the Future Evolution of Species." In Hoage 1985.

———. 1989. "A Major Extinction Spasm: Predictable and Inevitable?" In Western and Pearl 1989.

———. 1991a. "Biological Diversity and Global Stability." In Bormann and Kellert 1991.

———. 1991b. "Population, Environment, Development: The Emergent Synthesis." Lecture, Cornell University, 24 Oct.

———. 1992. "U.S. and the Global State." Lecture, Cornell University, 16 Oct.

———. 1993. *Ultimate Security: The Environmental Basis of Political Stability*. New York: Norton.

Nabhan, Gary Paul. 1995. "The Dangers of Reductionism in Biodiversity Conservation." *Conservation Biology* 9, 3: 479–81.

Naess, Arne. 1973. "The Shallow and the Deep, Long-Range Ecology Movement." *Inquiry* 16: 95–100.

———. 1986. "Intrinsic Value: Will the Defenders of Nature Please Rise?" In Soulé 1986.

———. 1990. "Deep Ecology and Conservation Biology." In *The Earth First! Reader: Ten Years of Radical Environmentalism*, ed. John Davis. Salt Lake City: Peregrine Smith Books.

Nash, Roderick. 1973. *Wilderness and the American Mind*. New Haven, Conn.: Yale University Press.

———. 1987. "Aldo Leopold's Intellectual Heritage." In Callicott 1987.

———. 1989. *The Rights of Nature: A History of Environmental Ethics*. Madison: University of Wisconsin Press.

National Academy of Sciences. 1991. *Policy Implications of Greenhouse Warming*. Washington, D.C.: National Academy of Sciences.

National Science Foundation. 1977. "The Biology of Aridity." *Mosaic* 8, 1: 28–35.

National Wildlife Federation. 1993. Program: Corporate Conservation Council Environmental Achievement Award Banquet. Mayflower Hotel, Washington, D.C., 26 Jan.

Nelkin, Dorothy. 1977. "Scientists and Professional Responsibility: The Experience of American Ecologists." *Social Studies of Science* 7: 75–95.

Nicholson, Marjorie Hope. 1959. *Mountain Gloom and Mountain Glory: The Development of the Aesthetics of the Infinite*. Ithaca, N.Y.: Cornell University Press.

Nilsson, Kate. 1992. "Prospecting for Green Gold." *Biomass Users Network News* 6, 6: 1–2.

Norton, Bryan G. 1986a. "On the Inherent Danger of Undervaluing Species." In Norton 1986b.

———. 1986b. *The Preservation of Species: The Value of Biological Diversity*, ed. Bryan G. Norton. Princeton, N.J.: Princeton University Press.

———. 1987. *Why Preserve Natural Variety?* Princeton, N.J.: Princeton University Press.

———. 1988. "Commodity, Amenity, and Morality: The Limits of Quantification in Valuing Biodiversity." In Wilson 1988a.

———. 1990. "Ethical Imperatives for Biodiversity." Conference, "Biodiversity and Landscapes: Human Challenges for Conservation in the Changing World," Penn State University, 23 Oct.

———. 1991. *Toward Unity among Environmentalists*. New York: Oxford University Press.

Noss, Reed F. 1983. "A Regional Landscape Approach to Maintain Diversity." *BioScience* 33, 11: 700–706.

———. 1989. "Who Will Speak for Biodiversity?" *Conservation Biology* 3, 2: 202–3.

———. 1990. "Indicators for Monitoring Biodiversity: A Hierarchical Approach." *Conservation Biology* 4, 4: 355–64.

———. 1991a. "From Endangered Species to Biodiversity." In Kohm 1991.

———. 1991b. "What Can Wilderness Do for Biodiversity?" *Wild Earth* 1, 2 (Summer): 51–56.

———. 1991–92. "Biologists, Biophiles, and Warriors." *Wild Earth* 1, 4: 56–60.

———. 1994–95. "Wilderness—Now More Than Ever: A Response to Callicott." *Wild Earth* 4, 4 (Winter): 60–63.

Noss, Reed F., and Larry D. Harris. 1986. "Nodes, Networks, and MVM's: Preserving Diversity at All Scales." *Environmental Management* 10, 3: 299–309.

Noss, Reed F., and Dennis D. Murphy. 1995. "Endangered Species Left Homeless in Sweet Home." *Conservation Biology* 9, 2: 229–31.

Nowak, Ronald M. 1992. "The Red Wolf Is Not a Hybrid." *Conservation Biology* 6, 4: 593–95.

Oelschlaeger, Max. 1991. *The Idea of Wilderness: From Prehistory to the Age of Ecology*. New Haven, Conn.: Yale University Press.

O'Neill, John. 1993. *Ecology, Policy, and Politics: Human Well-Being and the Natural World*. London: Routledge.

Orians, Gordon H. 1986. "An Ecological and Evolutionary Approach to Landscape Aesthetics." In *Landscape Meanings and Values*, ed. Edmund C. Penning-Rowsell and David Lowenthal. London: Allen & Unwin.

———. 1990. "Ecological Sustainability." *Environment* 32, 9: 10–15, 34–39.

Orr, David W. 1992. "For the Love of Life." *Conservation Biology* 6, 4: 486–87.

Ortega, Alfredo T. 1992. "¿Ha oido usted hablar de la biodiversidad?" *El Occidental* [Guadalajara, Jalisco], 6 Mar.

Ouspensky, P. D. 1982 [1920]. *Tertium Organum: A Key to the Enigmas of the World*. Random House, Vintage Books.

Pacchioli, David. 1991. "Life in the Balance." *Research/Penn State* 12, 3: 4–10.

Pennisi, Elizabeth. 1991. "Biodiversity Rides a Popular Wave." *The Scientist*, 15 Apr., 1, 8–11.

Peters, Robert L., and Thomas E. Lovejoy, eds. 1992. *Global Warming and Biological Diversity*. New Haven, Conn.: Yale University Press.

Petulla, Joseph. 1980. *American Environmentalism: Values, Tactics, Priorities.* College Station: Texas A & M Press.

Philip, Kavita. 1996. "The Role of Science in Colonial Discourses of Modernity: Anthropology, Forestry, and the Construction of 'Nature's' Resources in Madras Forests, 1858–1930." Ph.D. diss., Cornell University.

Pickett, Steward T. A., V. Thomas Parker, and Peggy L. Fiedler. 1992. "The New Paradigm in Ecology: Implications for Conservation Biology above the Species Level." In *Conservation Biology: The Theory and Practice of Nature Conservation Preservation and Management*, ed. Peggy L. Fiedler and Subodha K. Jain. New York: Chapman & Hall.

Pimentel, David. 1982. "Biological Diversity and Environmental Quality." In *Proceedings of the U.S. Strategy Conference on Biological Diversity, November 16–18, 1981*. Department of State Publication 9262. Washington, D.C.: Department of State.

Pimentel, David, Ulrich Stachow, David A. Takacs, Hans W. Brubaker, Amy R. Dumas, John J. Meaney, John A. S. O'Neil, Douglas E. Onsi, and David B. Corzilius. 1992. "Conserving Biological Diversity in Agricultural/Forestry Systems." *BioScience* 42, 5: 354–62.

Pimm, Stuart L. 1991. *The Balance of Nature? Ecological Issues in the Conservation of Species and Communities*. Chicago: University of Chicago Press.

Pitelka, Frank A. 1981. "The Condor Case: An Uphill Struggle in a Downhill Crush." *Auk* 98: 634–35.

Plotkin, Mark. 1988. "The Outlook for New Agricultural and Industrial Products from the Tropics." In Wilson 1988a.

Pollan, Michael. 1991. *Second Nature: A Gardener's Education*. New York: Delta.

Power, Alison G. 1989. "Agricultural Policies and the Environment: The Case of Costa Rica." *CUSLAR Newsletter*, Oct., 10–17.

Prescott-Allen, Robert, and Christine Prescott-Allen. 1990. "How Many Plants Feed the World?" *Conservation Biology* 4, 4: 365–74.

Price, Jennifer. 1993. "A Natural History of the Plastic Pink Flamingo." Conference, American Society for Environmental History, 6 Mar.

Raven, Peter H. 1981. Statement. Endangered Species Act Oversight. Hearings Before the Subcommittee on Environmental Pollution of the Committee on Environment and Public Works. U.S. Senate, 97th Cong., 1st sess., 10 Dec., 290–95.

———. 1987. "Biological Resources and Global Stability." In *Evolution and Coadaptation in Biotic Communities*, ed. Shoichi Kawano, Joseph H. Connell, and Toshitaka Hidaka. Tokyo: University of Tokyo Press.

———. 1990. "The Politics of Preserving Biodiversity." *BioScience* 40: 769–74.

———. 1994. "Defining Biodiversity." *Nature Conservancy*, Jan.–Feb., 10–15.

Ray, G. Carleton. 1988. "Ecological Diversity in Coastal Zones and Oceans." In Wilson 1988a.

Redford, Kent H., and Steven E. Sanderson. 1992. "The Brief, Barren Marriage of Biodiversity and Sustainability." *Bulletin of the Ecological Society of America* 73: 36–39.

Reid, Walter V. 1992. "The United States Needs a National Biodiversity Policy." *WRI Issues and Ideas*, Feb.

Reid, Walter, Charles Barber, and Kenton Miller. 1992. *Global Biodiversity Strategy: Guidelines for Action to Save, Study, and Use Earth's Biotic Wealth Sustainably and Equitably*. Washington, D.C.: World Resources Institute, World Conservation Union, and United Nations Environment Program.

Reid, Walter V., and Kenton R. Miller. 1989. *Keeping Options Alive: The Scientific Basis for Conserving Biodiversity*. Washington, D.C.: World Resources Institute.

Reid, Walter V., Sarah A. Laird, Carrie A. Meyer, Rodrigo Gámez, Ana Sittenfeld, Daniel H. Janzen, Michael A. Gollin, and Celestous Juma, eds. 1993. *Biodiversity Prospecting*. Washington, D.C.: World Resources Institute.

"RI, Costa Rica Agree in Biodiversity Cooperation." 1992. *Jakarta Post*, 3 Nov., 2.

Rohlf, Daniel J. 1991. "Six Biological Reasons Why the Endangered Species Act Doesn't Work—and What to Do about It." *Conservation Biology* 5, 3: 273–82.

Rojas, Martha. 1992. "The Species Problem and Conservation: What Are We Protecting?" *Conservation Biology* 6, 2: 170–78.

Rojas, Zaida. 1992. "En vías patentarse la biodiversidad nacional." *Esta Semana* (Costa Rica) 4, 169: 5.

Rolston, Holmes, III. 1985. "Duties to Endangered Species." *BioScience* 35, 11: 718–26.

———. 1991. "Environmental Ethics: Values in and Duties to the Natural World." In Bormann and Kellert 1991.

Royte, Elizabeth. 1990. "The Ant Man." *New York Times Magazine*, 22 July, 16–21, 38–39.

Rudwick, Martin J. S. 1985. *The Great Devonian Controversy: The Shaping of Scientific Knowledge among Gentlemanly Specialists*. Chicago: University of Chicago Press.

Ruse, Michael, and Edward O. Wilson. 1985. "The Evolution of Ethics." *New Scientist* 108 (17 Oct.): 50–52.

Sagoff, Mark. 1974. "On Preserving the Natural Environment." *Yale Law Journal* 81: 205–67.

———. 1985. "Fact and Value in Ecological Science." *Environmental Ethics* 7, 2: 99–116.

Salinas de Gortari, Carlos. 1992. Closing Speech. Conference, "Problemática del conocimiento y conservación de la biodiversidad," Yaxchilan, Mexico, 15 Feb.

Salwasser, Hal. 1991. "In Search of an Ecosystem Approach to Endangered Species Conservation." In Kohm 1991.

Santana, C. Eduardo, M. Rafael Gúzman, and P. Enrique Jardel. 1989. "The Sierra de Manantlan Biosphere Reserve: The Difficult Task of Becoming a Catalyst for Regional Sustained Development." In *Proceedings of the Symposium on Biosphere Reserves, Fourth Wilderness Congress, Sept. 14-17, 1987, YMCA at the Rockies, Estes Park, Colorado*, ed. William P. Gregg, Jr., Stanley L. Krugman, and James D. Wood, Jr. Atlanta: U.S. Department of the Interior, National Park Service.

Sawhill, John C. 1994. "The Nature Conservancy and Biodiversity." *Nature Conservancy*, Jan.–Feb., 5–9.

Scheuer, James D. 1991. Comments. National Biological Diversity Conservation and Environmental Research Act. Hearing Before the Subcommittee on Environmental Protection of the Committee on Environment and Public Works. U.S. Senate, 102d Cong., 1st sess., 26 July.

Schonewald-Cox, Christine M., Steven M. Chambers, Bruce MacBryde, and W. Lawrence Thomas, eds. 1983. *Genetics and Conservation: A Reference for Managing Wild Animal and Plant Populations*. Menlo Park, Calif.: Benjamin/Cummings.

Schrepfer, Susan R. 1983. *The Fight to Save the Redwoods: A History of Environmental Reform, 1917–1978*. Madison: University of Wisconsin Press.

Scott, J. Michael, Blair Csuti, James D. Jacobi, and John E. Estes. 1987. "Species Richness." *BioScience* 37, 11: 782–88.

Scott, J. Michael, Blair Csuti, Kent Smith, J. E. Estes, and Steve Caicco. 1991. Gap Analysis of Species Richness and Vegetation Cover: An Integrated Biodiversity and Conservation Strategy. In Kohm 1991.

Segerstrale, Ulrica. 1986. "Colleagues in Conflict: An 'in vivo' Analysis of the Sociobiology Controversy." *Biology and Philosophy* 1, 1: 53–87.

Seminario Universidad. 1993. Letter, 21 May, 9.

Shabecoff, Philip. 1990. "Loss of Tropical Forests Is Found Much Worse than Was Thought. *New York Times*, 8 June, A1, B6.

———. 1993. *A Fierce Green Fire: The American Environmental Movement*. New York: Hill & Wang.

Shen, Susan. 1987. "Biological Diversity and Public Policy." *BioScience* 37, 10: 709–12.

Sherman, Elizabeth J. 1991. "Champion of the Chain of Life." *The World and I*, Feb., 318–25.

Shiva, Vandana, Patrick Anderson, Heffa Schöcking, Andrew Gray, Larry Lohmann, and David Cooper. 1991. *Biodiversity: Social and Ecological Perspectives*. Penang, Malaysia: World Rainforest Movement.

Simberloff, Daniel. 1988. "The Contribution of Population and Community Biology to Conservation Science." *Annual Review of Ecology and Systematics* 19: 473–511.

Simon, Julian. 1986. "Disappearing Species, Deforestation, and Data." *New Scientist* 118, (15 May): 60–64.

Simon, Julian, and Aaron Wildavsky. 1993. "Facts, Not Species, Are Imperiled." *New York Times*, 13 May, A23.

Sismondo, Sergio. 1993. "Some Social Constructions." *Social Studies of Science* 23: 515–53.

———. 1996. *Science Without Myth: On Constructions, Reality and Social Knowledge*. Albany: State University of New York Press.

Sittenfeld, Ana, and Rodrigo Gámez. 1993. "Biodiversity Prospecting by INBio." In Reid et al. 1993.

Slack, Jennifer Daryl, and Laurie Anne Whitt. 1992. "Ethics and Cultural Studies." In *Cultural Studies*, ed. Lawrence Grossberg, Cary Nelson, and Paula A. Treichler. New York: Routledge.

Small, Meredith F. *What's Love Got to Do with It? The Evolution of Human Mating*. New York: Anchor Books.

Sober, Elliott. 1986. "Philosophical Problems for Environmentalism." In Norton 1986b.

"Societies Sound Alarm on Biodiversity." 1992. *Science* 257: 876.

Solbrig, Otto T. 1992. "Biodiversity, Global Change and Scientific Integrity." *Journal of Biogeography* 19: 1–2.

Soulé, Michael E. 1982. Comments. In *Proceedings of the U.S. Strategy Conference on Biological Diversity, November 16–18, 1981*. Department of State Publication 9262. Washington, D.C.: Department of State.

———. 1985. "What Is Conservation Biology?" *BioScience* 35: 727–34.

———, ed. 1986. *Conservation Biology: The Science of Scarcity and Diversity*. Sunderland, Mass: Sinauer Associates.

———. 1988. "Mind in the Biosphere; Mind of the Biosphere." In Wilson 1988a.

———. 1989. "Conservation Biology in the Twenty-First Century: Summary and Outlook." In Western and Pearl 1989.

———. 1991. "Conservation: Tactics for a Constant Crisis." *Science* 253: 744–50.

———. 1993. "Biophilia: Unanswered Questions." In Kellert and Wilson 1993.

———. 1995. "The Social Siege of Nature." In Soulé and Lease 1995.

Soulé, Michael W., and Gary Lease, eds. *Reinventing Nature? Responses to Postmodern Deconstruction*. Washington, D.C.: Island Press, 1995.

Stevens, William K. 1991. "Species Loss: Crisis or False Alarm?" *New York Times*, 20 Aug., C1, C8.

———. 1992. "Global Warming Threatens to Undo Decades of Conservation Efforts." *New York Times*, 25 Feb., C4.

———. 1993. "Study Undercuts Beliefs on Preserving Species." *New York Times*, 28 Sept., C4.

Strassmann, Beverly I. 1981. "Sexual Selection, Paternal Care, and Concealed Ovulation in Humans." *Ethology and Sociobiology* 2: 31–40.

Takacs, David. 1993. "'Biodiversity': An Idea as a Force of Nature." Conference, American Society for Environmental History, Pittsburgh. 6 Mar.

——— 1993. "'Biodiversity': At the Nexus of Biological Ideas, Environmental Values, and the Western Landscape." Conference, Western History Association, Albuquerque, 21 Oct.

———. 1994b. "Examining Biodiversity: Science and Technology Studies Meets Environmental History." Conference, "The Nature of Science Studies," Cornell University, 16 Apr.

Tallmadge, John. 1987. "Anatomy of a Classic." In Callicott 1987.

Tangley, Laura. 1986. "Biological Diversity Goes Public." *BioScience* 36: 708–11, 715.

Taylor, Kenneth J. 1988. "Deforestation and Indians in Brazilian Amazonia." In Wilson 1988a.

Taylor, Peter J. 1988. "Technocratic Optimism, H. T. Odum, and the Partial Transformation of Ecological Metaphor after World War II." *Journal of the History of Biology* 21: 213–44.

———. 1992. "Re/Constructing Socioecologies: System Dynamics Modeling of Nomadic Pastoralists in Sub-Saharan Africa." In *The Right Tools for the Job:*

At Work in Twentieth-Century Life Sciences, ed. Adele Clarke and Joan Fujimura. Princeton, N.J.: Princeton University Press.

Taylor, Peter J., and Raul García-Barrios. 1994. "The Social Analysis of Ecological Change—from Systems to Intersecting Processes." Conference, "The Nature of Science Studies," Cornell University, 16 Apr.

Thrupp, Lori Ann. 1990. "Politics of the Sustainable Development Crusade: From Elite Protectionism to Social Justice in Third World Resource Issues." *Environment, Technology, and Society* 58: 1–7.

Tierney, John. 1985. "Lonesome George of the Galapagos." *Science 85* 6, 5: 50–61.

Tilman, David, and John A. Downing. 1994. "Biodiversity and Stability in Grasslands." *Nature* 36: 363–65.

Tjøssem, Sara. 1993. "Preservation of Nature and Academic Respectability: Tensions in the Ecological Society of America, 1915–1979." Ph.D. diss., Cornell University.

Tobias, Michael, ed. 1985. *Deep Ecology.* San Diego: Avant Books.

Tudge, Colin. 1987. "Rembrandts in the Sky." *New Scientist* 116, 1580 (1 Oct.): 74–75.

Ugalde, Alvaro. 1989. "An Optimal Parks System." In Western and Pearl 1989.

Ulrich, Roger S. 1993. "Biophilia, Biophobia, and Natural Landscapes." In Kellert and Wilson 1993.

Umaña, Alvaro. 1990. "Museo Nacional 27 de Noviembre de 1987." In *La herencia de los recursos naturales.* San José, Costa Rica: Imprenta Nacional.

U.S. House. 1989. Committee on Science, Space, and Technology. Subcommittee on Natural Resources, Agricultural Research, and Environment. Hearing on the National Biological Diversity Conservation and Environmental Research Act. 101st Cong., 1st sess., 17 May.

Valerio Gutierrez, Carlos E. 1992. "INBio: A Pilot Project in Biodiversity." *Association of Systematics Collections Newsletter* 20, 2: 104–6.

Vargas Mena, Emilio. 1992. "El INBio y la Merck: El fin del colonialismo farmacologico?" *Ambien-Tico* 3 (Aug.): 3–4.

Western, David. 1989a. "Overview." In Western and Pearl 1989.

———. 1989b. "Population, Resources, and Environment in the Twenty-First Century." In Western and Pearl 1989.

———. 1989c. "Conservation Biology." In Western and Pearl 1989.

Western, David, and Mary C. Pearl, eds. 1989. *Conservation for the Twenty-First Century.* New York: Oxford University Press.

Western, David, Mary C. Pearl, Stuart L. Pimm, Brian Walker, Ian Atkinson, and David S. Woodruff. 1989. "An Agenda for Conservation Action." In Western and Pearl 1989.

Wetzler, Richard E. 1988. "The Maintenance of Biological Diversity: Review of *Biodiversity* and *Biodiversity: The Videotape.*" *Ecology* 69, 5: 1639–40.

"What Is INBio, the National Biodiversity Institute of Costa Rica?" 1993. Handout at INBio, dated Apr. 1993.

White, Richard. 1985. "Historiographical Essay; American Environmental History: The Development of a New historical Field." *Pacific Historical Review* 54: 297–335.

Wille, Chris. 1993. "The Booming Business of Biodiversity." *Outdoors Unlimited*, May, 8–9.

Willers, Bill. 1992. "Toward a Science of Letting Things Be." *Conservation Biology* 6, 4: 607.

Williams, C. B. 1964. *Patterns in the Balance of Nature.* New York: Academic Press.

Williams, J. Trevor. 1988. "Identifying and Protecting the Origins of Our Food Plants." In Wilson 1988a.

Williams, Raymond. 1976. "Nature." In *Keywords: A Vocabulary of Culture and Society.* New York: Oxford University Press.

———. 1980. "Ideas of Nature." In *Problems in Materialism and Culture.* London: Verso.

Wilson, Edward O. 1975. *Sociobiology: The New Synthesis.* Cambridge, Mass.: Harvard University Press, Belknap Press.

———. 1976. "Getting Back to Nature—Our Hope for the Future; A Provocative Interview with Sociobiologist Edward O. Wilson." *House and Garden*, Feb., 65.

———. 1978. *On Human Nature.* Cambridge, Mass.: Harvard University Press.

———. 1979. "Biophilia." Advertisement: "The Column: Capital Ideas from People Who Publish with Harvard." *New York Times Book Review*, 14 Jan., 43.

———. 1980. "Resolutions for the 80's." *Harvard Magazine*, Jan.–Feb., 22–26.

———. 1981. Statement. Endangered Species Act Oversight. Hearings Before the Subcommittee on Environmental Pollution of the Committee on Environment and Public Works. U.S. Senate, 97th Cong., 1st sess., 10 Dec., 288–90.

———. 1984. *Biophilia.* Cambridge, Mass.: Harvard University Press.

———. 1985a. "The Biological Diversity Crisis: A Challenge to Science." *Issues in Science and Technology* 2: 20–29.

———. 1985b. "In the Queendom of the Ants: A Brief Autobiography." In *Leaders in the Study of Animal Behavior: Autobiographical Perspectives*, ed. Donald A. Dewsbury. Cranbury, N.J.: Bucknell University Press, 1985.

———. 1987. "The Little Things That Run the World (the Importance and Conservation of Vertebrates)." *Conservation Biology* 1: 344–46.

———, ed. 1988a. *BioDiversity.* Washington, D.C.: National Academy Press.

———. 1988b. "The Current State of Biological Diversity." In Wilson 1988a.

———. 1989. "Conservation: The Next Hundred Years." In Western and Pearl 1989.

———. 1991a. "Arousing Biophilia: A Conversation with E. O. Wilson." *Orion* 10 (Winter): 9–15.

———. 1991b. "Ants." *Wings* 16, 2: 4–13.

———. 1991c. "Biodiversity, Prosperity, and Value." In Bormann and Kellert 1991.

———. 1992a. *The Diversity of Life.* Cambridge, Mass.: Harvard University Press.

———. 1992b. "The Return to Natural Philosophy." *Harvard Divinity Bulletin* 21, 3: 12–15.

———. 1993. "Biophilia and the Conservation Ethic." In Kellert and Wilson 1993.

———. 1994. *Naturalist.* Washington, D.C.: Island Press.

Windle, Phyllis. 1992. "The Ecology of Grief." *BioScience* 42, 5: 363–66.

Winner, Langdon. 1986. *The Whale and the Reactor: A Search for Limits in an Age of High Technology.* Chicago: University of Chicago Press.

Wise, James. 1993. "Saving Depleted Forests: Wave of Future?" *Oregonian,* 12 Jan., D9.

Woodruff, David. 1989. "The Problems of Conserving Genes and Species." In Western and Pearl 1989.

Worster, Donald. 1985. *Nature's Economy: A History of Ecological Ideas.* Cambridge: Cambridge University Press.

———, ed. 1988a. *The Ends of the Earth: Perspectives on Modern Environmental History.* Cambridge: Cambridge University Press.

———. 1988b. "Appendix: Doing Environmental History." In Worster 1988a.

———. 1990a. "Seeing Beyond Culture." *Journal of American History* 76, 4: 1142–47.

———. 1990b. "The Ecology of Order and Chaos." *Environmental History Review* 14: 1–18.

———. 1995. "Nature and the Disorder of History." In Soulé and Lease 1995.

Yearley, Steven. 1989a. "Environmentalism: Science and a Social Movement." *Social Studies of Science* 19: 343–55.

———. 1989b. "Bog Standards: Science and Conservation at a Public Inquiry." *Social Studies of Science* 19: 421–38.

———. 1991. *The Green Case: A Sociology of Environmental Issues, Arguments, and Politics.* London: Harper Collins.

———. 1995. "The Environmental Challenge to Science Studies." In Jasanoff et al. 1995.

Yoon, Carol Kaesuk. 1991. "Do Earth's Species Face Death?" *Oregonian,* 15 Aug., E1–2.

———. 1993. "Rain Forests Seen as Shaped by Human Hand." *New York Times,* 27 July, C1, C10.

———. 1995. "Monumental Inventory of Costa Rican Forest's Insects Is Under Way." *New York Times,* 11 July, C4.

Zuniga, José. 1995. "Call for Endangered Species Protection." *Washington Blade,* 17 Mar., 41.

INDEX

Library of Congress Cataloging-in-Publication Data

Takacs, David.
    The idea of biodiversity : philosophies of paradise / David Takacs.
        p.   cm.
    Includes bibliographical references (p.   ) and index.
    ISBN 0-8018-5400-8
    1. Biological diversity conservation—Philosophy.   I. Title.
QH75.T235   1996
333.95'16'01—dc20   96-15924